Springer Series in Measurement Science and Technology

The Springer Series in Measurement Science and Technology comprehensively covers the science and technology of measurement, addressing all aspects of the subject from the fundamental principles through to the state-of-the-art in applied and industrial metrology, as well as in the social sciences. Volumes published in the series cover theoretical developments, experimental techniques and measurement best practice, devices and technology, data analysis, uncertainty, and standards, with application to physics, chemistry, materials science, engineering and the life and social sciences.

More information about this series at http://www.springer.com/series/13337

Kyung-Young Jhang · Cliff J. Lissenden ·
Igor Solodov · Yoshikazu Ohara ·
Vitalyi Gusev
Editors

Measurement of Nonlinear Ultrasonic Characteristics

Springer

Editors
Kyung-Young Jhang
School of Mechanical Engineering
Hanyang University
Seoul, Korea (Republic of)

Cliff J. Lissenden
Engineering Science and Mechanics
Pennsylvania State University
University Park, PA, USA

Igor Solodov
Zerstörungsfreie Prüfung
Institut für Kunststofftechnik
University of Stuttgart
Stuttgart, Germany

Yoshikazu Ohara
Department of Materials Processing
Tohoku University
Sendai, Japan

Vitalyi Gusev
Laboratoire d'Acoustique de
l'Université du Mans
Le Mans Université
Le Mans, France

ISSN 2198-7807 ISSN 2198-7815 (electronic)
Springer Series in Measurement Science and Technology
ISBN 978-981-15-1460-9 ISBN 978-981-15-1461-6 (eBook)
https://doi.org/10.1007/978-981-15-1461-6

This Springer imprint is published by the registered company Springer Nature Singapore Pte Ltd.
The registered company address is: 152 Beach Road, #21-01/04 Gateway East, Singapore 189721, Singapore

Preface

The evaluation of accumulated damage or degradation in the early stages of fracture is important in refinery plants, nuclear power plants, or aircraft parts to ensure their structural safety. The ultrasonic method is the most powerful nondestructive technique for evaluating material degradation, as the characteristics of ultrasonic wave propagation are directly related to the mechanical properties of the material.

Traditional ultrasonic methods are based on linear theory, which generally relies on the measurement of a few specific parameters such as the sound velocity, attenuation, and transmission or reflection amplitudes to determine the elastic properties of a material or detect defects. The sound velocity depends on the elastic constants, while the attenuation is related to the microstructure. The presence of defects changes the phase and/or amplitude of the transmitted or reflected waves. However, the methods employing ultrasonic characteristics in the linear elastic region are only sensitive to gross defects or open cracks, while being less sensitive to the evenly distributed micro-cracks or degradation.

Nonlinear ultrasonic characteristics have been studied to overcome the above-mentioned limitation. The principal difference between linear and nonlinear ultrasonic methods is that in the latter, the existence and characteristics of the defects are often related to an acoustic signal whose frequency differs from that of the input signal. This is related to the radiation and propagation of finite amplitude (particularly high power) ultrasound and its interaction with discontinuities such as cracks, interfaces, and microstructures. As material failure or degradation is usually preceded by some type of nonlinear mechanical behavior before significant plastic deformation or material damage occurs, recent studies have shown considerable focus on the application of nonlinear ultrasonic methods.

Typical nonlinear ultrasonic characteristics include higher harmonic generation, sub-harmonic generation, resonance frequency shifts (nonlinear resonance), or mixed frequency response (nonlinear frequency mixing). Additionally, these phenomena occur not only in the longitudinal waves, but also in the surface waves and guided waves. Thus, many researchers have focused on the development of various methodologies and applied technologies through theoretical and experimental studies. Furthermore, the advances in measuring instruments have led to numerous studies on techniques that facilitate the measurement of nonlinear ultrasonic characteristics. Nevertheless, a lot of experience and high proficiency are required to obtain significant experimental data, as these nonlinear ultrasonic characteristics are extremely minute as compared to linear phenomena. Thus, many researchers, particularly beginners including graduate school students, still face difficulties while measuring the nonlinear ultrasonic characteristics.

In this book, the major nonlinear ultrasonic methodologies can be covered at a glance, and the key technological know-how/knowledge of each methodology has been provided by the world's best experts with years of experience in this field, which will help researchers to obtain good quality data. An important feature of this book is that it provides the key technological know-how/knowledge required to quickly learn the nonlinear ultrasonic measurement skills and many tips to acquire significant data on the nonlinear ultrasonic characteristics. Thus, this book will serve as an essential guidebook for the measurement of nonlinear ultrasonic characteristics. We hope that this book will be useful for various research investigations.

The main experts invited to edit this book include Prof. Igor Solodov (Stuttgart University, Germany), Prof. Cliff J. Lissenden (Penn State, USA), Prof. Vitalyi Gusev (Le Mans Université, France), and Prof. Yoshikazu Ohara (Tohuku University, Japan). They have joined as the respective authors of each chapter, and I would like to express my sincere gratitude to them. In addition, I would also like to express my gratitude to the many professionals who helped draft each chapter.

Seoul, Korea (Republic of) Kyung-Young Jhang

Contents

Chapter 1
Overview—Nonlinear Ultrasonic Characteristics

Kyung-Young Jhang

Abstract Nonlinear ultrasonic characteristics are the features that result from the nonlinear interactions between the propagating ultrasonic waves and the material or defects, which typically include higher harmonic generation, sub-harmonic generation, nonlinear resonance and nonlinear frequency-mixing. In this book, such nonlinear interactions are divided into two categories, namely the material elastic nonlinearity and the contact acoustic nonlinearity, and the methodologies to measure the nonlinear ultrasonic characteristics associated with each category will be introduced with the corresponding applications.

The evaluation of accumulated damage or degradation in the material properties and detection of micro-defects in the early stages of fracture are important to ensure the safety of various industrial structures. The ultrasonic method is the most powerful nondestructive technique, as the characteristics of ultrasonic wave propagation are directly related to the mechanical properties of the material. Traditional ultrasonic nondestructive evaluation (NDE) is based on linear acoustic theory that is associated with the propagation of vibrations through a medium. Deviations from the equilibrium state of a medium caused by these vibrations are assumed to be small; i.e. the propagating wave is assumed to have a small amplitude or low intensity and maintain constant wave velocity. This type of linear ultrasonic technique generally relies on the measurement of a few specific parameters such as the sound velocity, attenuation, and reflectivity. The sound velocity depends on the elastic constants, while the attenuation is related to microstructural features such as the grain size. Furthermore, the presence of defects changes the phase and/or amplitude of the output signal [1]. However, this type of technique is less sensitive to the evenly distributed micro-defects or degradation.

An alternative technique to overcome the above-mentioned limitation is to employ the nonlinear ultrasonic characteristics. The principal difference between linear and non-linear ultrasonic NDE is that in the latter, the propagating wave is assumed to

K.-Y. Jhang (✉)
Hanyang University, Seongdong-gu, Seoul, Republic of Korea
e-mail: kyjhang@hanyang.ac.kr

K.-Y. Jhang et al. (eds.), *Measurement of Nonlinear Ultrasonic Characteristics*, Springer Series in Measurement Science and Technology,
https://doi.org/10.1007/978-981-15-1461-6_1

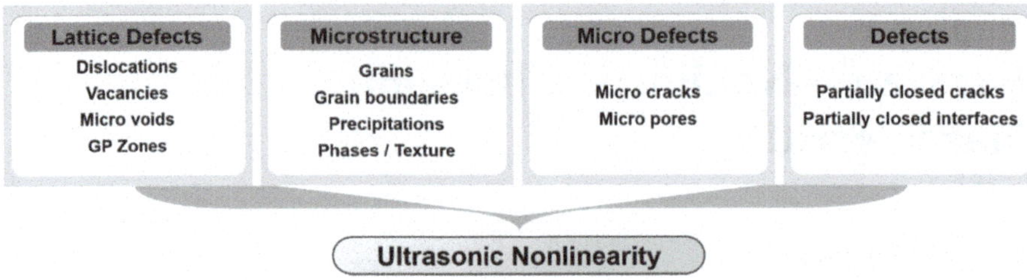

Fig. 1.1 The microstructure and defects affecting the nonlinear ultrasonic characteristics

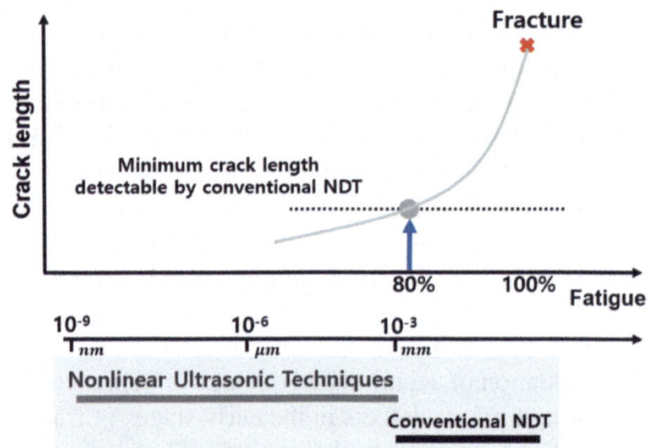

Fig. 1.2 An example of approximate crack progression due to fatigue

have a finite amplitude and is accompanied by numerous effects, whose magnitudes depend on the vibration amplitude. For example, the wave velocity varies with the vibration amplitude, and the frequency of the wave signal differs from that of the input signal. This is related to the radiation and propagation of finite amplitude (particularly high power) ultrasound and its nonlinear mechanical interaction with the microstructure or defects. This includes lattice defects such as dislocations and vacancies, microstructure such as slip bands, grains, precipitates, and micro voids, and micro-defects such as micro-cracks and micro-pores, as shown in Fig. 1.1 [2]. Partially closed cracks and interfaces are also important types of anomalously high nonlinearities if the crack-opening distance or gap between the contact interfaces is less than the amplitude displacement of the ultrasonic waves.

As material failure or degradation is usually preceded by some type of nonlinear mechanical behavior before significant plastic deformation or material damage occurs, recent studies have shown considerable focus on the application of nonlinear ultrasonics [3–5]. For example, Fig. 1.2 shows the approximate crack progression due to fatigue. The conventional linear ultrasonic technique (LUT) detects the crack only after a macro crack is initiated. The detectable size is usually greater than 1 mm, which corresponds to the damage stage that is over 80% of the fatigue life. Furthermore, the time to fracture occurrence might be shorter than expected, as the

Table 1.1 Nonlinear ultrasonic characteristics resulting from the nonlinear interactions of the ultrasonic waves

Nonlinear interactions	Nonlinear ultrasonic characteristics
Material elastic nonlinearity	Higher harmonics Resonance frequency shift (nonlinear resonance) Mixed frequency (nonlinear frequency-mixing)
Contact acoustic nonlinearity	Higher harmonics Sub-harmonics Resonance frequency shift (nonlinear resonance) Mixed frequency (nonlinear frequency-mixing)

crack propagation speeds up significantly from this point onwards. Thus, a crack that was not found during the previous periodic inspection might cause a fracture before the next inspection. In contrast, the nonlinear ultrasonic technique (NUT) can be used to evaluate the micro-damage prior to macro crack initiation.

Nonlinear ultrasonic characteristics are the features that result from the nonlinear interactions between the propagating ultrasonic waves and the material or defects. Such nonlinear interactions can be divided into two categories, namely the material elastic nonlinearity and contact acoustic nonlinearity. Table 1.1 summarizes the nonlinear ultrasonic characteristics resulting from the nonlinear interactions of the ultrasonic waves.

Material elastic nonlinearity is based on the nonlinear relationship between the stress and strain. This type of interaction causes higher harmonic generation, resonant frequency shifting (nonlinear resonance), and mixed frequency response (nonlinear frequency mixing). These phenomena are strongly influenced by the structure and interaction of solids, thus allowing the application of ultrasonic waves in material characterization. Moreover, these effects are significant in damaged materials but are nearly unmeasurable in undamaged materials, indicating their usefulness for the evaluation of material degradation. Meanwhile, elastic nonlinearity inherently exists in all materials. Thus, it is important to measure its change from the existing nonlinearity. The amount of variation in the nonlinear characteristics might be small and should therefore be measured properly. Nonetheless, it is important to note that the change in the nonlinear characteristics is greater as compared to the linear characteristics when the material properties change.

Contact acoustic nonlinearity originates when repeated collisions occur between the two contact interfaces induced by the incident ultrasonic waves. This type of interaction causes a similar phenomenon as that seen in material elastic nonlinearity. However, a peculiar phenomenon in this case is the occurrence of subharmonics. Subharmonic generation is only a result of contact acoustic nonlinearity and not material elastic nonlinearity. These phenomena can be used to detect partially closed cracks or evaluate imperfect bonding. The crack-opening distance or gap between the contact interfaces is an important factor in contact acoustic nonlinearity. Even if the length or

Table 1.2 Typical measurement parameters in the linear and nonlinear ultrasonic techniques

	Linear ultrasonic characteristics	Nonlinear ultrasonic characteristics
Measurement parameters	Sound speed (or time of flight), Attenuation, Scattering, Dispersion, Reflectivity	Higher harmonics, Sub-harmonics, Resonance frequency (nonlinear resonance) Mixed frequency (nonlinear frequency-mixing)
To evaluate or detect	Elastic modulus Thickness (or thickness loss) Anisotropy, Porosity, Cracks, Delamination, Voids, Inclusions	Ultrasonic nonlinear parameter, Nonlinear elastic modulus, Degradation, Partially closed crack, Insufficient bonding (interfacial stiffness)

size of the defect is macroscopic, contact acoustic nonlinearity occurs when the crack-opening distance or gap between the contact interfaces is less than the displacement amplitude of the ultrasonic waves. This type of nonlinearity is known to be much larger than elastic nonlinearity. Thus, contact acoustic nonlinearity measurements are relatively easier than elastic nonlinearity measurements. However, to activate contact acoustic nonlinearity, it is necessary to input an ultrasonic wave with a very large displacement amplitude or cause resonance. In general, the displacement amplitude of the ultrasonic wave is <1 nm, thus the interface with a smaller gap can be detected. The larger the amplitude of the ultrasonic waves, the defects with larger interfacial gaps can be detected.

Table 1.2 summarizes the differences between the LUT and NUT measurement parameters. In the LUT, the sound speed, attenuation, dispersion, or backscattering are generally measured to evaluate the elastic modulus, thickness, anisotropy, grain size, or porosity, while the reflectivity or transmittance can be measured to detect the crack, delamination, void, or inclusion. In the NUT, the higher harmonics, resonance frequency, or mixed frequency can be measured to evaluate the nonlinear elastic modulus, which can be extended to evaluate the hardness or strength and detect the partially closed crack or insufficient bonding. The sub-harmonics are measured only to detect cracks or interfaces. It can be observed that the signal amplitude in the time domain is generally used in the LUT, while the signal amplitude in the frequency domain is used in the NUT. However, the amplitude in the frequency domain can be used in special LUT applications, while the amplitude in the time domain might be used in special NUT applications.

Additionally, the nonlinear ultrasonic characteristics induced by the material elastic nonlinearity or contact acoustic nonlinearity can appear in all types of waves, bulk waves (longitudinal or shear waves), surface waves, or guided waves. As different types of nonlinear interactions can occur with various types of ultrasound, various methodologies have been developed to measure these nonlinear ultrasonic characteristics.

This book covers the typical nonlinear ultrasonic characteristics induced in solids. The methodologies that employ each nonlinear ultrasonic characteristic will be introduced with the corresponding applications, which have been categorized into two parts: Part I covers the measurements of the nonlinear ultrasonic characteristics associated with material elastic nonlinearity, while Part II covers the measurements of the nonlinear ultrasonic characteristics associated with contact acoustic nonlinearity.

Part I consists of two chapters: Chaps. 2 and 3. Chapter 2 describes the measurement of the nonlinear ultrasonic parameters from the higher harmonics, which quantitatively represent the nonlinear elastic properties. The use of longitudinal waves is mainly dealt. Chapter 3 describes secondary wave generation, including harmonic generation of various guided waves such as Lamb waves and shear horizontal waves.

Part II consists of three chapters: Chaps. 4, 5, and 6. Chapter 4 gives an overview of various nonlinear acoustic techniques, and covers different vibration-based nonlinear resonance techniques, particularly for measuring contact nonlinearities in defects. Chapter 5 introduces the subharmonic measurement technique used for the closed crack interfaces, particularly the array technique employed for imaging. Chapter 6 covers the nonlinear frequency mixing effect in crack, and introduces measurement techniques based on laser ultrasonic techniques.

References

1. A.S. Birks, in *Nondestructive Testing Handbook 7: Ultrasonic Testing*. ASNT Handbook (1991)
2. N.G.H. Meyendorf, P.B. Nagy, S.I. Rokhlin (eds.), *Nondestructive Materials Characterization with Applications to Aerospace Materials*. Springer Series in Materials Science, vol. 67 (2004), p. 4
3. K.Y. Jhang, Applications of nonlinear ultrasonics to the NDE of material degradation. IEEE-UFFC **47**(3), 540–548 (2000)
4. H. Jeong, S.H. Nahm, K.Y. Jhang, Y.H. Nam, A nondestructive method for estimation of the fracture toughness of CrMoV rotor steels based on ultrasonic nonlinearity. Ultrasonics **41**(7), 543–549 (2003)
5. K.Y. Jhang, Nonlinear ultrasonic techniques for non-destructive assessment of micro damage in material: A review. Int. J. Precis. Eng. Manuf. **10**(1), 123–135 (2009)

Part I
Measurements of Nonlinear Ultrasonic Characteristics Related with Material Elastic Nonlinearity

Chapter 2
Measurement of Nonlinear Ultrasonic Parameters from Higher Harmonics

Kyung-Young Jhang, Sungho Choi and Jongbeom Kim

Abstract Higher harmonic generation is a classic phenomenon of nonlinear ultrasonic interaction, which can occur in both material elastic and contact acoustic nonlinearities. However, this chapter covers only material elastic nonlinearity. The nonlinear ultrasonic parameters quantitatively represent the nonlinear elastic properties of a material and can be measured from the higher harmonics generated during ultrasonic wave propagation in the material. In this chapter, we describe the measurement process of nonlinear ultrasonic parameters. Furthermore, we introduce the configuration of the experimental apparatus and the signal processing techniques in detail. In particular, we focused on the error-inducing factors. Methods that use longitudinal waves are mainly considered; however, a method that uses surface waves is also introduced in brief.

Nomenclature

$A(\omega)$	displacement amplitude
A_1	displacement amplitude of the fundamental frequency component
A'_1	detected signal amplitude of the fundamental frequency component
$A'_{1,0}$	detected signal amplitude of the fundamental frequency component for the reference material
A_2	displacement amplitude of the second-order harmonic frequency component
A'_2	detected signal amplitude of the second-order frequency component

K.-Y. Jhang (✉) · S. Choi
Hanyang University, Seongdong-gu, Seoul, Republic of Korea
e-mail: kyjhang@hanyang.ac.kr

S. Choi
e-mail: sunghochoi@hanyang.ac.kr

J. Kim
Korea Atomic Energy Research Institute, Daejeon, Republic of Korea
e-mail: jkim8@kaeri.re.kr

K.-Y. Jhang et al. (eds.), *Measurement of Nonlinear Ultrasonic Characteristics*, Springer Series in Measurement Science and Technology,
https://doi.org/10.1007/978-981-15-1461-6_2

$A'_{2,0}$	detected signal amplitude of the second-order frequency component for the reference material
$\mathrm{A_V}(\omega)$	amplitude of a detected voltage signal
$\mathrm{C_i}(\omega)$	response function of couplant layer function at position i
C_{IJ}	second-order elastic constant
C_{IJK}	third-order elastic constant
$\mathrm{D}(\omega)$	amplitude composition of an electrical signal sent to a transmitting transducer
$\mathrm{G}(\omega)$	response function of an amplifier
$H(\omega)$	calibration function
$I'_{in}(\omega)$	input current in the calibration
$I_{out}(\omega)$	current measured at the receiving part in the nonlinear measurement
$I'_{out}(\omega)$	output current in the calibration
$K(\omega)$	transfer function
M_{IJ}	second-order Huang coefficient
M_{IJK}	third-order Huang coefficient
N	window length
N_p	number density of precipitates
$P_{A,in}(\omega)$	monochromatic acoustic power at the transmitting transducer
$P_{A,out}(\omega)$	acoustic power at the receiving transducer
$P_{\mathrm{E,cal-in}}(\omega)$	generated signal in the calibration
$P_{E,cal-out}(\omega)$	received signal in the calibration
$P_{E,in}(\omega)$	monochromatic electrical power at the transmitting transducer
$P_{E,out}(\omega)$	electrical power at the receiving transducer
$\mathrm{R}(\omega)$	response function of a receiving transducer
$\mathrm{T}(\omega)$	response function of transmitting transducer
$V'_{in}(\omega)$	input voltage in the calibration
$V'_{\mathrm{out}}(\omega)$	output voltage in the calibration
X	propagation direction
$Z(\omega)$	electrical impedance
a	area of the receiving transducer
d	line beam spacing
f_{fund}	fundamental wave frequency determined by selection of line beam spacing
k	wave number
k'	wave number compensation factor
k_0	wave number for the reference material
n	data number
q	quantum voltage
r_p	average radius of the precipitates
t	time
u	displacement
u_1	first-order perturbation solution
u_2	second-order perturbation solution

$u_{(0^\circ)}$	displacement of ultrasonic wave when the phase of the incident ultrasound is 0°
$u_{(180^\circ)}$	displacement of ultrasonic wave when the phase of the incident ultrasound is 180°
v	particle velocity
w	line beam width,
x	propagation distance
$\Delta\beta$	variation in the nonlinear ultrasonic parameter
Λ	dislocation density
α'	displacement proportionality compensation factor
α_1	displacement proportionality coefficient for the fundamental frequency
$\alpha_{1,0}$	displacement proportionality coefficient for the fundamental frequency for the reference material
α_2	displacement proportionality coefficient for the second-order harmonic frequency
$\alpha_{2,0}$	displacement proportionality coefficient for the second-order harmonic frequency for the reference material
α_i	displacement proportionality coefficient
β	second-order nonlinear ultrasonic parameter
β'	relative nonlinear ultrasonic parameter
β_0	nonlinear ultrasonic parameter for the reference material
β_e	estimated nonlinear ultrasonic parameter
β_p	plate-shaped precipitates
$\beta_p{}'$	rod-shaped precipitates
β_p''	needle-shaped precipitates
ε	strain
ε_J	strain tensor
ε_K	strain tensor
ε_q	quantization error
κ	coefficient that determines the profile of the Tukey window
λ	wavelength
ρ	density
σ	stress
σ_I	stress tensor
σ_y	yield strength
ω	angular frequency
ω_1	fundamental angular frequency
ω_2	second-order harmonic angular frequency

2.1 Higher Harmonic Generation

Higher harmonic generation is a classic phenomenon wherein the incident wave waveform is distorted by the nonlinear elastic response of the medium to the incident wave and higher harmonic waves are generated in the transmitted wave as shown in Fig. 2.1.

Harmonic generation related with the elastic nonlinearity is based on the fact that the relationship between stress and strain is inherently nonlinear even in the elastic range, as shown in Fig. 2.2. This elastic nonlinearity in an isotropic material can be represented as Eq. (2.1) for a one-dimensional case, which is called the nonlinear

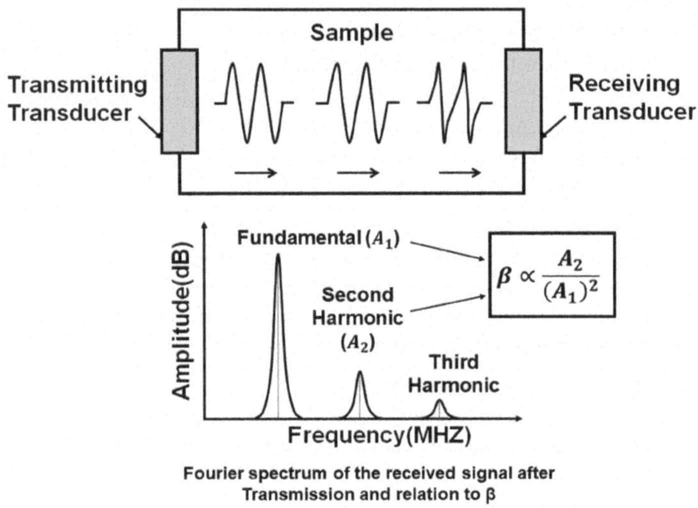

Fig. 2.1 Distortion in the waveform during propagation by nonlinear elasticity and higher harmonic generation. $\boldsymbol{\beta}$ is a nonlinear parameter defined by the amplitudes of the fundamental and second–order harmonic components

Fig. 2.2 Nonlinear relationship between stress and strain

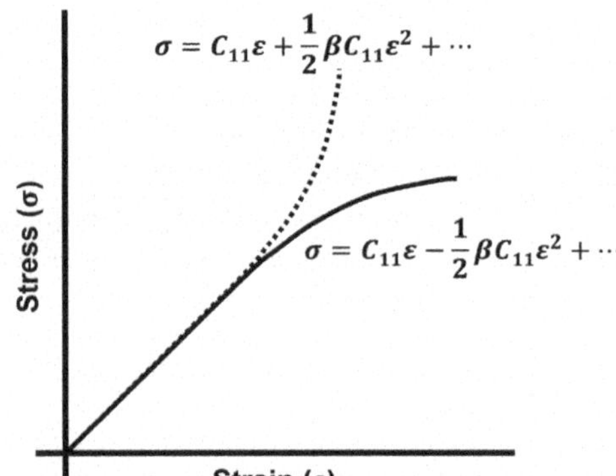

Hooke's equation [1–3].

$$\sigma = C_{11}\varepsilon \pm \frac{1}{2}\beta C_{11}\varepsilon^2 + \cdots \tag{2.1}$$

where σ, ε, C_{11}, and β represent stress, strain, elastic constant, and second–order nonlinear parameter, respectively.

The sign of the nonlinear term, i.e., the second term on the right-hand side of Eq. (2.1), depends on the curvature direction of the curve. In most metallic materials, the sign becomes negative; however, it is positive for soft materials such as rubber.

Harmonic generation can be easily explained by the relationship between the oscillating force and displacement in a single spring, as shown in Fig. 2.3. Consider the displacement y as an output when an oscillating force F is applied as an input. If the stiffness of the spring is perfectly linear, the waveform of the displacement output becomes exactly the same as that of the input, as shown in Fig. 2.4a. However, if the spring has nonlinear stiffness, the output displacement will be distorted because of the asymmetry between the positive and negative phases, as shown in Fig. 2.4b. This distortion generates harmonics in the spectral domain. As the input amplitude increases, the distortion in the output waveform increases, and the magnitudes of the harmonics increase.

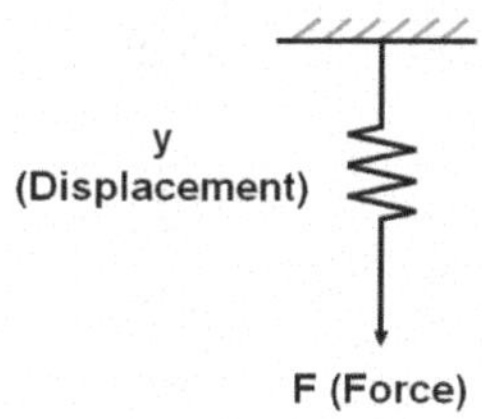

Fig. 2.3 Single spring model to represent the oscillating force and displacement

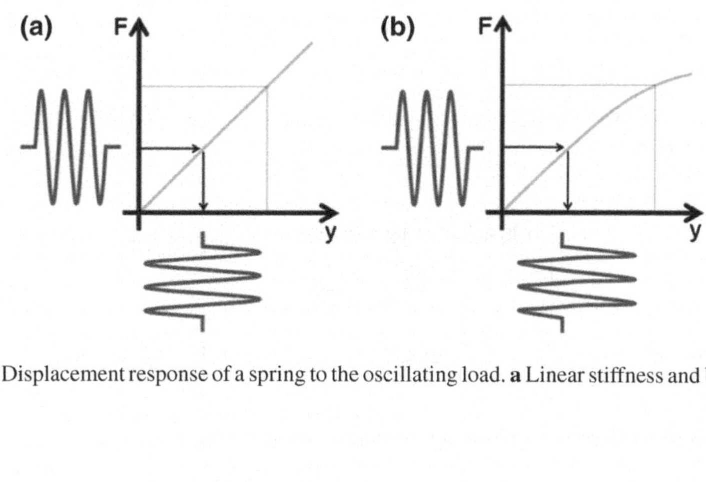

Fig. 2.4 Displacement response of a spring to the oscillating load. **a** Linear stiffness and **b** nonlinear stiffness

Similarly, when an ultrasonic sinusoidal wave of a given frequency is introduced in a nonlinear elastic material, the ultrasonic waveform propagating through the material responds to the nonlinear stress–strain relationship. Considering the stress (or sound pressure) and strain as force and displacement, respectively, the nonlinear response of ultrasonic wave is the same as that in the spring shown above. If the sound pressure amplitude of the ultrasonic wave is small, the received ultrasonic waveform remains similar to the incident waveform because the stress–strain relation is almost linear. However, if the sound pressure amplitude of the ultrasonic wave is large, the incident waveform may get distorted as it propagates. Note that this does not mean that the nonlinear phenomenon appears only when the sound pressure is high. Nonlinearity exists inherently irrespective of sound pressure, and whether it is measurable or not is the key.

Figure 2.5 shows the waveform and spectrum changes of the received ultrasonic wave signal in water when the incident wave amplitude is changed. The incident wave signal is 3 MHz tone burst. As we can see, when the incident wave amplitude is small the waveform remains sinusoidal and almost no harmonic frequency components are generated. However, when the incident amplitude becomes larger, the received signal waveform gradually becomes distorted and the harmonic component grows.

Furthermore, higher harmonic generation can be understood simply by considering the response of a nonlinear system. Let us consider a second–order nonlinear system (Fig. 2.6) with a sine wave as input. The output of this system will have a second–order higher harmonic wave component. When we consider third or higher

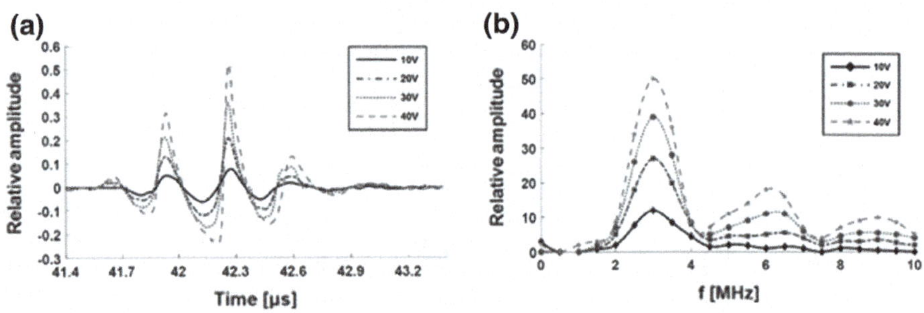

Fig. 2.5 **a** Waveform and **b** spectrum of the ultrasonic wave when the incident wave amplitude is increased

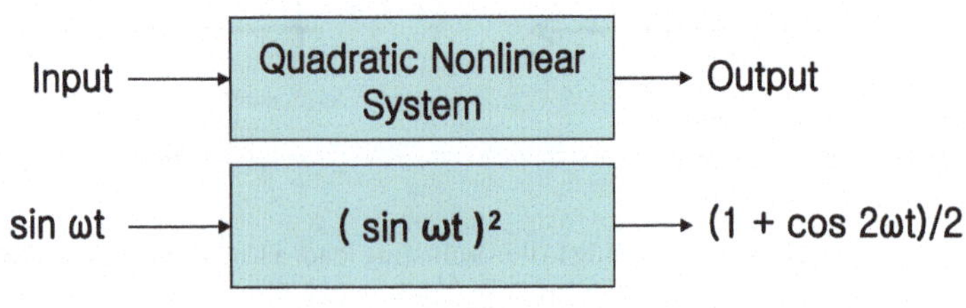

Fig. 2.6 Second harmonic generation in a quadratic nonlinear system

order nonlinearity, then the system output will have third or higher harmonics. However, usually the nonlinear effect of third or higher order nonlinearity is much smaller than that of second–order. Therefore, we often only consider up to the second–order nonlinearity.

2.2 Nonlinear Ultrasonic Parameters

The three-dimensional nonlinear stress–strain relationship, i.e., up to a second-order approximation for a solid material, can be described as follows [4–8]:

$$\sigma_I = M_{IJ}\varepsilon_J + \frac{1}{2}M_{IJK}\varepsilon_J\varepsilon_K \tag{2.2}$$

where σ_I is the stress tensor, ε_J and ε_K are the strain tensors, and M_{IJ} and M_{IJK} are the second- and third-order Huang coefficients related to the second- and third-order elastic constants C_{IJ} and C_{IJK} [4, 5], respectively.

To explain the generation of second-order harmonic waves, consider a one-dimensional longitudinal wave propagating in an isotropic material. Further, the nonlinear stress–strain relationship can be described as follows:

$$\sigma = M_{11}\varepsilon + \frac{1}{2}M_{111}\varepsilon^2 \tag{2.3}$$

where the second- and third-order Huang coefficients M_{11} and M_{111} can be expressed as the second- and third-order elastic constants as follows:

$$M_{11} = C_{11} \tag{2.4}$$

$$M_{111} = 3C_{11} + C_{111} \tag{2.5}$$

where C_{11} and C_{111} are the second- and third-order elastic constants, respectively. Further, from the equality of Eqs. (2.1) and (2.3), the nonlinear ultrasonic parameter β can be expressed as Huang coefficients and elastic constants as follows [7–9]:

$$\beta = -\frac{M_{111}}{M_{11}} = -\frac{3C_{11} + C_{111}}{C_{11}} \tag{2.6}$$

Meanwhile, the nonlinear wave equation for a longitudinal wave propagating in the direction of X can be expressed as follows [7–9]:

$$\rho\frac{\partial^2 u}{\partial t^2} = C_{11}\frac{\partial^2 u}{\partial t^2}\left(1 + \beta\frac{\partial u}{\partial X}\right) \tag{2.7}$$

where ρ is the density, u is the displacement in the X direction, and t is the time.

The perturbation theory is applied to obtain the solution of this equation. For this purpose, the perturbation solution of displacement u is assumed as follows:

$$u = u_1 + u_2 \tag{2.8}$$

where u_1 is the first-order perturbation solution, which is the displacement of the fundamental frequency component, and u_2 is the second-order perturbation solution, which is the displacement of the second-order harmonic frequency component. The resulting solutions can be expressed as follows:

$$u_1 = A_1 \cos(kx - \omega t) \tag{2.9}$$

$$u_2 = A_2 \cos 2(kx - \omega t) = \frac{1}{8}\beta k^2 A_1^2 x \cos 2(kx - \omega t) \tag{2.10}$$

where A_1 is the displacement amplitude of the fundamental frequency component; A_2 is the displacement amplitude of the second-order harmonic frequency component, which is dependent on the amplitude of A_1, wavenumber k, propagation distance x, and nonlinear ultrasonic parameter β. By rearranging Eq. (2.10), the nonlinear ultrasonic parameter can be expressed as follows:

$$\beta = \frac{8A_2}{A_1^2 k^2 x} \tag{2.11}$$

This parameter is generally referred to as the absolute nonlinear ultrasonic parameter. However, this absolute parameter is difficult to measure, because the displacement amplitude of the second-order harmonic component is very small (less than 1 nm). Thus, many researchers use the relative parameter, which is defined by the detected signal amplitude, regardless of the displacement. The relative parameter β' can be expressed as follows [10]:

$$\beta' = \frac{A_2'}{A_1'^2} \tag{2.12}$$

where A_1' and A_2' are the detected signal amplitudes of the fundamental and second-order frequency components, respectively. This parameter can be measured more easily than the absolute parameter, because it is determined using the voltage output signal detected with a conventional piezoelectric transducer. However, the relative parameter β' is only effective for comparing the degrees of degradation of the intact and damaged materials. Note that, when measuring the relative parameters to compare the damage, the measurement conditions, such as the equipment, signal pre- and post-processing, transducer, and couplant, as well as the wave frequency and the thickness of specimens, should be constant. This is to ensure that even if extra nonlinearity from the measurement system is incorporated, it is maintained at a constant value so that there is no problem in relative comparison. Nevertheless, the absolute parameter is still valuable for quantitative assessment of material degradation because the relative parameter cannot assess material degradation by single measurement. However, the absolute and relative parameters can be directly correlated under certain conditions, which are discussed in Sect. 2.3.3.

2.3 Measurement of Nonlinear Ultrasonic Parameters

2.3.1 Absolute Nonlinear Ultrasonic Parameter (Piezoelectric Method)

Only a few techniques have been designed to measure the absolute nonlinear ultrasonic parameter. Some common methods for measuring the absolute nonlinear ultrasonic parameter are capacitive detection [11], laser interferometry [12], and piezoelectric detection [13, 14]. In the capacitive detection method, the gap between the specimen and the capacitive detector is measured. The sensitivity of this method is very high; however, it requires that an optically flat and parallel specimen should be placed over the detector and the gap should be very small (only a few microns) [15]. The laser interferometric method has various advantages, such as noncontact measurement and broad-bandwidth frequency sensitivity. However, the interferometer uses the specular reflection of light; thus, the surface of the tested material should be like a mirror. Furthermore, it is practically difficult to use these methods for measurements of the absolute nonlinear ultrasonic parameter, because the displacement amplitude of the high-frequency harmonic component is very small.

The piezoelectric detection method, which was developed by G. E. Dace et al. [13, 14], measures the displacement amplitude using a contact piezoelectric transducer in a roundabout way—that is, through calibration, which converts a measured electrical current into acoustic displacement, as shown in Fig. 2.7. In this method, the signal-to-noise ratio is high, and the measurement results are less affected by the surface roughness of the material, compared with other methods [16]. Thus, piezoelectric methods have been widely adopted to measure the absolute nonlinear ultrasonic parameter despite the inconvenience of the additional processes required for calibration.

The piezoelectric method consists of two steps, i.e., nonlinear measurement and calibration, as shown in Figure 2.8. In the nonlinear measurement shown in Fig. 2.8a, the monochromatic electrical power $\boldsymbol{P}_{E,in}(\omega)$ is converted into the monochromatic acoustic power $\boldsymbol{P}_{A,in}(\omega)$ at the transmitting transducer. As the acoustic power propagates through the material, a harmonic component is generated. Then, the acoustic power, which includes the harmonic frequency component, $\boldsymbol{P}_{A,out}(\omega)$ is converted into the electrical power $\boldsymbol{P}_{E,out}(\omega)$ at the receiving transducer. This mechanism at the receiving transducer can be expressed using the transfer function $\boldsymbol{K}(\omega)$ as follows:

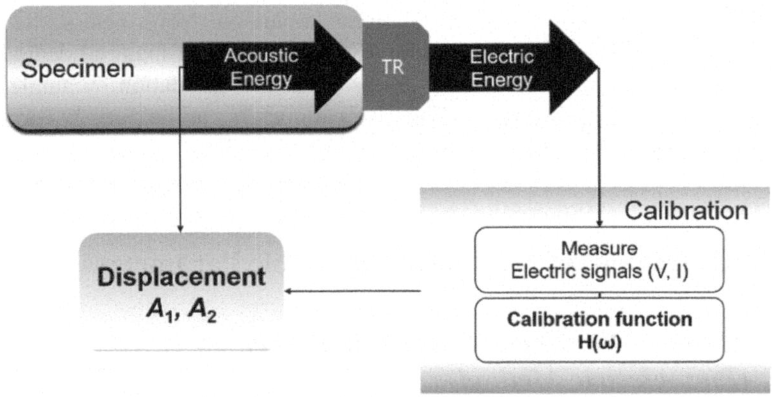

Fig. 2.7 Schematic of the piezoelectric method

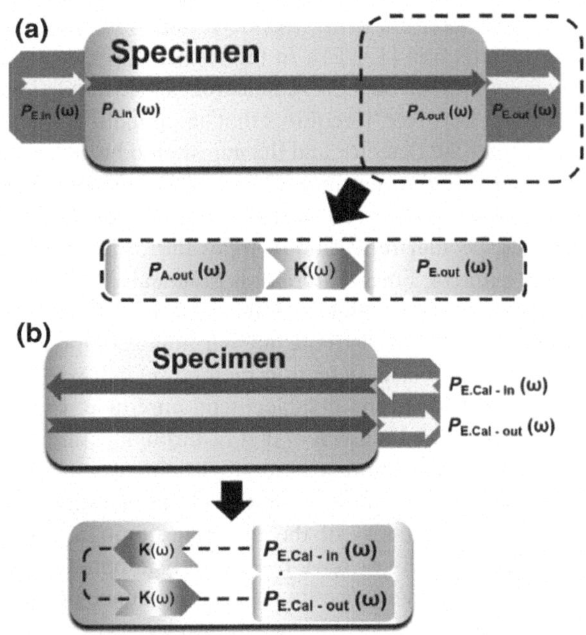

Fig. 2.8 Experimental design of the piezoelectric method for **a** nonlinear measurement and **b** calibration

$$P_{E,out}(\omega) = K(\omega)P_{A,out}(\omega) \tag{2.13}$$

where ω is the angular frequency. Then, Eq. (2.13) can be expressed in terms of the displacement amplitude $A(\omega)$ as follows:

$$\frac{1}{2}|I_{out}(\omega)|^2\mathrm{Re}(Z(\omega)) = K(\omega)\frac{1}{2}|\omega A(\omega)|^2\rho va \tag{2.14}$$

where $I_{out}(\omega)$ is the current measured at the receiving part, $Z(\omega)$ is the electrical impedance, ρ is the density of material, v is the particle velocity, and a is the area of the receiving transducer. The displacement amplitude $A(\omega)$ can be obtained by rearranging Eq. (2.14) as follows:

$$|A(\omega)| = \sqrt{\frac{\mathrm{Re}\left(Z_{(\omega)}\right)}{\omega^2 \rho v a K(\omega)}} |I_{out}(\omega)| \tag{2.15}$$

where the calibration function $H(\omega)$, which converts the detected electric energy into acoustic energy, can be defined as

$$|H(\omega)| = \sqrt{\frac{\mathrm{Re}\left(Z_{(\omega)}\right)}{\omega^2 \rho v a K(\omega)}} \tag{2.16}$$

Thus, Eq. (2.15) can be expressed as Eq. (2.17)

$$|A(\omega)| = |H(\omega)||I_{out}(\omega)| \tag{2.17}$$

In the calibration shown in Fig. 2.8b, a broadband pulse signal is reflected from the back-wall of the material and received. This mechanism can be expressed using the transfer function $\boldsymbol{K}(\boldsymbol{\omega})$ as follows:

$$P_{E,cal-in}(\omega) = K(\omega)^2 P_{E,cal-out}(\omega) \tag{2.18}$$

where $P_{\mathrm{E,cal-in}}(\omega)$ and $P_{E,cal-out}(\omega)$ are the generated and received signals in calibration. This mechanism can be formularized by two-port transducer modeling. Then, the transfer function $K(\omega)$ can be expressed using the input current and voltage, $I'_{in}(\omega)$ and $V'_{in}(\omega)$, and the output current and voltage, $I'_{out}(\omega)$ and $V'_{\mathrm{out}}(\omega)$, as follows:

$$K(\omega) = \sqrt{\frac{2\left|I'_{out}(\omega)\right|\mathrm{Re}\left(\frac{V'_{out}(\omega)}{I'_{out}(\omega)}\right)}{\left|I'_{in}(\omega)\left(\frac{V'_{out}(\omega)}{I'_{out}(\omega)}\right) + V'_{in}(\omega)\right|}} \tag{2.19}$$

Then, the calibration function $H(\omega)$ can be obtained by substituting Eq. (2.19) into (2.16).

$$H(\omega) = \sqrt{\frac{\left|I'_{in}(\omega)\left(\frac{V'_{out}(\omega)}{I'_{out}(\omega)}\right) + V'_{in}(\omega)\right|}{2\omega^2 \rho\, \upsilon\, a\left|I'_{out}(\omega)\right|}} \tag{2.20}$$

In Eqs. (2.19) and (2.20), the prime sign denotes the measured parameter during the calibration process.

To evaluate the accuracy of the absolute nonlinear ultrasonic measurement setup and procedure, measurements were performed on Al 6061-T6 whose nonlinear ultrasonic parameter has been reported in the literature [17, 18].

First, a calibration process was conducted, as shown Fig. 2.9a, to obtain the calibration function for converting the electrical output of the receiving transducer into the displacement amplitude. The calibration function was obtained using the broadband pulse-echo technique [13, 14]. Only the 10 MHz $LiNbO_3$ transducer was calibrated, which was used as the receiver in the nonlinear measurement process. A broadband pulse signal generated by a pulser–receiver (Panamertric, PR5072, USA) drives the transducer to input a broadband pulse wave into the specimen. The incident wave is reflected from the back-wall of the specimen and received by the same transducer. During the calibration process, the input current I'_{in} and voltage V'_{in} applied to the transducer are detected, and the output current I'_{out} and voltage V'_{out} of the back-wall echo signal received by the transducer are detected using current (Lecroy, CP030, USA) and voltage probes (Tektronix, P2220, USA), respectively. The signal was acquired by a digital oscilloscope (Lecroy, Wavesurfer452, USA). The acquired signal was Fourier-transformed by using fast Fourier transform (FFT) after applying the Hanning window to minimize the influence of the side lobes [19]. The calibration function $H(\omega)$ was calculated from the final spectrum, as shown in Fig. 2.10.

Further, nonlinear measurement was performed using the through-transmission technique, as shown in Fig. 2.9b. The transmitter was a 5 MHz $LiNbO_3$ transducer. An input tone-burst signal with 20 cycles was generated from a high-power gated amplifier (RITEC, RAM-5000 SNAP). It was important to keep the contact condition of the receiver undisturbed, i.e., exactly the same as that in calibration. Accordingly, we developed a pneumatic device, which applied a consistent pressure on the contact transducers and observed that the repeatability was significantly improved. A constant contact pressure of 0.4 MPa was applied to both transducers using the pneumatic device to minimize the variations in the contact condition during repeated measurements. The current output $I_{out}(\omega)$ was detected using the current probe used in calibration. The output current signals were obtained for increasing input voltage levels. Figure 2.11a shows a typical output current signal. More than 100 signals

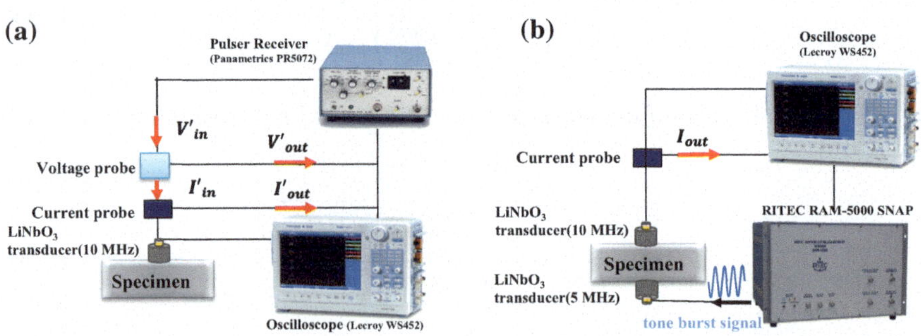

Fig. 2.9 Experimental setup for absolute measurements. **a** Calibration and **b** nonlinear measurement

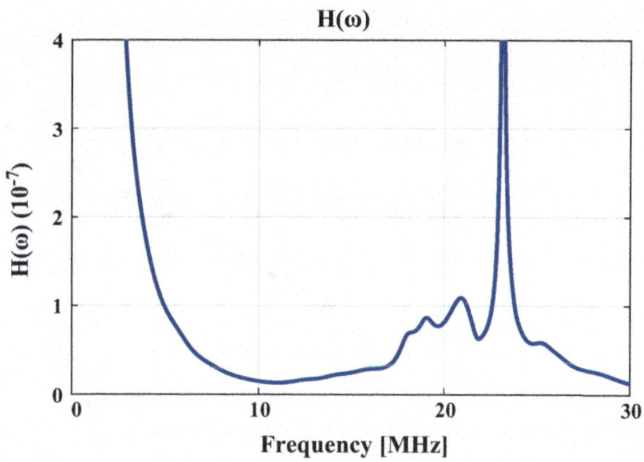

Fig. 2.10 Calibration function obtained by absolute measurement in Al 6061-T6

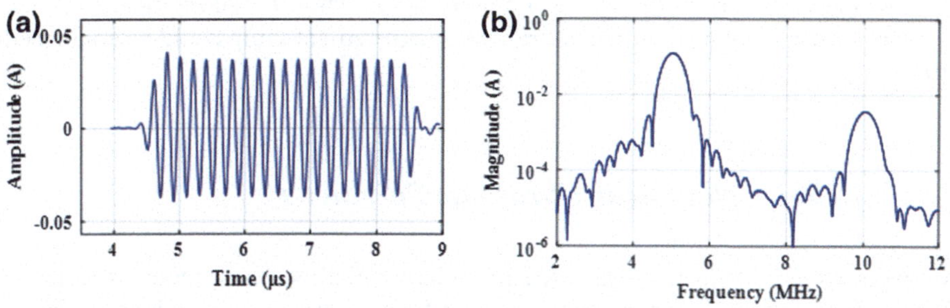

Fig. 2.11 **a** Received current output signal by the oscilloscope and **b** FFT result in Al 6061-T6

were averaged to improve the signal-to-noise ratio. The frequency spectrum of the averaged signal was calculated through FFT after applying the Hanning window, and the power loss caused by windowing was compensated by rescaling the magnitude of the frequency spectrum. The signal processing is described in detail in Sect. 2.4.4. Figure 2.11b shows the frequency spectrum of the signal shown in Fig. 2.11a, from which the magnitudes of the fundamental and second-order harmonic frequency components were determined. The particle displacement amplitude spectrum $A(\omega)$ was calculated from the transfer function $H(\omega)$ and current output spectrum $I_{out}(\omega)$ using Eq. (2.17).

Figure 2.12 shows a good linear relationship (with a correlation coefficient of 0.99) between the square of the fundamental frequency component A_1^2 and the amplitude of the second-order harmonic component A_2. Further, the absolute nonlinear ultrasonic parameter β is determined from the slope of the fitted line (A_1^2 versus A_2) with wavenumber k and thickness of specimen, x.

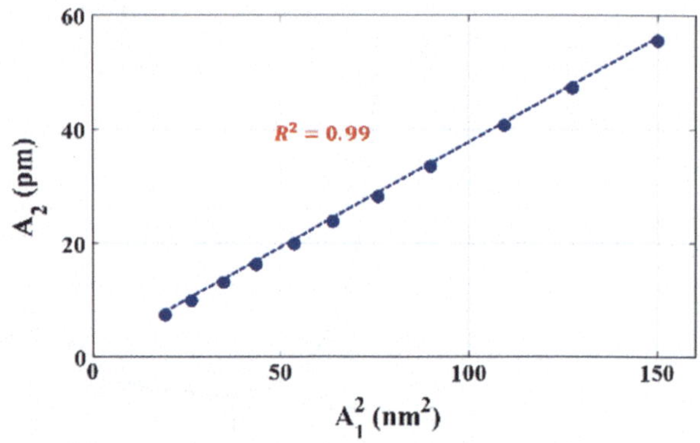

Fig. 2.12 Relationship between A_1^2 and A_2 in Al 6061-T6

The value of the measured absolute nonlinear ultrasonic parameter for Al 6061-T6 is 5.41. This value is in excellent agreement with the reference values (4.5–5.69) [17, 18], and the variations in the values are quite small. Thus, based on this result, the accuracy of the absolute nonlinear ultrasonic measurement setup could be confirmed [10].

2.3.2 *Relative Nonlinear Ultrasonic Parameter*

An experimental setup for measuring the relative nonlinear parameter is same as that employed in the nonlinear measurement for measuring the absolute nonlinear ultrasonic parameter. The only difference is that the relative nonlinear ultrasonic parameter is directly obtained from the transducer output voltage signal, as shown in Fig. 2.13.

Figure 2.14a shows the signal received from the Al 6061-T6 specimen. A stable 15-cycle tone-burst signal of the received signal is acquired and subsequently processed using FFT after applying the Hanning window [19]. The magnitudes of the fundamental $\left(A_1'\right)$ and second-order $\left(A_2'\right)$ harmonic frequency components are determined from the FFT results, as shown in Fig. 2.14b. To measure the relative nonlinear ultrasonic parameter β', the measurements are repeated by increasing the input power of the high-power pulse generator. Figure 2.15 shows the relationship between $A_1'^2$ and A_2'. A clear linear relationship is observed between the experimentally measured values, with the correlation coefficient as high as 0.99. Finally, the relative nonlinear ultrasonic parameter β' is determined from the slope of the fitted line of the linear relationship.

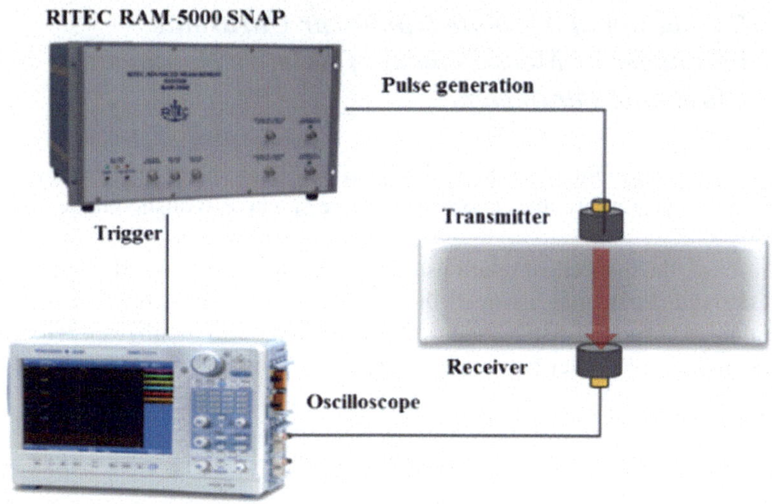

Fig. 2.13 Experimental setup to measure the relative nonlinear ultrasonic parameter

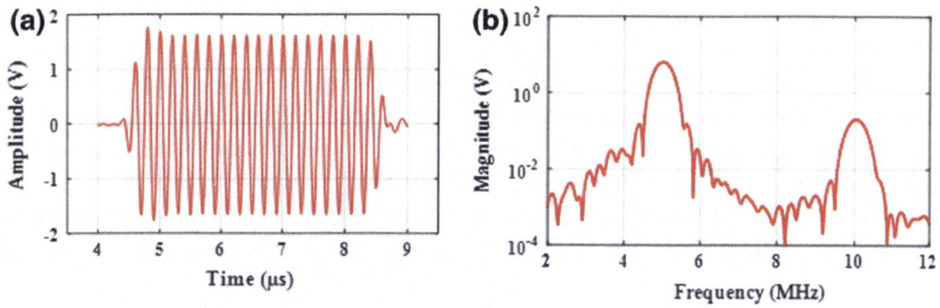

Fig. 2.14 **a** Received signal by the oscilloscope and **b** FFT result of the received signal

Fig. 2.15 Linear relationship between experimentally determined $A_1'^2$ and A_2'

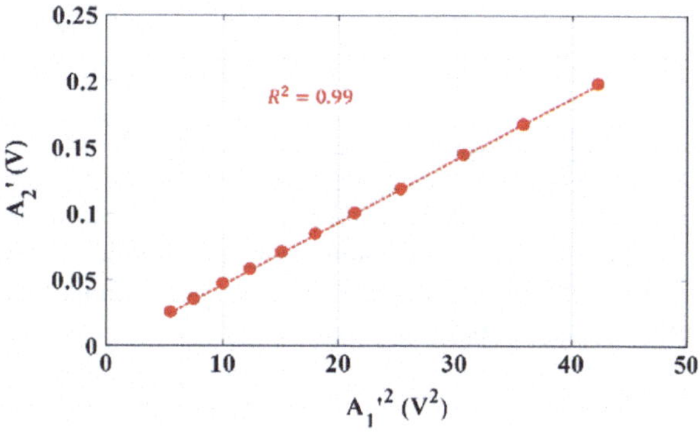

2.3.3 *Estimation of Absolute Nonlinear Ultrasonic Parameter by Measurement of Relative Nonlinear Ultrasonic Parameters*

As mentioned earlier, the absolute and relative nonlinear ultrasonic parameters are correlated; however, more direct correspondence can be established under specified conditions.

In this case, the theoretical relationship between the absolute and relative parameters is derived. First, it is assumed that the detected signal amplitude is linearly proportional to the actual displacement amplitude of the wave with respect to the fundamental and second-order harmonic components, which is represented as follows [20]:

$$
\begin{aligned}
A_1 &= A'_1 \cdot \alpha_1 \\
A_2 &= A'_2 \cdot \alpha_2
\end{aligned}
\tag{2.21}
$$

where α_i is the displacement proportionality coefficient of the detected signal amplitude ($i = 1$ for the fundamental frequency and $i = 2$ for the second-order harmonic frequency), which is related to the frequency-dependent sensitivity of the transducer (the efficiency of conversion of acoustic energy into electrical energy). Piezoelectric transducers (PZTs) produce a voltage output proportional to the displacement of an acoustic wave at a fixed frequency; thus, the above assumption is reasonable for PZTs. Note that the effects of attenuation and diffraction were neglected.

Next, let us consider the fractional ratio of the nonlinear ultrasonic parameters between the reference and test materials, assuming that the thickness of specimen x remains constant. From Eq. (2.21), the relationship between the fractional ratio of the absolute and relative parameters can be derived as follows [10, 20]:

$$
\frac{\beta}{\beta_0} = \frac{\frac{1}{k} \cdot \frac{A_2}{A_1^2}}{\frac{1}{k_0} \cdot \frac{A_{2,0}}{A_{1,0}^2}} = \frac{\frac{1}{k} \cdot \frac{\alpha_2 \cdot A'_2}{\alpha_1^2 \cdot A_1'^2}}{\frac{1}{k_0} \cdot \frac{\alpha_{2,0} \cdot A'_{2,0}}{\alpha_{1,0}^2 \cdot A_{1,0}'^2}} = k' \cdot \alpha' \cdot \frac{\frac{A'_2}{A_1'^2}}{\frac{A'_{2,0}}{A_{1,0}'^2}} = k' \cdot \alpha' \cdot \frac{\beta'}{\beta'_0}
\tag{2.22}
$$

where the subscript "0" indicates the value for the reference material, $k' = \frac{k_0}{k}$, and $\alpha' = \frac{\alpha_2/\alpha_1^2}{\alpha_{2,0}/\alpha_{1,0}^2} = \left(\frac{\alpha_2}{\alpha_{2,0}}\right)\left(\frac{\alpha_{1,0}}{\alpha_1}\right)^2$.

Equation (2.22) represents how the fractional ratio of the absolute nonlinear parameter between two materials is related to that of the relative nonlinear parameter. If the absolute nonlinear parameter of the reference material is known, this relationship can be used to estimate the absolute nonlinear parameter of a test material from the relative measurements of both materials. The estimation process is described in two cases. One is a general case wherein two materials are completely different, and the other is a special case wherein two materials are similar.

CASE (I): Different materials

If two materials are different, the wavenumber is different even if the frequency is same, and the displacement proportionality coefficient is also different because of the difference in acoustic impedance even if the measurement conditions are kept consistent. Therefore, to obtain the fractional ratio of the absolute parameters by measuring the relative parameters, k' and α' should be compensated using Eq. (2.22). The transmission of the acoustic energy from the surface of the material to the transducer may affect the amplitude of the detected signal, which is related to the difference in the acoustic impedances of the material and the transducer. In such cases, the compensation factor α' includes the effect of difference in acoustic impedances of the materials.

From this, the absolute parameter of the test material can be estimated as follows:

$$\beta = k' \cdot \alpha' \frac{\beta'}{\beta_0'} \times \beta_0 \tag{2.23}$$

Therefore, once the absolute parameter of the reference material is known, the absolute parameter of the test material can be obtained by measuring the fractional ratio of the relative parameters after compensating the wavenumber and detection sensitivity.

CASE (II): Similar materials

If two materials are the same but one is intact and the other is damaged, we can consider those two materials are similar. In this case, the wavenumbers and detection sensitivities will be similar also. Thus, the values of k' and α' can be considered as unity, and Eq. (2.22) can be simplified as follows [10]:

$$\frac{\beta}{\beta_0} = \frac{\beta'}{\beta_0'} \tag{2.24}$$

In this case, the fractional ratio of the relative parameters of the reference and test materials is identical to that of the absolute parameters, and the quantitative ratio of the absolute nonlinear ultrasonic parameters can be obtained by measuring the relative parameters. Then, the absolute parameter of the damaged material can be estimated as follows:

$$\beta = \frac{\beta'}{\beta_0'} \times \beta_0 \tag{2.25}$$

Therefore, once the absolute parameter of the material in intact condition is known, the absolute parameter of the material after damage can be obtained by measuring the fractional ratio of the relative parameters. To apply this method to the health monitoring of a material in operation, one must preserve an intact specimen of the same material and measure the relative parameters of both the material in operation and the intact specimen. As mentioned earlier, when measuring the ratio of the

Table 2.1 Results of absolute and relative measurements in fused silica and Al 6061-T6

	Fused silica	Al 6061-T6	Ratio ($\frac{Al6061-T6}{fused\ silica}$)
Absolute parameter (β)	10.06 ± 0.46	5.41 ± 0.37	0.54
Relative parameter (β')	0.01061 ± 0.000345	0.00509 ± 0.000308	0.48 (0.55 after compensation)
Literature data of β	8.67–14.0 [9, 14, 17, 21, 22]	4.5–5.69 [17, 18]	0.32–0.66

relative parameters, it is important to keep the measurement conditions consistent. Thus, it is preferred that the relative parameter of the intact specimen is measured each time the relative parameter of the material in operation is measured.

The followings are experimental verification of the above theory. Note that the absolute parameter estimated by Eqs. (2.23) and (2.25) will be expressed as β_e .

CASE I: Different Materials

To experimentally verify the relationship between the absolute and relative nonlinear ultrasonic parameters in different materials, the fractional ratio of the absolute parameter of Al 6061-T6 to that in fused silica was compared with the compensated ratio of the relative parameter of Al 6061-T6 to that in fused silica. Then, the feasibility of the proposed method for estimating the absolute parameter using relative parameters was verified.

The experimental results are listed in Table 2.1, which compares the absolute parameters with the relative parameters. The ratio of the absolute parameter of Al 6061-T6 to that of fused silica is 0.54, which is within the range of the literature data. Moreover, the ratio of the relative parameters is 0.48 before compensation, which is slightly different from that of the absolute parameters.

Further, the compensation factors k' and α', shown in Eq. (2.22), were considered. To obtain the displacement-proportionality coefficient α for fused silica and Al 6061-T6, the transmitting transducer was fixed, and the displacement amplitude at the receiving side was detected using a laser detector [23]. This amplitude was compared with the voltage amplitude detected using the piezoelectric contact transducer at a constant input power. This process was repeated after increasing the input power. The results are shown in Fig. 2.16, wherein the slopes correspond to the displacement proportionality coefficients of fused silica and Al 6061-T6. In this figure, the horizontal axis is the voltage amplitude detected using the piezoelectric transducer, and the vertical axis is the displacement amplitude detected using the laser detector. The values along both axes were normalized by their respective initial values. The displacement proportionality coefficient α, is almost the same in fused silica and Al 6061-T6, regardless of the measuring frequency; Therefore, the compensation factor α' in Eq. (2.22) can be ignored.

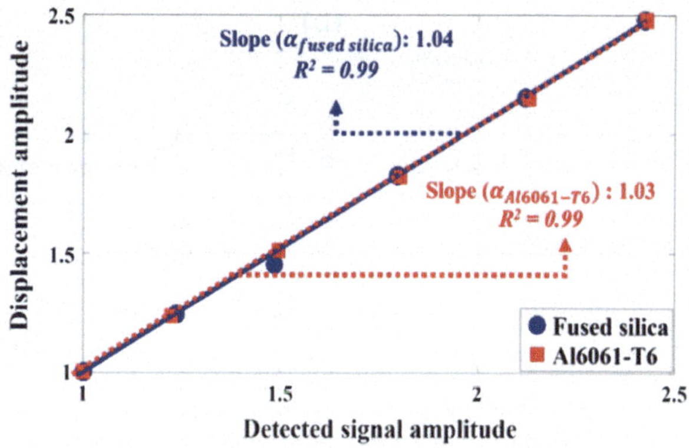

Fig. 2.16 Comparison of the detected signal and displacement amplitudes for estimating the compensation factor α'

Alternatively, the above-described result can be explained by rewriting the relationship between the displacement and detected current signal amplitudes as follows:

$$|A(\omega)| = |H(\omega)||I_{\text{out}}(\omega)| = \frac{|H(\omega)|}{|Z(\omega)|}|V_{\text{out}}(\omega)| = \frac{|H(\omega)|}{|Z(\omega)|}\left|A'(\omega)\right| \quad (2.26)$$

where $|V_{\text{out}}(\omega)|$ is the output voltage signal for relative measurement, which is detected from the signal amplitude $\left|A'(\omega)\right|$. That is, the displacement proportionality coefficient α corresponds to the ratio of the calibration function $|H(\omega)|$ to the electrical impedance $|Z(\omega)|$.

The experimental setup for absolute measurement was the same as that for relative measurement, where the only difference was that the current was measured instead of voltage to obtain the absolute nonlinear ultrasonic parameter. The electrical impedance (about 50 Ω) remained constant during the measurements at the measurement frequencies (5 and 10 MHz), as shown in Fig. 2.17. Furthermore, the calibration function for fused silica and Al 6061-T6 was almost the same, as shown in the previous result. Therefore, the displacement-proportionality coefficient α was similar for both materials.

In contrast, the wavenumbers were different, because the longitudinal velocities for fused silica and Al 6061-T6 are quite different. Therefore, the compensation factor k' cannot be ignored. The value of factor k' for these materials was 1.15. After multiplying this factor by the ratio of the relative parameters, as shown in Eq. (2.23), the ratio of the relative parameters could be obtained, which was 0.55. This is similar to the fractional ratio of the absolute parameters, as presented in Table 2.1. These results satisfy Eq. (2.22).

Finally, according to Eq. (2.23), the absolute parameter for Al 6061-T6 is estimated from the fractional ratio of the relative parameters measured in Al 6061-T6 and fused silica $\left(\beta'/\beta'_0\right)$ and the absolute parameter of fused silica (β_0), which is a

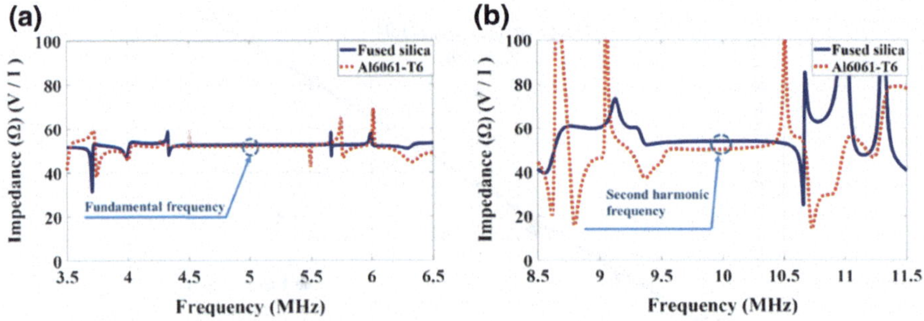

Fig. 2.17 Output electrical impedance measured in the nonlinear measurement process for absolute measurement at interesting frequencies: **a** 3.5–6.5 MHz and **b** 8.5–11.5 MHz

reference material after compensating for the wavenumber and detection sensitivity. Table 2.2 presents the comparison of the measured and estimated absolute nonlinear ultrasonic parameters for Al 6061-T6. The values of both parameters are similar with a difference of only 2.3%.

These experimental results demonstrate that if the absolute nonlinear ultrasonic parameter of the reference specimen β_0 is known, then the absolute parameter β for the test material can be obtained from the fractional ratio of the relative parameters $\left(\beta'/\beta'_0\right)$ after compensating for the wavenumber and detection sensitivity (displacement-proportionality coefficient) depending on the material.

CASE II: Similar materials (Degradation of a material)

To experimentally verify the relationship between the absolute and relative nonlinear ultrasonic parameters in degradation of a material, the Al 6061-T6 alloy specimens were prepared by heat treatment at a constant temperature of 220 °C for different aging times (0, 60, 120, 600, and 6000 min). The absolute and relative parameters were measured for the Al 6061-T6 specimens, and the variation of the absolute parameter was compared with that of the relative parameter as a function of aging time. Then, the values of the absolute parameter estimated using the proposed method were compared with those obtained through direct measurements.

Figure 2.18 shows the measured absolute parameter β and relative parameter β' as a function of the aging time. Initially, both the parameters increase with the aging time (the absolute parameter by 4.8% and the relative parameter by 2.7%) to reach the first maximum peak at 60 min, and then decrease (the absolute parameter by 9.2% and the relative parameter by 6.2%) to reach the first minimum peak at 120 min. With further aging, both parameters again increase and decrease during the period

Table 2.2 Comparison of the measured absolute parameter β and estimated absolute parameter β_e in Al 6061-T6

	Measured	Estimated	Literatures
Absolute β	5.41 ± 0.37	5.53 ± 0.47	4.5–5.69 [17, 18]

spanning 120–60,000 min. The fluctuations in both parameters as a function of the aging time can be explained by the precipitation sequences in Sect. 2.5 [24–32].

These experimental results show that the fluctuations in the two parameters as a function of aging time are very similar. In this case, the wavenumber and detection sensitivity were not compensated because they are scarcely changed as a function of aging time.

Further, the absolute parameter is estimated from the relative parameter using Eq. (2.25). Figure 2.19 shows the comparison between the measured absolute parameter β and the estimated absolute parameter β_e as a function of aging time. The fluctuations in the estimated absolute parameter β_e and the measured absolute parameter β are similar. Note that the deviations in the results obtained from repeated measurements are very small, with an overall deviation rate of 1.5%, and the ranges of deviations in β and β_e overlap each other throughout the aging time.

These experimental results demonstrate that if the absolute nonlinear ultrasonic parameter of the intact material β_0 is already known, the absolute parameter β for the degraded material can be obtained from the fractional ratio of the relative parameters $\left(\beta'/\beta'_0\right)$. Consequently, the absolute measurement can be replaced by relative measurement to evaluate the material degradation.

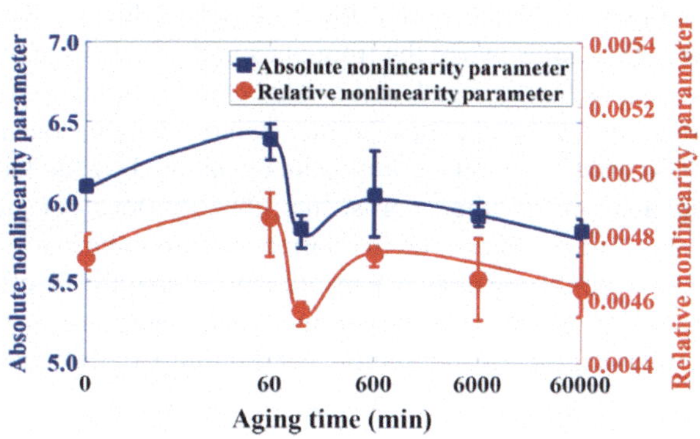

Fig. 2.18 Experimental results for the absolute parameter $\boldsymbol{\beta}$ and the relative parameter $\boldsymbol{\beta}'$ as a function of aging time

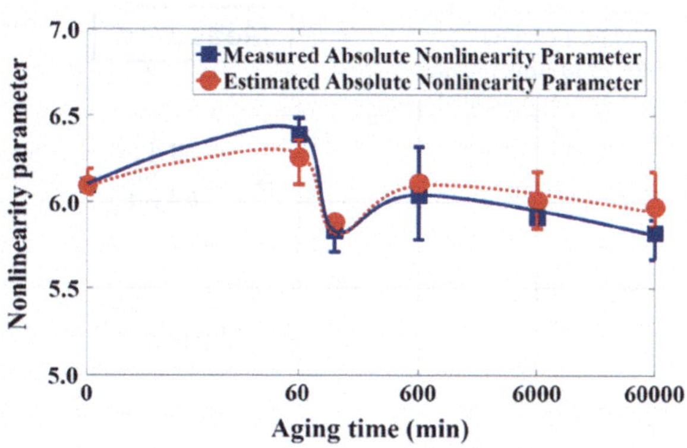

Fig. 2.19 Comparison of the measured absolute parameter $\boldsymbol{\beta}$ and the estimated absolute parameter $\boldsymbol{\beta}_e$ as a function of aging time

2.3.4 Measurement Reliability Test

In actual measurement, the following points should be considered when determining nonlinear ultrasonic parameters.

The measured A_2 (or A_2') contains the second–order harmonic component generated by the nonlinearity of the material as well as the second–order harmonic component generated within the instrument. Unless the latter is removed, the measured A_1 and A_2 do not give the true nonlinear ultrasonic parameter of the material. Therefore, in absolute measurement, the second–order harmonic component generated within the instrument should be eliminated. In contrast, in relative measurement this problem does not occur. This issue is described in the next section (Sect. 2.4).

Even if the harmonic components generated within the instrument are negligible or have been removed in an appropriate manner, we need to make sure that the nonlinear parameters have been measured successfully. This procedure is essential for both absolute and relative measurements. From the definition of nonlinear parameter, it can be seen that there are three input variables, i.e., wave number, propagation distance, and input power. This means that β should be constant in a material even if these three input variables are different. However, the frequency is fixed in most measurements; thus, there are two methods of varying the input power and propagation distance.

Figure 2.20 shows that there must be a linear relationship between A_1 squared and A_2. That is, when the frequency and propagation distance are fixed, and only the input power is changed, A_2 should be proportional to A_1 squared. This relationship is the same for relative measurements. Note that this linear relationship also holds when there exists harmonic components generated by the instrument or when there exists attenuation. In other words, this linearity does not guarantee the reliability of the measurement. However, in relative measurements, at least the nonlinearity generated by the measuring device does not matter, so only the effects of attenuation need to be considered. If the relative nonlinear parameters are compared between materials with similar degrees of attenuation, the reliability of the measurement can be confirmed by

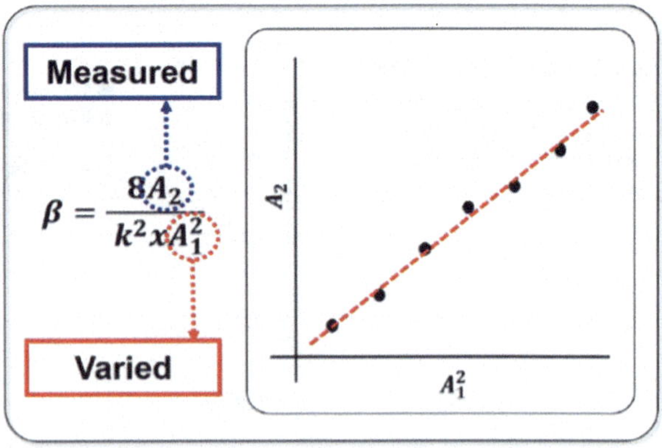

Fig. 2.20 Dependence of second–order harmonic amplitude A_2 on the amplitude of fundamental frequency component A_1

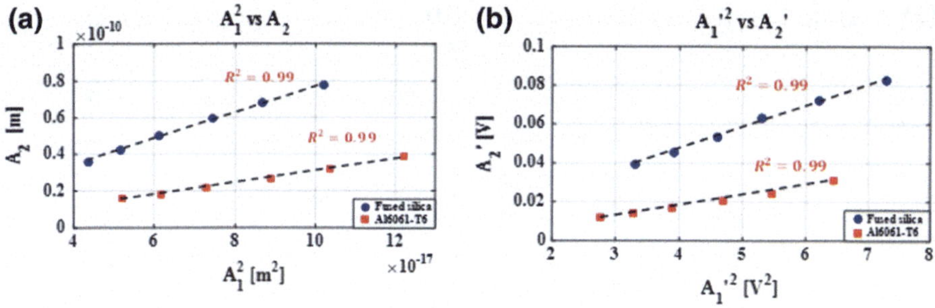

Fig. 2.21 Relationship between A_1^2 and A_2 in fused silica (circle) and Al 6061-T6 (square). **a** Absolute measurement and **b** relative measurement

confirming this linear relationship. However, additional confirmations are necessary if we do not know the attenuation of the two materials to compare.

Figure 2.21 shows the experimental data of fused silica and Al 6061-T6. Figure 2.21a shows the relationship obtained during absolute measurement, wherein the unit of A_1 and A_2 is the absolute displacement amplitude (m). Figure 2.21b shows the relationship obtained during relative measurement, wherein the unit of A_1' and A_2' is the voltage amplitude (V). The results show a clearly linear relationship between the square of the fundamental frequency component A_1^2 (or $A_1'^2$) and the amplitude of the second-order harmonic component A_2 (or A_2') with a high correlation coefficient of 0.99. Furthermore, the absolute parameter β is determined from the slope of the fitted line obtained in absolute measurement, with wavenumber k and thickness of specimen x, using Eq. (2.11). The relative parameter β' is determined from the slope of the fitted line obtained from relative measurement using Eq. (2.12).

Figure 2.22 shows that there must be a linear relationship between A_2 and the propagation distance. That is, when the frequency and the input power are fixed and only the distance is changed, A_2 should be proportional to the propagated distance. This relationship is the same for relative measurements.

Fig. 2.22 Dependence of second–order harmonic amplitude A_2 on the propagation distance

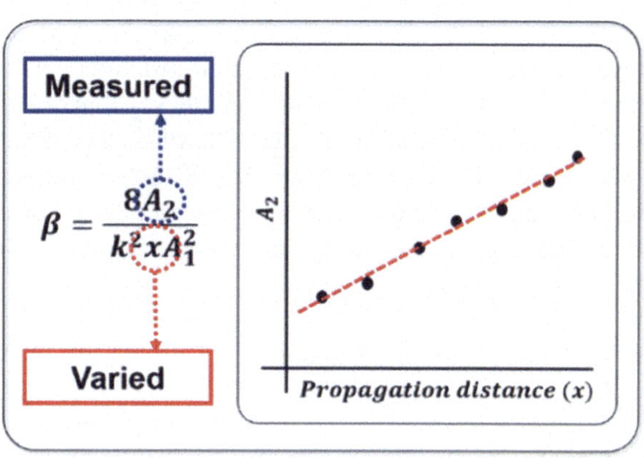

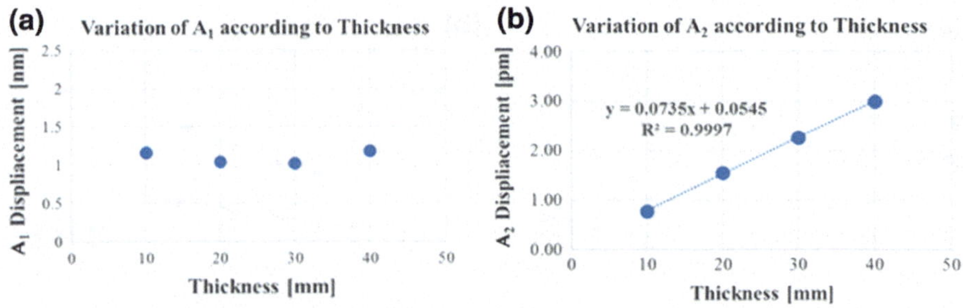

Fig. 2.23 Variation in the amplitude of **a** fundamental frequency component ($\boldsymbol{A}_1$) and **b** second–order harmonic component ($\boldsymbol{A}_2$) according to the difference of thickness of fused silica

However, in actual measurements, the amplitude of the fundamental wave may change by diffraction (because of the finite size of the transducer) and material attenuation. In this case, the second–order harmonic component does not grow proportionally to distance and the straightforward use of Eq. (2.11) can yield an incorrect nonlinear ultrasonic parameter. For example, the measurements of the same material but with different thicknesses can produce different results. If we can ignore diffraction and attenuation, or if two materials of similar material are compared; for example, when comparing before and after degradation of a particular material; then this will not be a concern. However, if we cannot ignore diffraction and attenuation, then appropriate correction is required. Research on this problem is being conducted [33].

Figure 2.23 shows the experimental results for fused silica of different thickness. Four different thicknesses were tested, i.e., 10, 20, 30, and 40 mm. Figure 2.23a shows the amplitude of the fundamental frequency component (A_1) and Fig. 2.23b shows the amplitude of the second–order harmonic frequency component (A_2). These are the results of absolute measurements. In the experiments, a 7 MHz low-pass filter was placed between the high-power pulser and the transmitting transducer to suppress the second–order harmonic component generated within the pulser. Furthermore, experiments were conducted using $LiNbO_3$ transducers because $LiNbO_3$ has small nonlinearity. Therefore, we are sure that the nonlinearity from the instruments is insignificant. We can see that A_1 is almost constant and A_2 grows proportionally to the thickness. From these results, we can say that our experimental setup satisfied the perturbation approximation and it is hardly affected by the diffraction and attenuation up to 40 mm. If we compensate the diffraction and attenuation effects, then we can measure the nonlinear parameter more accurately. In addition, it is worth noting that the y-intercept of the fitting line shown in Fig. 2.23b, which is corresponding to the system nonlinearity generated by the instruments, is negligible compared with the material nonlinearity. This reconfirms that nonlinearity from the instruments is insignificant in this experiment.

2.4 Factors Affecting Measurement Reliability

Let us recall the measurement procedure of the nonlinear ultrasonic parameter. An electrical signal generated by a high-power electric signal generator is sent to a transmitting transducer, ultrasonic waves excited by the transmitter pass through a couplant layer, and then these waves are incident on a specimen. While the ultrasonic waves propagate through the specimen, the ultrasonic waveform is distorted by nonlinear elasticity. Further, the distorted ultrasonic waves pass through another couplant layer and are converted into a voltage signal from ultrasonic energy at a receiving transducer. The received signal is generally amplified using a pre-amplifier and then is converted into a digital signal through an analog-to-digital converter. Finally, spectral analysis is performed on the digitally converted signal to analyze the amplitudes of the fundamental and second-order harmonic components to calculate the nonlinear ultrasonic parameter. Meanwhile, the digital voltage signal in the last step can include not only nonlinear elasticity to be measured but also system nonlinearity that can occur in various measurement system elements. This system nonlinearity can result in errors in the measurement of the nonlinear ultrasonic parameter. To improve measurement reliability for the nonlinear ultrasonic parameter, it is necessary to analyze the system nonlinearity factors and to suppress them as much as possible.

The system components that can affect the measurement of a nonlinear ultrasonic parameter are systematically shown in Fig. 2.24, where $D(\omega)$, $T(\omega)$, $R(\omega)$, $C_i(\omega)$, $G(\omega)$, and $A_V(\omega)$ represent the amplitude composition of an electrical signal sent to a transmitting transducer as a function of ω, the response function of a transmitting transducer, the response function of a receiving transducer, the couplant response function at position i, the response function of an amplifier, and the amplitude of a detected voltage signal at ω, respectively. First, the presence of the uncertain initial harmonic components in the electrical signal generated by a high-power electric signal generator can significantly affect the measurement reliability for the nonlinear ultrasonic parameter, which should be suppressed. Assuming that there are no initial harmonic components in the ultrasonic waves incident on the specimen, the amplitudes of the fundamental and second-order harmonic components to be measured can be expressed by [34, 35]

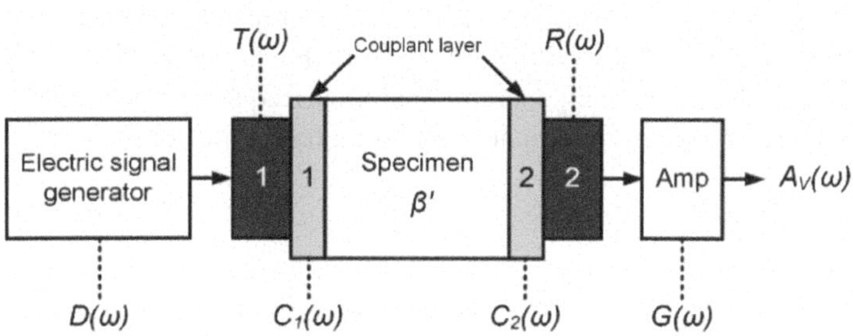

Fig. 2.24 System components that can affect the measurement of the nonlinear ultrasonic parameter

$$A_V(\omega_1) = D(\omega_1)T(\omega_1)C_1(\omega_1)C_2(\omega_1)R(\omega_1)G(\omega_1) \quad (2.27)$$

$$A_V(\omega_2) = \beta'(D(\omega_1)T(\omega_1)C_1(\omega_1))^2 C_2(\omega_2)R(\omega_2)G(\omega_2), \quad (2.28)$$

where ω_1 and ω_2 are the fundamental and second-order harmonic angular frequencies, respectively. Further, the relative nonlinear ultrasonic parameter to be measured, β'_m, is expressed as

$$\beta'_m = \frac{A_V(\omega_2)}{(A_V(\omega_1))^2} = \beta' \frac{C_2(\omega_2)R(\omega_2)G(\omega_2)}{(C_2(\omega_1)R(\omega_1)G(\omega_1))^2}. \quad (2.29)$$

Meanwhile, it is relatively easy to keep the response characteristics of the receiver R(ω) and the amplifier G(ω) consistent during a series of measurements of the nonlinear ultrasonic parameter, by not replacing the receiver and the amplifier, and by keeping their performance constant during the measurements. Therefore, the major factor that affects the measurement reliability would then be the initial harmonics included in D(ω); the response characteristic of the couplant layer C(ω), which depends on the contact pressure of the transducer to the specimen; and the type and thickness of the couplant. The processes of analog-to-digital conversion and digital signal processing may also affect the measurement reliability; however, it is not shown in Fig. 2.24. In the following sections, we describe the major factors that affect measurement reliability and suggest effective solutions to suppress them.

2.4.1 Uncertain Initial Harmonics

Ideally, in the measurement of a nonlinear ultrasonic parameter, it is desired that an electrical signal with a single fundamental frequency is sent to a transmitting transducer, and the ultrasonic waves with a single frequency are incident on a specimen. Therefore, the uncertain initial harmonic components that can be contained in the incident electrical signal should be suppressed before being sent to the transmitter. An effective solution for the suppression is to filter out the initial harmonic components using a low-pass filter with a cutoff frequency below the second-order harmonic frequency. As an example, the electrical signals before and after filtering with a 7 MHz low-pass filter on the 5 MHz tone-burst electrical signal generated by a high-power electric signal generator, and their frequency spectra, are shown in Fig. 2.25. The low-pass filter eliminates the initial harmonic component of 10 MHz.

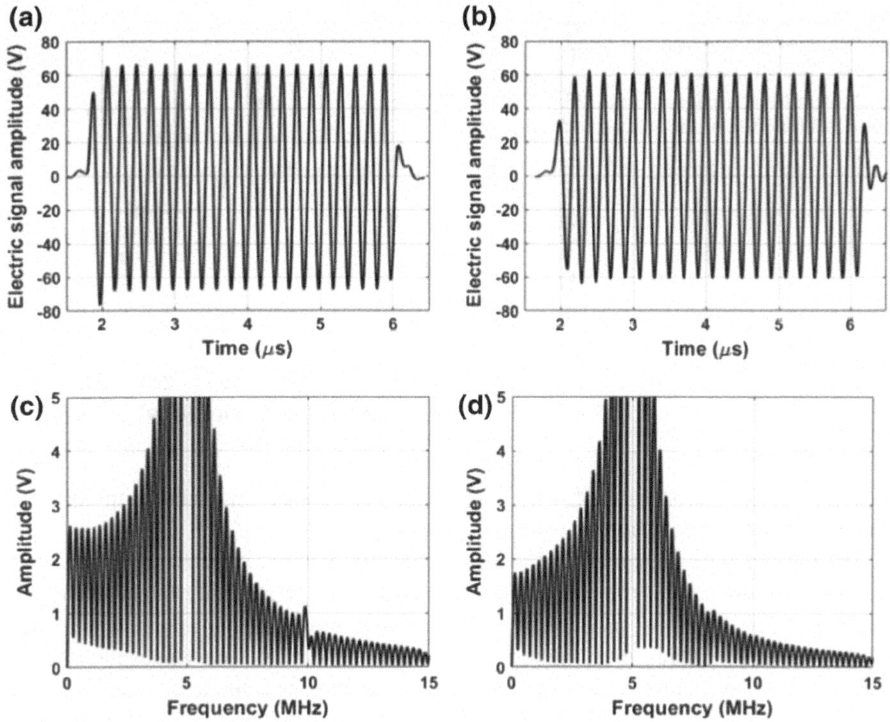

Fig. 2.25 Electrical signals **a** before and **b** after filtering with a 7 MHz low-pass filter on the 5 MHz tone-burst electrical signal generated by a high-power electric signal generator, **c** and **d** frequency spectra of **a** and **b**, respectively

2.4.2 Couplant Layer

In contact ultrasonic measurements, a couplant layer is normally used to transfer ultrasonic waves between a transducer and a specimen, and the influence of the couplant layer on nonlinear ultrasonic measurements should be noted. The response characteristic of the layer depends on various experimental conditions including the contact pressure of the transducer to the specimen, and the type and thickness of the couplant, where the most dominant factor is the contact pressure.

Let us consider the influence of the contact pressure of a transducer on a specimen. Figure 2.26 shows the peak-to-peak amplitudes of the received voltage signals measured in the through-transmission mode as a function of the contact pressure [36], where the specimen was aluminum and general 5 and 10 MHz PZTs were used with a gel-type couplant. The voltage values exhibit significant amplitude fluctuations and deviations at low contact pressures of 160 kPa or below. This is because the condition of the couplant layer was unstable at low pressures, even when a consistent pressure was applied to the transducer. As the contact pressure increases over 180 kPa, the

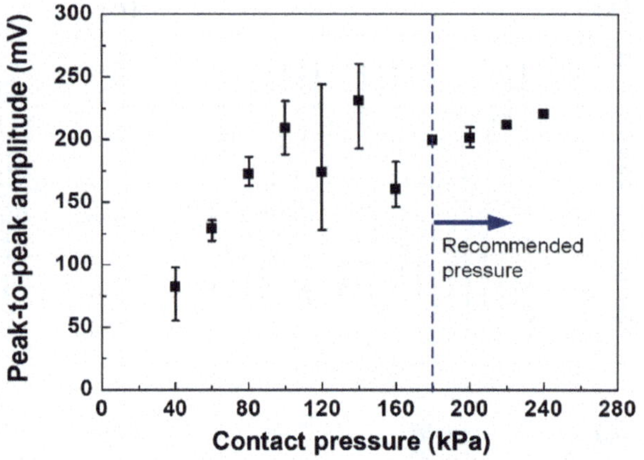

Fig. 2.26 Peak-to-peak amplitudes of the received voltage signals measured in the through-transmission mode as a function of the contact pressure

measured values are stabilized, which means that the condition of the couplant layer became almost constant at these contact pressures.

The tendency of the measured values to become stable above a specific contact pressure also appears in the measurement results of the relative nonlinear ultrasonic parameter, as shown in Fig. 2.27 [36]. The relative nonlinear parameters measured at low contact pressures are relatively large because the measured values include not only the nonlinear elasticity component but also system nonlinearity caused by the unstable couplant layer. Meanwhile, the contact pressures above 180 kPa provide consistent measurements of the nonlinear ultrasonic parameter. The contact pressure above this threshold value is recommended for general nonlinear ultrasonic measurements. However, the threshold pressure may vary slightly depending on various experimental conditions, such as the transducer size, couplant type, and specimen. Thus, it is recommended to analyze the influence of the contact pressure in advance from the preceding experiments and to select the appropriate contact pressure suitable for the experimental system before performing a series of nonlinear ultrasonic

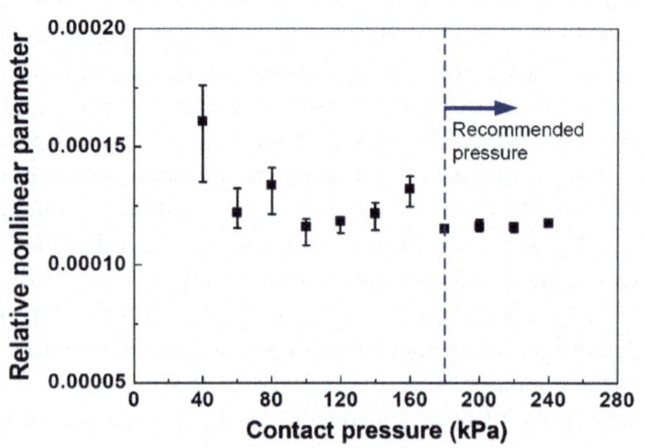

Fig. 2.27 Relative nonlinear parameters as a function of the contact pressure

measurements. Note that excessive pressure may cause specimen deformation or fatal damage to the transducer.

To maintain consistent contact conditions between the transducer and the specimen, it is recommended to use a pneumatic control system that can apply a constant pressure on the transducer. Figure 2.28 shows the experimental system for nonlinear ultrasonic measurements using a pneumatic control system. Together with the pneumatic system, the use of a holder fixture to secure the transducer makes it possible to keep the measurement position constant and to maintain the alignment of the transmitter and the receiver consistently, thereby enabling reliable nonlinear ultrasonic measurements. In some cases, a holder fixture with flexible suspension, as shown in Fig. 2.29, may be effective to align the transducer perpendicular to the specimen surface.

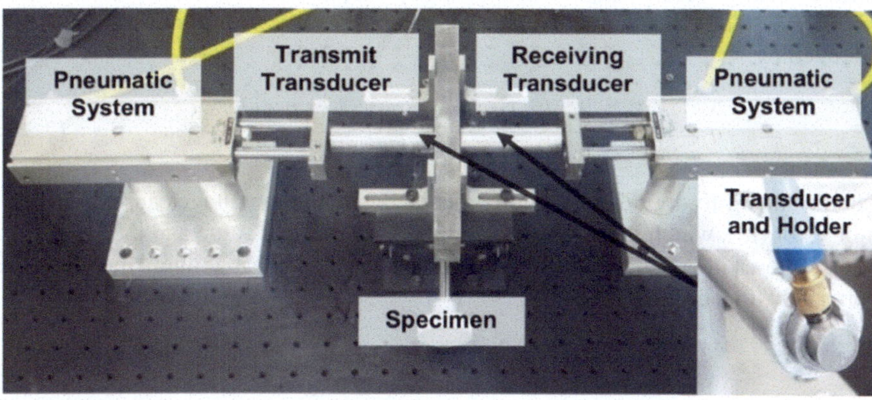

Fig. 2.28 Experimental system for nonlinear ultrasonic measurements using the pneumatic control system [37]

Fig. 2.29 Holder fixture with flexible suspension to align the transducer perpendicular to the specimen surface [34]

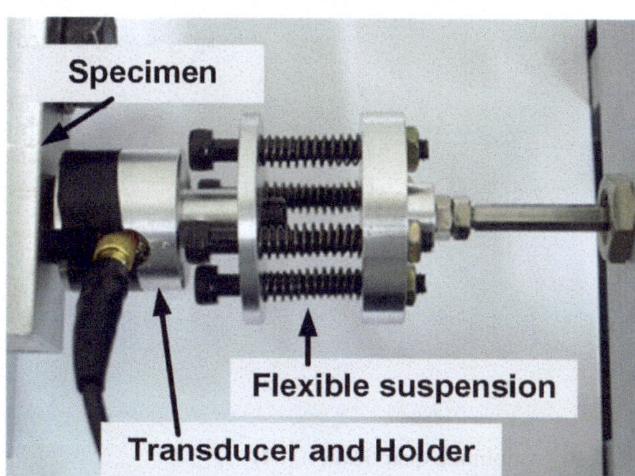

2.4.3 Analog-to-Digital Conversion

The amplitude of the second-order harmonic component is very small as a few % of the fundamental component; thus, careful attention should be paid to the error that can occur in the analog-to-digital conversion process of the received voltage signal. As an example, an original single-cycle sinusoidal signal and its digitally converted signal are shown in Fig. 2.30 as black and blue solid lines, respectively. This analog-to-digital conversion process is called quantization, and the quantization interval is called a quantum. The voltage corresponding to the quantum, called quantum voltage (q), is equivalent to the maximum measurement range (Full Scale (FS)) divided by the resolution of n-bit in an analog-to-digital converter. When a digital oscilloscope is used to display and record the voltage signal detected by a receiving transducer, the maximum measurement range is determined by the product of the vertical resolution (Volt/Division) and the number of divisions. Consequently, the quantum voltage can be defined by

$$q = \frac{FS}{2^n} = \frac{Volt/Div. \times Number\ of\ divisions}{2^n}. \tag{2.30}$$

In the analog-to-digital conversion process, the quantum voltage causes a quantization error (ε_q), which is the difference between the actual value and the converted value, as shown in Fig. 2.30 as a red solid line. In this case, the maximum quantization error can be defined as

$$\varepsilon_{q}\_\text{max} = \frac{\pm q}{2} = \pm\frac{FS}{2^{n+1}}. \tag{2.31}$$

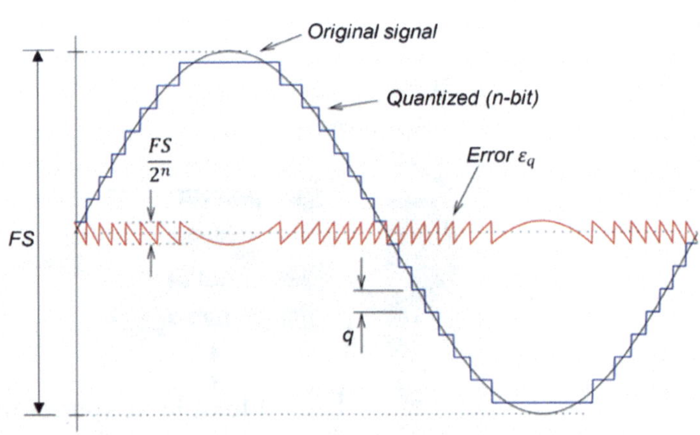

Fig. 2.30 Amplitude quantization [34]

If the amplitude of the second-order harmonic component contained in the received voltage signal is less than or equal to the quantum voltage, the quantization error can significantly affect the measurement reliability for the nonlinear ultrasonic parameter.

Let us examine the influence of quantization on the analysis of the second-order harmonic component using the following nonlinear ultrasonic time-domain signal:

$$V(t) = 1 \times \sin(2\pi \cdot 5\,\text{MHz} \cdot t) + 0.01 \times \sin(2\pi \cdot 10\,\text{MHz} \cdot t), \tag{2.32}$$

where the frequencies of the fundamental and second-order harmonic components are 5 and 10 MHz, and their amplitudes are 1 and 0.01 V, respectively. The amplitude of the second-order harmonic component is analyzed as the quantum voltage changes from 1/10 to 3 times the second-order harmonic amplitude.

The measured amplitudes of the second-order harmonic as a function of the quantum voltage at a sampling time of 0.5 ns are shown in Fig. 2.31. When the quantum voltage is smaller than the harmonic amplitude ($q < A_2' = 0.01$ V), the quantization error is within 5%. However, as the quantum voltage becomes larger than the harmonic amplitude ($q > A_2' = 0.01$ V), the quantization error increases to about 10%. Furthermore, this quantization error can be further increased with the increase in sampling time, as shown in Fig. 2.32. As the sampling time increases to 5 ns, even if the quantum voltage is smaller than the harmonic amplitude, the quantization error increases to about 15%. In particular, when the quantum voltage is greater than the harmonic amplitude, the quantum error further increases to about 60%. The analog signal is quantized at sampling time intervals with respect to the time-axis; thus, the sampling time is also a major factor that affects the quantization error in addition to the quantum voltage.

This quantization error can also be confirmed experimentally. In the experiment, a 60 mm thick aluminum specimen was used. The experimental setup was based on the through-transmission mode using a transmitting transducer with a 5 MHz

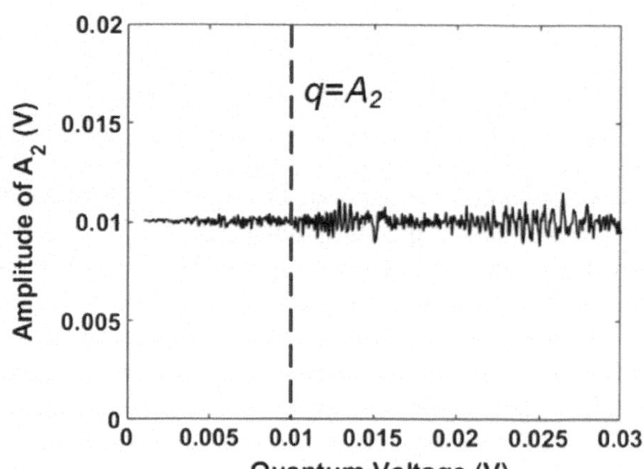

Fig. 2.31 Measured amplitudes of the second-order harmonic as a function of the quantum voltage at a sampling time of 0.5 ns

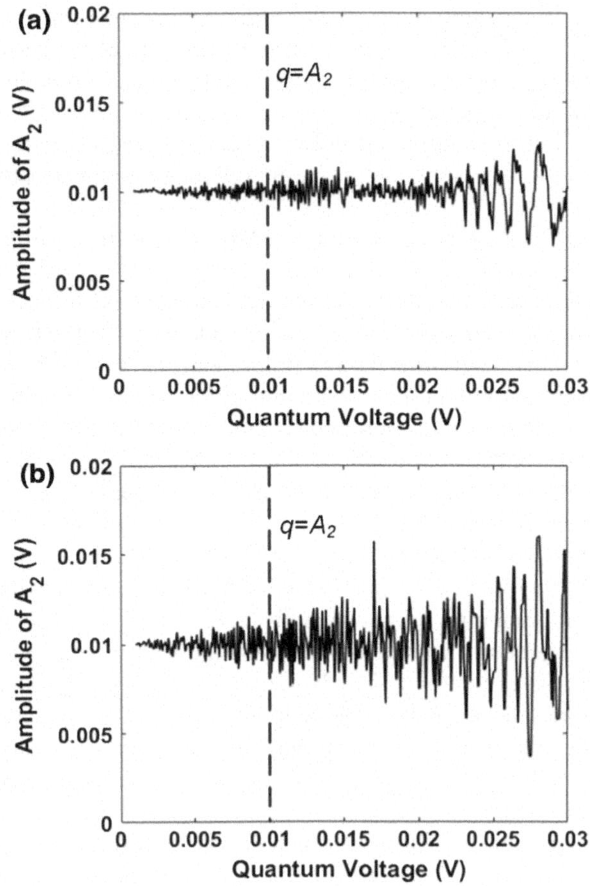

Fig. 2.32 Measured amplitudes of the second-order harmonic as a function of the quantum voltage at sampling times of **a** 1 ns and **b** 5 ns

center frequency and a receiving transducer with a 10 MHz center frequency. The variations of the second-order harmonic and the nonlinear ultrasonic parameter were analyzed while reducing the voltage of the input electric signal sent to the transmitting transducer by 2 dB. The digital oscilloscope used in this experiment provided an 8-bit resolution and a 0.5 ns sampling time. The signals detected by the receiving transducer were averaged 128 times. Here, two cases were compared. In CASE 1, the vertical resolution (Volt/Division) of the oscilloscope was kept constant at 200 mV to have a constant quantum voltage. In CASE 2, the vertical resolution was finely adjusted in each measurement so that the received signals were within the maximum measurement range, to minimize the quantum voltage at each measurement.

The measured amplitudes of the second-order harmonic and the quantum voltage as a function of the voltage of the input electric signal are shown in Fig. 2.33. For CASE 1, the quantum voltage is constant irrespective of the input voltage because the vertical resolution of the oscilloscope was kept constant. The amplitude of the second-order harmonic varies approximately in proportion to the square of the input

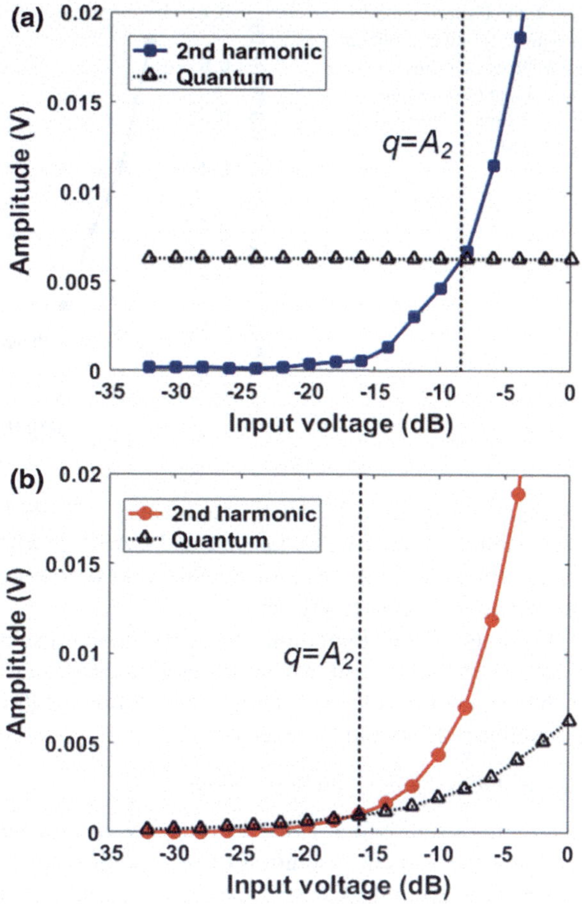

Fig. 2.33 Experimentally measured amplitudes of the second-order harmonic and the quantum voltage as a function of the voltage of the input electric signal for **a** CASE 1 and **b** CASE 2

voltage, and its amplitude is similar to the quantum voltage when the input voltage was reduced by −8 dB. For CASE 2, the quantum voltages are generally smaller than that in CASE 1 because the quantum voltage was minimized by adjusting the vertical resolution at each measurement. Therefore, the input voltage at which the amplitude of the second-order harmonic was similar to the quantum voltage was lowered to −16 dB.

Figure 2.34 shows the measured nonlinear ultrasonic parameters as a function of the input voltage. Ideally, if the quantum voltage is negligible compared with the amplitude of the second-order harmonic, the nonlinear ultrasonic parameter should have a constant value irrespective of the input voltage. However, in CASE 1, the amplitude of the second-order harmonic was similar to the quantum voltage when the input voltage decreased to −8 dB; thus, the nonlinear ultrasonic parameters measured below the critical input voltage fluctuated because of the quantization error. Meanwhile, in CASE 2, the measured parameters were almost constant until

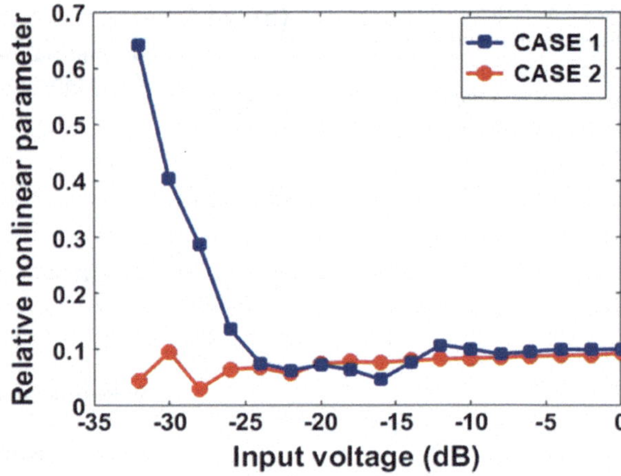

Fig. 2.34 Experimentally measured relative nonlinear parameters as a function of the voltage of the input electric signal for CASE 1 and CASE 2

the input voltage decreased to approximately −20 dB. By comparing CASE 1, where the quantum voltage was constant, and CASE 2, where it was minimized in each measurement, the influence of quantization on the nonlinear ultrasonic parameter measurement is clearly visible.

From the above simulation and experimental results, it is confirmed that reliable measurements are possible when the amplitude of the second-order harmonic is larger than the quantum voltage. Therefore, the following three methods are recommended to minimize the quantization error.

(a) Minimizing the quantum voltage

If a digital oscilloscope is used to display and record the received voltage signal, reducing the vertical resolution of the oscilloscope can reduce the quantum voltage, as verified in the CASE 2 example, but only until the received signal does not exceed the maximum measurement range. When the maximum measurement range is completely matched with the peak-to-peak voltage of the received signal, n-bit analog-to-digital conversion will produce the quantum voltage defined in Eq. (2.30). Under the matched condition, an 8-bit resolution will produce a quantum voltage that is 1/128 times the maximum amplitude of the received signal.

(b) Using a narrowband receiving transducer

The second-order harmonic can be quantized with higher resolution if the system response characteristics are adjusted to increase the amplitude ratio ($A_V(\omega_2)/A_V(\omega_1)$) by increasing the harmonic amplitude while reducing the fundamental amplitude. From Eqs. (2.27) and (2.28), the amplitude ratio can be expressed as

$$\frac{A_V(\omega_2)}{A_V(\omega_1)} = \beta' D(\omega_1)T(\omega_1)C_1(\omega_1)\frac{C_2(\omega_2)R(\omega_2)G(\omega_2)}{C_2(\omega_1)R(\omega_1)G(\omega_1)}. \tag{2.33}$$

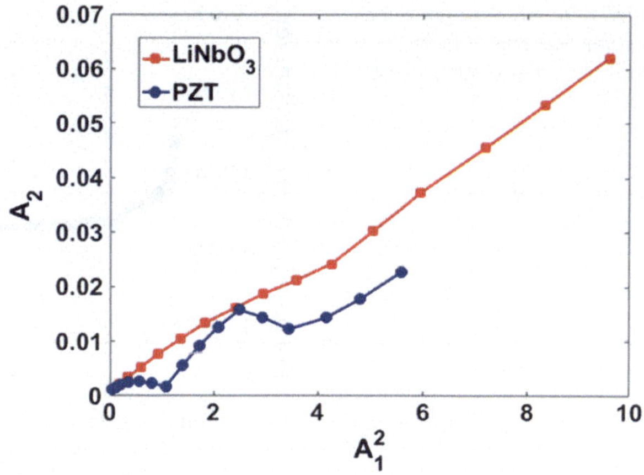

Fig. 2.35 Relationships between $A_1'^2$ and A_2' as a function of the increasing input voltage level when general PZTs and $LiNbO_3$ transducers were used

The ratio depends on the response characteristics of various system components. One way of increasing this ratio is to increase the $R(\omega_2)/R(\omega_1)$ term, corresponding to the response function of a receiving transducer, by using a receiving transducer, whose response at the second-order harmonic frequency is higher than at the fundamental frequency. Generally, for nonlinear ultrasonic measurements, the center frequency of a receiving transducer is set to be twice the fundamental frequency to measure the second-order harmonic more sensitively. This is equivalent to increasing the value of the $R(\omega_2)/R(\omega_1)$ term. The same effect can also be obtained by using an additional high-pass filter that passes harmonic frequency components.

In addition to matching the receiver's center frequency to the second-order harmonic frequency, using a narrowband receiving transducer will increase the $R(\omega_2)/R(\omega_1)$ term further. As an example, Fig. 2.35 shows the experimentally measured relationships between $A_1'^2$ and A_2' as a function of the increasing input voltage level when general PZTs were used and when $LiNbO_3$ transducers that provide a narrower bandwidth than the PZTs were used. This experiment was based on the through-transmission mode in a 20 mm thick aluminum specimen, and the fundamental and receiver's center frequencies were 5 and 10 MHz, respectively. The bandwidth of each transducer was 5.13 MHz for both 5 and 10 MHz PZTs, and 2.76 and 3.93 MHz for 5 and 10 MHz $LiNbO_3$ transducers, respectively. As shown in Fig. 2.35, A_2 measured by PZTs fluctuates because of the quantization error when A_1^2 is less than 3.5. Only when $A_1'^2$ is above that value, there exists a good linear relationship between $A_1'^2$ and A_2'. Meanwhile, the relationship measured by the $LiNbO_3$ transducers is linear. This is because the $R(\omega_2)/R(\omega_1)$ term in Eq. (2.33) increased by using the narrowband $LiNbO_3$ transducers, which resulted in reducing the quantum voltage and quantization error.

(c) Increasing the incident ultrasonic power

Increasing the amplitude of the ultrasonic wave incident on a specimen can increase the amplitude of the second-order harmonic, because the second-order harmonic amplitude is proportional to the square of the incident wave amplitude. This can be

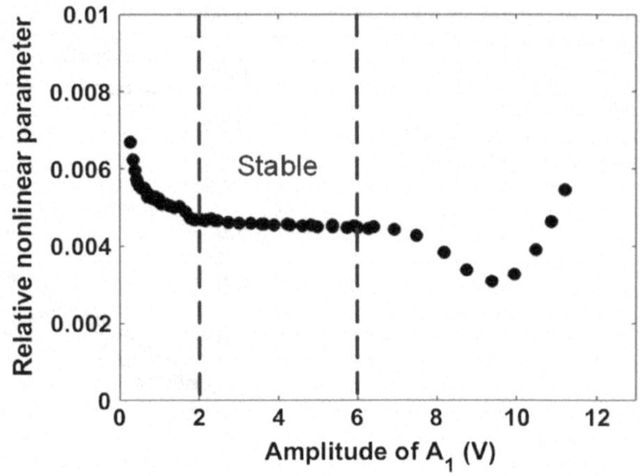

Fig. 2.36 Relative nonlinear parameters as a function of the increasing input voltage level

achieved by increasing the voltage of the input electric signal sent to a transmitting transducer, using a high-efficiency transmitting transducer, and improving the contact condition of the incident side. This means increasing the $D(\omega_1)T(\omega_1)C_1(\omega_1)$ term in Eq. (2.33). However, it should be noted that too much input voltage can cause breakage of the transducer and can also cause the nonlinear response in the transducer itself. Therefore, it is recommended to select the appropriate input voltage after selecting the input voltage range where the nonlinear ultrasonic parameter can be measured stably. As an example, Fig. 2.36 shows the experimentally measured relative nonlinear parameters as a function of increasing the input voltage level by using 5 and 10 MHz $LiNbO_3$ transducers in an aluminum specimen. Severe fluctuations of the nonlinear parameter appear at the received A_1 range lower than 2 V and higher than 6 V. The measured nonlinear parameters are stable only for the received amplitude range of 2–6 V.

2.4.4 Digital Signal Processing

The nonlinear ultrasonic measurement generally requires spectral analysis to analyze their amplitudes in the received ultrasonic wave signals, and thereafter to calculate the nonlinear ultrasonic parameter. Accordingly, tone-burst ultrasonic waves with a narrow bandwidth are widely used so that the fundamental frequency component and the second-order harmonic component can be clearly separated in the frequency domain. For spectral analysis, an FFT that performs a discrete Fourier transform at high speed is normally used. Here, careful attention is required to guarantee the measurement reliability includes selecting the section to perform FFT on the received ultrasonic signal and using a suitable FFT window.

The tone-burst ultrasonic signal received for the nonlinear ultrasonic measurement can generally be divided into three parts, as shown in Fig. 2.37; the front-transient

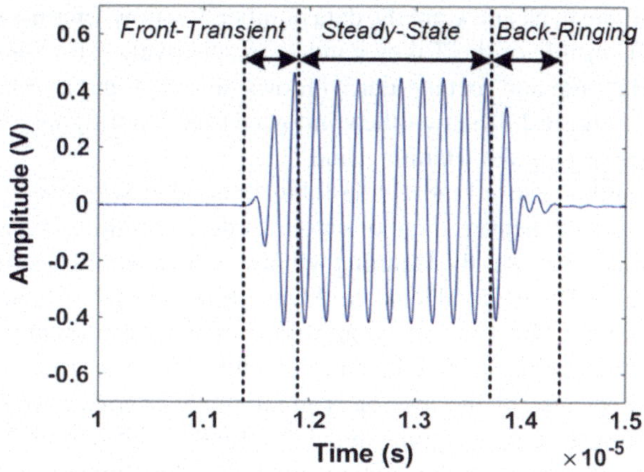

Fig. 2.37 Typical tone-burst ultrasonic signal received for the nonlinear ultrasonic measurement

response part caused by the frequency response characteristic of a receiving transducer, the steady-state part, and the back-ringing effect part. If the unstable front-transient response and back-ringing effect parts are included in the section to perform the FFT, those unstable components can yield inconsistent results. Therefore, it is desirable that spectral analysis is performed only for the steady-state part. The use of windowing, which is a process of taking a finite specific signal in an entire signal, makes it possible to perform the FFT only for the specific steady-state section. Meanwhile, windowing with the discontinuity between the endpoints of the windowed section cause spectral leakage that results in side-lobes in frequency domain. The side-lobes can overlap with the second-order harmonic component in the frequency domain; thus, the side-lobes should be minimized as much as possible. The reduction of side-lobes is possible using a window with an amplitude that varies smoothly and gradually toward zero at the endpoints.

A window function that can be effectively used for nonlinear ultrasonic measurements includes the Hanning and Tukey windows. These windows are defined as follows:

$$w_{Hanning}(n) = \begin{cases} 0.5\left(1 - \cos\left(\frac{2\pi n}{N-1}\right)\right), & 0 \le n \le N-1 \\ 0, & otherwise \end{cases} \tag{2.34}$$

$$w_{Tukey}(n) = \begin{cases} 0.5\left[1 + \cos\left(\pi\left(\frac{2n}{\kappa(N-1)} - 1\right)\right)\right], & 0 \le n \le \frac{\kappa(N-1)}{2} \\ 1, & \frac{\kappa(N-1)}{2} \le n \le (N-1)\left(1 - \frac{\kappa}{2}\right) \\ 0.5\left[1 + \cos\left(\pi\left(\frac{2n}{\kappa(N-1)} - \frac{2}{\kappa} + 1\right)\right)\right], & (N-1)\left(1 - \frac{\kappa}{2}\right) \le n \le (N-1) \\ 0, & otherwise \end{cases} \tag{2.35}$$

where n, N, and κ are the data number, window length, and coefficient that determines the profile of the Tukey window, respectively. The Tukey window is identical to the Hanning and rectangular windows, at $\kappa = 1$ and $\kappa = 0$, respectively.

Figure 2.38 shows the windowed tone-burst signals and their frequency spectra for the rectangular, Hanning, and Tukey windows ($\kappa = 0.5$), respectively. The rectangular window caused more side-lobes than other windows, which is disadvantageous for accurate measurement of second-order harmonic. The window that provides the least side-lobes was the Hanning window, which would be effective in minimizing the error caused by spectral leakage. Meanwhile, when the Hanning and Tukey windows were applied, the analyzed amplitudes of the fundamental and second-order harmonic components were reduced to 50 and 75% of original amplitudes, respectively. This is because of the loss of original signal energy, depending on the window profile. To restore the original signal amplitudes, the scaling factors of 2 and 4/3 should be compensated to the FFT results using the Hanning and Tukey windows ($\kappa = 0.5$), respectively.

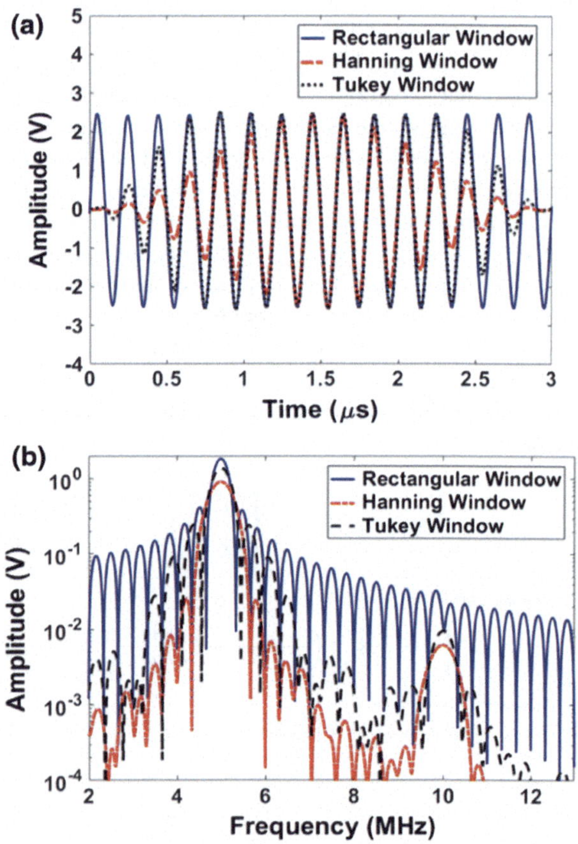

Fig. 2.38 **a** Windowed tone-burst signals and **b** their frequency spectra for the rectangular, Hanning, and Tukey windows ($\kappa = 0.5$) [37]

To minimize the effect of side-lobes on the analysis of second-order harmonic, it is recommended that a null point between side-lobes be located at the second-order harmonic frequency in the frequency domain, which is achieved by setting the window length to an integral multiple of the period of the fundamental component. In addition, the use of zero-padding will improve the frequency resolution in the FFT.

2.5 Application to Assess Thermal Aging in Al Alloy

This section introduces the thermal aging assessment of aluminum alloy as an example of the application of the techniques described above. In the case of thermal aging, the precipitates that produce local strain fields may also produce dislocations around them [38], and interact with dislocations [39, 40]. These interactions cause changes in the nonlinear ultrasonic parameter. In this study, we investigate the correlation between the absolute nonlinear ultrasonic parameter and the mechanical strength varied by the nucleation and growth of the precipitates because of thermal aging in Al 6061-T6 alloy. First, the Al 6061-T6 alloy specimens were heat-treated at a constant temperature of 220 °C for different exposure times: 20, 40, 60, 120, 600, and 6000 min. Further, ultrasonic measurements were conducted using the contact piezo-electric detection method [13, 14] (introduced in Sect. 2.3.1) to determine the absolute nonlinear ultrasonic parameter β as a function of aging time. Note that the same result can be obtained even if the absolute nonlinearity parameter is estimated from the relative measurements introduced in Sect. 2.3.3 Case II.

After ultrasonic measurements, the tensile and micro-Vickers hardness tests were conducted to investigate the correlation of the measured parameter β with the yield strength (and the hardness). Furthermore, microscopy using OM and TEM was performed to observe changes in microstructure and to support the nonlinear ultrasonic results.

Seven Al 6061-T6 alloy specimens with dimensions 40 × 20 × 200 mm were prepared by cutting from an aluminum plate. All specimens were solution-heat-treated for 240 min at 540 °C and quenched in water for 60 min. Further, each specimen was heat-treated at a constant temperature of 220 °C for different exposure times: 20, 40, 60, 120, 600, and 6000 min. The ultrasonic measurement is first performed for each specimen, followed by the tensile test, micro-Vickers hardness test, and microstructural observations using OM and TEM. Further, the tensile test specimens were taken from these heat-treated specimens and had dimensions 33 mm × 6 mm × 2 mm according to ASTM E8 M standard. The tensile tests were performed at the head speed of 1 mm/min at room temperature using a universal testing machine (Instron, MTS793). The hardness tests were carried out according to ASTM E384 standard using a micro-Vickers hardness tester (Shimadzu, HMV-2T) with 1 kg load and a dwell time of 10 s. The test was repeated over ten different locations to obtain an average hardness value. For the microscopy, the specimens are etched with 55 mL

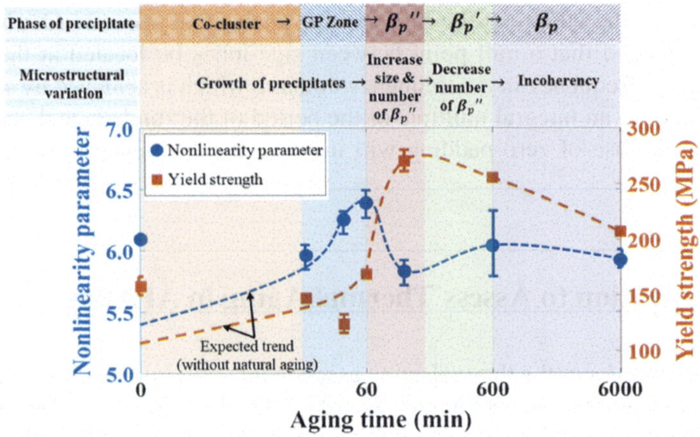

Fig. 2.39 Experimental results of the nonlinear ultrasonic parameter β and yield strength σ_y as a function of aging time (the marked points are measured data. The dashed lines are the trend of expected change of the nonlinear ultrasonic parameter and the yield strength without natural aging effect)

Keller's solution (60% nitric acid (HNO3) 1 mL + 35–37% hydrochloric acid (HCl) 2 mL + 49–52% hydrofluoric acid (HF) 2 mL + diwater 50 mL) for 30 s.

Figure 2.39 shows the changes in the absolute nonlinear ultrasonic parameter β and yield strength σ_y as a function of aging time. The marked points are measured data and the error bar indicates the range of the maximum and minimum values. The dashed lines are the trend of expected change in the nonlinear ultrasonic parameter and the yield strength as a function of the aging time without natural aging effect. The differences appeared between the measured data and dashed lines at the beginning of aging within 60 min because of the natural aging effect. In the experimental results, both the nonlinear ultrasonic parameter and yield strength decrease slightly during the first 20 min. From 20 to 120 min these parameters vary considerably.

At 120 min, the nonlinear ultrasonic parameter reaches a negative peak and the yield strength reaches a positive peak. With further aging, the nonlinear ultrasonic parameter slightly increases whereas the yield strength decreases. Overall, the nonlinear ultrasonic parameter β and the yield strength σ_y showed a good correlation over the aging time.

The variations in the nonlinear ultrasonic parameter and the yield strength are caused by the nucleation and growth of the precipitates, which can be well described by the precipitation sequence [12, 24–27, 29–32, 38, 41, 42].

Super-saturated solid solution ⟶ Mg-Si co-clusters ⟶ Guinier–Preston zones (GP zones): spherical, coherent with the aluminum matrix ⟶ Needle-shaped precipitates β_p'': coherent with the aluminum matrix ⟶ Rod-shaped precipitates β_p': semi-coherent with the aluminum matrix ⟶ Plate-shaped precipitates β_p: incoherent with the aluminum matrix.

At the start of the aging, the number of super-saturated solute atoms in aluminum matrix is gradually reduced, and Mg-Si co-clusters are formed. Generally, the co-clusters affect to increase the nonlinear ultrasonic parameter and yield strength. As the aging time increases, the co-clusters gradually grow and GP zones are formed between 20–60 min. The GP zones have a spherical shape with 2–5 nm diameter and are fully coherent with the matrix [31]. The growth of the GP zones causes a positive coherency strain, which increases the lattice irregularity. This irregularity contributes to the distortion of the propagating ultrasonic waves; thus, the nonlinear ultrasonic parameter increases. The physical mechanism is the vibration of the dislocations pinned by precipitates, which can be expressed by $\Delta\beta \propto \Lambda r_p^3/N_p^{\frac{1}{3}}$, where $\Delta\beta$ is the variation in the nonlinear ultrasonic parameter, Λ is the dislocation density, r_p is the average radius of the precipitates, and N_p is the number density of precipitates [2, 43]. Further growth of the GP zones can also impede the movement of dislocations through the material resulting in an increase in the material hardness [12] that is expected proportional yield strength. That is, the GP zones generally lead to an increase in the nonlinear ultrasonic parameter and yield strength. However, in the experimental results shown in Fig. 2.39, the nonlinear ultrasonic parameter and the yield strength slightly decrease until 20 and 40 min, respectively, which is attributed to the reversion of some GP zones formed because of natural aging [41]. These measurements are conducted after a few days of artificial aging; thus, the specimens are affected not only by the artificial aging but also the natural aging. Miao and Laughlin [41] observed a similar trend in the hardness in Al 6022 alloy during artificial aging at 175 °C after natural aging. They saw the hardness increase from 55 to 72 only because of natural aging and decrease initially during artificial aging. The detrimental effect of natural aging is crucial for short aging time until 60 min at 220 °C. To ensure that these initial drops are because of natural aging, hardness measurements were conducted on the specimens immediately after the heat treatment. Figure 2.40 shows that the hardness monotonically increases until 120 min are passed; this trend is quite similar to other's results, for example, those described in [27]. As the hardness and

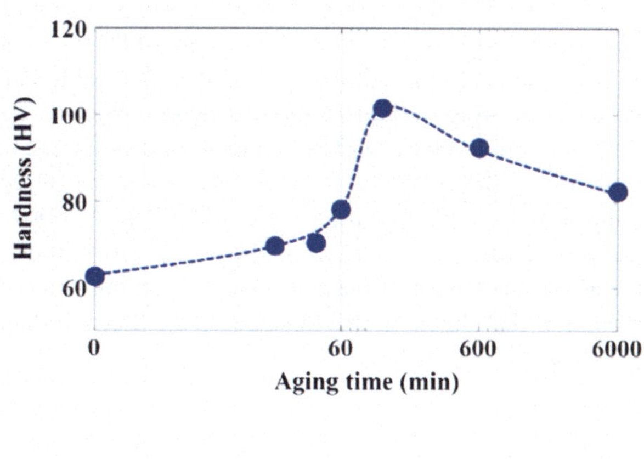

Fig. 2.40 Experimental results of micro-vickers hardness as a function of aging time

yield strength in the aged Al 6016-T6 material has a simple proportional relationship [27], the hardness shows a trend similar to the trend of expected change of the yield strength shown in Fig. 2.39 during artificial aging [44]. This is also the reason why the nonlinear ultrasonic parameter of naturally aged Al 6061-T6 specimen (6.1, the initial value shown in Fig. 2.39) is larger than that of specimen without natural aging (5.4 in Table 2.1). The effect of natural aging on both nonlinear ultrasonic parameter and hardness will be reported in some other study.

Further aging, a reversion of the GP zones occurs back into the matrix and needle-shaped precipitates β_p'' are formed. These precipitates are aligned in the Al <100> direction and coherent with the matrix. The length and diameter of these precipitates β_p'' are about 20–40 and 10 nm [31]. Both the size of the precipitates and the number density of the precipitates increase significantly in this period. The nonlinear ultrasonic parameter decreases according to the following equation: $\Delta\beta \propto \Lambda r_p^3/N_p^{\frac{1}{3}}$ [2, 32]. The formed β_p'' also impedes the motion of dislocations, which increases the yield strength. The result shows that β_p'' is the most effective hardening phase [27, 32], i.e., the yield strength increases significantly with the formation of β_p'' [27, 32, 45]. Consequently, the yield strength dramatically increases and reaches a positive peak while the nonlinear ultrasonic parameter decreases and reaches a negative peak at 120 min.

With aging time extending further between 120 and 600 min, the precipitates grow further, thereby transforming from the needle-shaped β_p'' precipitates to rod-shaped β_p' precipitates aligned in the Al <100> direction and semi-coherent with the matrix. The length and diameter of these β_p' precipitates are about 30–100 nm and 10–20 nm, respectively [31]. The decrease in the number density of the needle-shaped precipitates causes a decrease in the yield strength. With the decrease in the density of β_p'', the nonlinear ultrasonic parameter increases. Consequently, the nonlinear ultrasonic parameter increases, whereas the yield strength decreases between 120 and 600 min. Further, the β_p' precipitates transform to incoherent plate-shaped equilibrium precipitates β_p. The precipitates coarsen beyond the critical size; thus, the nonlinear ultrasonic parameter and yield strength decrease with the aging time because of the loss of coherency (over-aging). This loss is associated with an increase in the dislocation loop length and mobility caused by the coarsening precipitates [12].

The experimental results showed that the nonlinear ultrasonic parameter β and the yield strength σ_y had an interesting correlation. The variations of the nonlinear ultrasonic parameter β and the yield strength σ_y were explained by the precipitation sequence in the Al 6061-T6 alloy. Based on the observed correlation, the optimal aging time to reach maximum strength of the Al 6061-T6 alloy can be nondestructively evaluated by monitoring the variation of the nonlinear ultrasonic parameter. Moreover, the nonlinear ultrasonic parameter is also useful for evaluating the change in the elastic properties caused by thermal degradation and fatigue damage in industrial applications.

2.6 Associated Methodologies

2.6.1 *Phase Inversion Technique*

The phase inversion technique is a digital signal processing technique that can be effectively used in the measurement of a nonlinear ultrasonic parameter, for measuring a second-order harmonic component that is much smaller than a fundamental component [46–50]. This technique superposes two ultrasonic signals measured at 180° out-of-phase inputs to extract only even harmonics by canceling out the odd harmonics. When the phase of the incident ultrasound is 0° or 180°, the plane wave displacement solutions of the nonlinear wave equation can be expressed by the following equations:

$$u_{(0^\circ)} = A_1 \cos(kx - \omega t) + A_2 \cos 2(kx - \omega t) \tag{2.36}$$

$$\begin{aligned} u_{(180^\circ)} &= A_1 \cos(kx - \omega t + \pi) + A_2 \cos 2(kx - \omega t + \pi) \\ &= -A_1 \cos(kx - \omega t) + A_2 \cos 2(kx - \omega t) \end{aligned}. \tag{2.37}$$

The sum of the two displacements is as follows:

$$u_{(0^\circ)} + u_{(180^\circ)} = 2A_2 \cos 2(kx - \omega t) \tag{2.38}$$

which results in the second-order harmonic term to be doubled and the fundamental components to cancel out each other.

As an example, two typical time-domain signals measured with 180° out-of-phase inputs are shown in Fig. 2.41a. The two signals are symmetric with respect to the time axis. Figure 2.41b shows the superposed signal of the two signals shown in Fig. 2.41a. The amplitude of the superposed signal is small (about 10% of that of the raw signals)

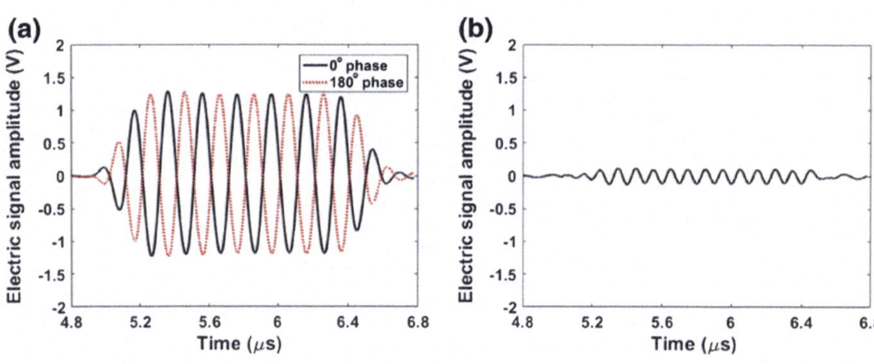

Fig. 2.41 **a** Typical two time-domain signals measured with 180° out-of-phase inputs and **b** the superposed signal of the two signals shown in **a** [34]

and its period is short in half. This is because the fundamental components in the raw signals cancel out each other, and only the second-order harmonic components remain, as expressed in Eq. (2.38). These results can be clearly confirmed in the spectral analysis results shown in Fig. 2.42. The superposed signal has only the second-order harmonic component, which is twice as large as that of the raw signal.

In particular, the use of this technique is strongly recommended when measuring the nonlinear ultrasonic parameter using an ultrasonic signal with a broad bandwidth, such as a pulsed or a single-cycle signal, or on thin plates. If traditional spectral analysis is applied to broadband signals, it would be difficult to measure the second-order harmonic component, which is less than a few percentage of the amplitude of the fundamental component. This is because the second-order harmonic component can be buried in the side-lobe of the fundamental component, as shown in Fig. 2.43 [50]. On the other hand, if the phase inversion technique is used, only the second-order

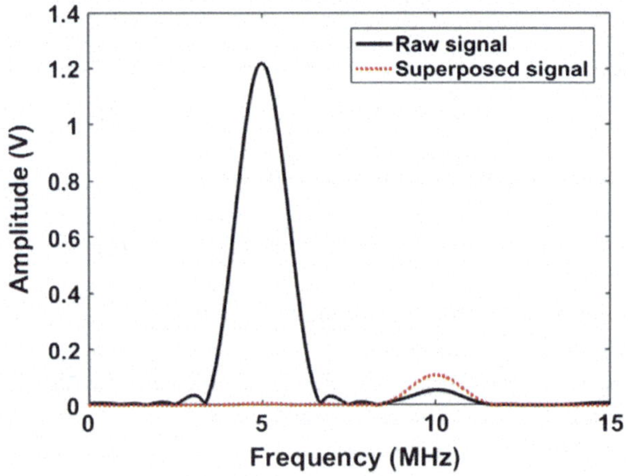

Fig. 2.42 Frequency spectra of the raw and superposed signals [34]

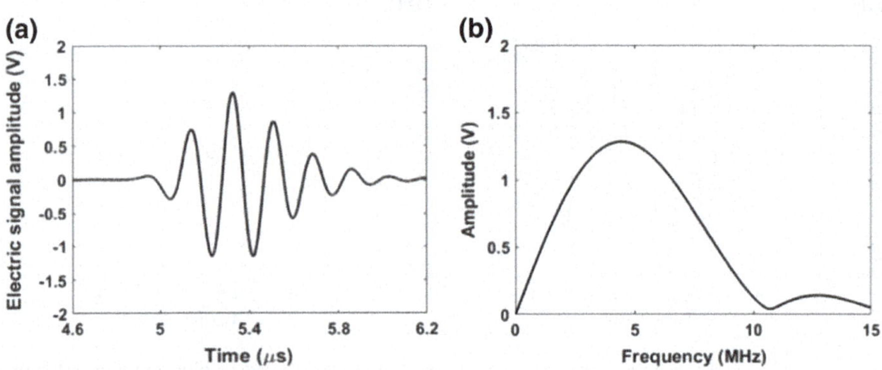

Fig. 2.43 **a** Typical broadband ultrasonic signal and **b** its frequency spectrum

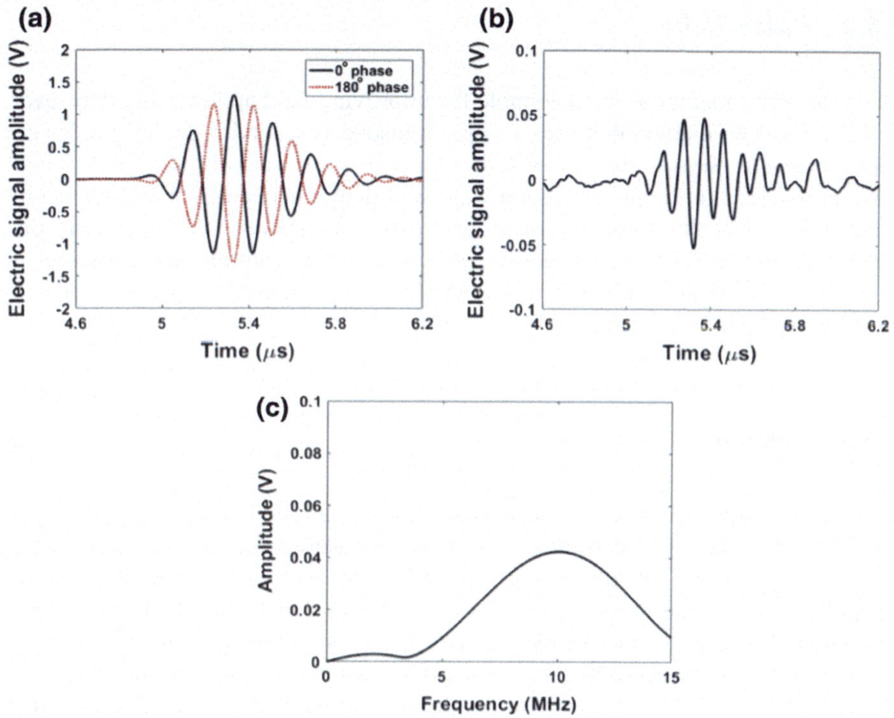

Fig. 2.44 **a** Typical two broadband ultrasonic signals measured with 180° out-of-phase inputs, **b** superposed signal of the two signals shown in **a**, and **c** frequency spectrum of the superposed signal shown in **b** [50]

harmonic component can be extracted clearly, and its amplitude can be accurately measured, as shown in Fig. 2.44 [50]. This application is particularly useful for local measurement of the nonlinear ultrasonic parameter in thin plates.

Meanwhile, this technique requires the use of an electric signal generator that can precisely control the phase of the input electrical signal. If ideal 180° out-of-phase inputs are not provided, the two measured signals would be slightly asymmetric with respect to the time axis, which can cause the fundamental components to still remain in the superposed signal. In the case that it is not possible to provide ideal 180° out-of-phase inputs systematically, it is highly recommended to use additional post-signal processing or filtering to remove the remaining components. If not, the error factors caused by the phase difference can reduce the measurement reliability for the nonlinear ultrasonic parameter.

2.6.2 *Pulse-Echo*

The pulse-echo method is most suitable for improving field applicability. However, in the pulse-echo method that uses a single transducer, second-order harmonics are hardly received because the second-order harmonic generated by the fundamental wave propagated in the pre-reflection stage and that generated by the fundamental wave reflected at the boundary cause destructive interference. To overcome this problem, a technique has been developed wherein a transmitter and a receiver in close proximity or a dual element transducer that have a coaxially separated type are used. Refer the literatures for details [51, 52].

2.6.3 *V-Scan*

In all the techniques discussed so far, the through-transmission mode was used. This is because the second–order harmonic is difficult to measure when using the echo mode. However, if only one side of the object is accessible, the transmission mode cannot be applied. In this case, the V-scan method can be considered as an alternative. In this technique, the beam path looks like a V-line when two transducers for transmitting and receiving are located opposite to each other within one skip distance space, as shown in Fig. 2.45. Of course, as in the pulse echo method, the disadvantage that the second-order harmonic is detected smaller than actual due to the destructive interference between the forward propagation and the reverse propagation is still inevitable.

To verify the validity of this technique, the results of the normal through-transmission mode and the V-scan method were compared for a stainless-steel alloy specimen with concentrated fatigue damage at the center [53]. Figure 2.46

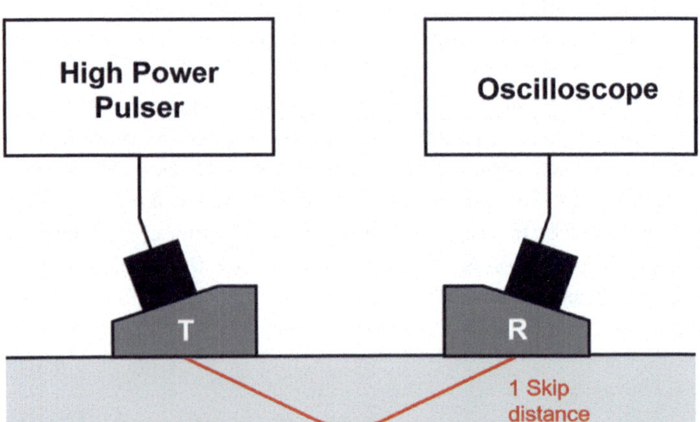

Fig. 2.45 Schematic diagram of the V-Scan technique for measuring ultrasonic nonlinearity

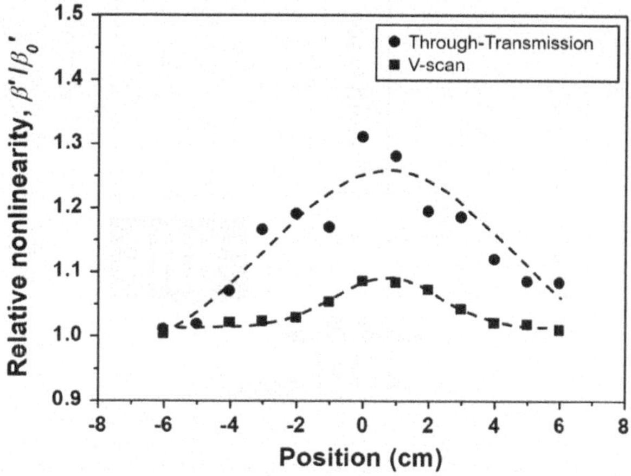

Fig. 2.46 Comparison of relative nonlinearity measured by the normal through-transmission technique and the V-scan method as a function of distance from the center of the stainless-steel alloy specimen with concentrated fatigue damage at the center

shows the relative ultrasonic nonlinearities, β'/β'_0, measured by the normal through-transmission technique and the V-scan method as a function of distance from the center (0 position). Note that β'_0 is the mean value in the nonlinear parameter obtained from the grip region (far from the center). Both results show a bell-shaped curve of the ultrasonic nonlinearity as a function of position, which indicates that the fatigue damage in the test specimen was localized to the center. Due to the disadvantages of the pulse echo method and the difference in experimental conditions in two techniques, the absolute increase in ultrasonic nonlinearity is different each other, but we have to note that both results show the same tendency.

2.6.4 Laser-Ultrasonic Surface Acoustic Waves

The laser-ultrasonic technique has various advantages, such as non-contact generation and detection of ultrasonic waves without couplant, ability to operate on curved or rough surfaces, and increase in scanning speed [54–57]. Numerous achievements have been reported with the use of the laser-ultrasonic technique; however, most are related to crack detection based on linear theory [58–62], thereby leaving the necessary challenge of applying it to the measurement of ultrasonic nonlinearity pending.

In this section, the laser-ultrasonic technique to measure the nonlinearity of surface acoustic waves is introduced in brief [63]. There are several advantages of using surface waves. First, measurements are made only on one side of target, which is advantageous for field applications. Second, the degradation of a material usually starts from the surface, and surface acoustic wave is advantageous for evaluation of surface damage. Furthermore, since it is easy to change the propagation distance, the measurement reliability test (refer the Sect. 2.3.4) based on the dependence of second-order harmonic amplitude on the propagation distance is convenient.

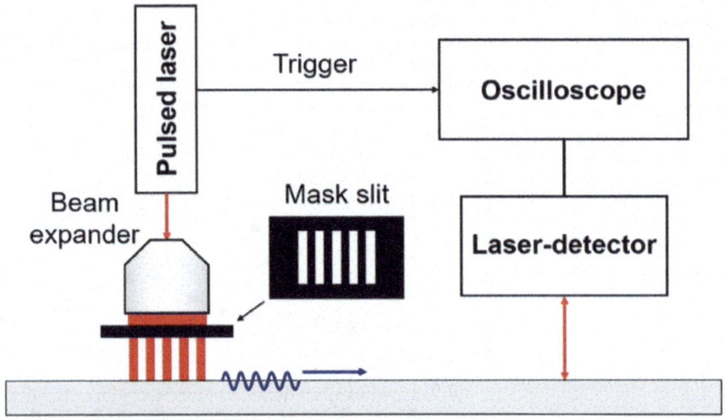

Fig. 2.47 Schematic of the experimental setup used for measuring the acoustic nonlinearity of laser-generated surface waves

Nonlinearities in surface waves are slightly different from those in longitudinal waves; however, the use of higher harmonics are basically the same. A schematic of the experimental setup for measuring the ultrasonic nonlinearity of the laser-generated surface waves is shown in Fig. 2.47. A pulsed laser was used to excite the surface waves, and a laser ultrasonic detector was employed to receive the laser-generated surface waves. Considering the harmonic frequency magnitude for the ultrasonic nonlinearity test, a narrow bandwidth wave signal is much more effective for analyzing the frequency characteristics of the ultrasonic waves. The line-arrayed laser beams can be used to produce a tone-burst signal of surface waves. In particular, using a line-arrayed slit mask with an expanded laser beam is an effective and easy method of producing line-arrayed illumination sources to excite a narrowband surface wave [54, 55]. A slit mask was attached to the specimen surface to negate beam diffraction. As a detector, an interferometer can be used basically, which should ensure sufficient sensitivity in the ultrasonic bandwidth. Currently, the two-wave mixing photorefractive interferometers or the laser doppler interferometers are used.

Note that, in this technique, to test the measurement reliability mentioned in Sect. 2.3.4, both the method of changing the intensity of the laser pulse for the generation of surface acoustic waves and the method of changing the transmitting/receiving interval are applicable.

Acknowledgements A major portion of this work was financially supported by the National Research Foundation of Korea (NRF) Grant funded by the Korean Government (NRF-2013M2A2A9043241).

We would like to thank all of the ISNDE Lab (Intelligent Sensing & Nondestructive Evaluation Lab, Department of Mechanical Convergence Engineering, Hanyang University) graduates over the past 27 years for their research efforts, discussions, and contributions to the field of nonlinear ultrasonics. A special tribute to Dr. Kyung-Cho Kim, the first Ph.D. graduate of ISNDE to pioneer nonlinear ultrasonics research in solid materials, Dr. Taehun Lee, developed fundamental technologies for the measurement of nonlinear ultrasonic parameters, and Dr. Chung-Seok Kim, who worked as a research Professor at Hanyang University from 2010 to 2012 (currently a Professor at Chosun University) on various applications of nonlinear ultrasonics for material characterization.

Many thanks to Dr. Yob Ha, Dr. Hyunmook Kim, and Dr. Hongjoon Kim, who developed miscellaneous methods for the measurement of nonlinear ultrasonic characteristics, including laser ultrasonic techniques, and Dr. Hogeon Seo who pioneered the measurement of contact acoustic nonlinearity and its synthetic imaging.

Special thanks to the students currently working in the ISNDE Lab: Ph.D. students Dong-Gi Song, Jihyun Jun, Seong-Hyun Park, and Master's degree students Juyoung Ryu, Youngchang Lee, Seunghoon Lee, Jungyean Hong, who have improved the measurement systems, developed various application techniques, and contributed to the editing and data organization of this chapter.

References

1. J.K. Na, J.H. Cantrell, W.T. Yost, Linear and nonlinear ultrasonic properties of fatigued 410Cb stainless steel. Rev. Prog. Quant. Nondestr. Eval. **15**, 1347–1352 (1996)
2. A. Hikata, B.B. Chick, C. Elbaum, Dislocation contribution to the second harmonic generation of ultrasonic waves. J. Appl. Phys. **36**, 229–236 (1965)
3. A. Hikata, B.B. Chick, C. Elbaum, Effect of dislocations on finite amplitude ultrasonic waves in aluminum. Appl. Phys. Lett. **3**, 195–197 (1963)
4. M. Hong, Z. Su, Q. Wang, L. Cheng, X. Qing, Modeling nonlinearities of ultrasonic waves for fatigue damage characterization: theory, simulation, and experimental validation. Ultrasonics **54**, 770–778 (2014)
5. J.-Y. Kim, L.J. Jacobs, J. Qu, J.W. Littles, Experimental characterization of fatigue damage in a nickel-base superalloy using nonlinear ultrasonic waves. J. Acoust. Soc. Am. **120**, 1266–1273 (2006)
6. K.-Y. Jhang, Application of nonlinear ultrasonic to the NDE of material degradation. IEEE Trans. Ultrason. Ferroelectr. Freq. Control **47**, 540–548 (2000)
7. K.H. Matlack, J.-Y. Kim, L.J. Jacobs, J. Qu, Review of second harmonic generation measurement techniques for material state determination in metals. J. Nondestr. Eval. **34**, 1–23 (2014)
8. G. Gutiérrez-Vargas, A. Ruiz, J.-Y. Kim, L.J. Jacobs, Characterization of thermal embrittlement in 2507 super duplex stainless steel using nonlinear acoustic effects. NDT E Int. **94**, 101–108 (2018)
9. R.B. Thompson, O. Buck, D.O. Thompson, Higher harmonics of finite amplitude ultrasonic waves in solids. J. Acoust. Soc. Am. **59**, 1087–1094 (1976)
10. J. Kim, D.-G. Song, K.-Y. Jhang, Absolute measurement and relative measurement of ultrasonic nonlinear parameters. Res. Nondestr. Eval. **28**, 211–225 (2017)
11. W.B. Gauster, M.A. Breazeale, Detector for measurement of ultrasonic strain amplitudes in solids. Rev. Sci. Instrum. **37**, 1544–1548 (1966)
12. J.H. Cantrell, W.T. Yost, Effect of precipitate coherency strains on acoustic harmonic generation. J. Appl. Phys. **81**, 2957–2962 (1997)
13. G.E. Dace, R.B. Thompson, L.J.H. Brasche, D.K. Rehbein, O. Buck, Nonlinear acoustics, a technique to determine microstructural changes in material. Rev. Prog. Quant. Nondestr. Eval. **10B**, 1685–1692 (1991)
14. G.E. Dace, R.B. Thompson, O. Buck, Measurement of the acoustic harmonic generation for materials characterization using contact transducers. Rev. Prog. Quant. Nondestr. Eval. **11B**, 2069–2076 (1992)
15. D.J. Barnard, G.E. Dace, O. Buck, Acoustic harmonic generation due to thermal embrittlement of Inconel 718. J. Nondestr. Eval. **16**, 67–75 (1997)

16. D.J. Barnard, Variation of nonlinearity parameter at low fundamental amplitudes. Appl. Phys. Lett. **74**, 2447–2449 (1999)
17. P. Li, W.P. Winfree, W.T. Yost, J.H. Cantrell, Observation of collinear beam-mixing by an amplitude modulated ultrasonic wave in a solid, in *Ultrasonics Symposium* (1983), pp. 1152–1156
18. P. Li, W.T. Yost, J.H. Cantrell, K. Salama, Dependence of acoustic nonlinearity parameter on second phase precipitates of aluminum alloys, in *IEEE 1985 Ultrasonics Symposium* (1985), pp. 1113–1115
19. K.-J. Lee, J. Kim, D.-G. Song, K.-Y. Jhang, Effect of window function for measurement of ultrasonic nonlinear parameter using fast Fourier Transform of tone-burst signal. J. Korean Soc. Nondestr. Test. **35**, 251–257 (2015)
20. G. Ren, J. Kim, K.-Y. Jhang, Relationship between second- and third-order acoustic nonlinear parameters in relative measurement. Ultrasonics **56**, 539–544 (2015)
21. D.C. Hurley, C.M. Fortunko, Determination of the nonlinear ultrasonic parameter using a michelson interferometer. Meas. Sci. Technol. **8**, 634–642 (1997)
22. T. Kang, T. Lee, S.-J. Song, H.-J. Kim, Measurement of ultrasonic nonlinearity parameter of fused silica and Al2024-T4. J. Korean Soc. Nondestr. Test. **33**, 14–19 (2013)
23. J. Kim, K.-Y. Jhang, Measurement of ultrasonic nonlinear parameter by using non-contact ultrasonic receiver. Trans. Korean Soc. Mech. Eng. A **38**, 1133–1137 (2014)
24. H. Demir, S. Gündüz, The effects of aging on machinability of 6061 aluminum alloy. Mater. Des. **30**, 1480–1483 (2009)
25. L.P. Troeger, E.A.S. Jr, Microstructural and mechanical characterization of a superplastic 6xxx aluminum alloy. Mater. Sci. Eng. A **277**, 102–113 (2000)
26. S. Rajasekaran, N.K. Udayashankar, J. Nayak, T4 and T6 treatment of 6061 Al-15 vol. % SiCP composite. ISRN Mater. Sci. **2012**, 1–5 (2012)
27. F. Ozturk, A. Sisman, S. Toros, S. Kilic, R.C. Picu, Influence of aging treatment on mechanical properties of 6061 aluminum alloy. Mater. Des. **31**, 972–975 (2010)
28. W.F. Miao, D.E. Laughlin, Precipitation hardening in aluminum alloy 6022. Scripta Mater. **40**, 873–878 (1999)
29. G. Mrówka-Nowotnik, Influence of chemical composition variation and heat treatment on microstructure and mechanical properties of 6xxx alloys. Arch. Mater. Sci. Eng. **46**, 98–107 (2010)
30. J. Buha, R.N. Lumley, A.G. Crosky, K. Hono, Secondary precipitation in an Al-Mg-Si-Cu alloy. Acta Mater. **55**, 3015–3024 (2007)
31. X. Fang, M. Song, K. Li, Y. Du, Precipitation sequence of an aged Al-Mg-Si alloy. J. Min. Metall. Sect. B. **46**, 171–180 (2010)
32. G.A. Edwards, K. Stiller, G.L. Dunlop, M.J. Couper, The precipitation sequence in Al-Mg-Si alloys. Acta Mater. **46**, 3893–3904 (1998)
33. H. Jeong, S. Zhang, S. Cho, X. Li, Development of explicit diffraction corrections for absolute measurements of acoustic nonlinearity parameters in the quasilinear regime. Ultrasonics **70**, 199–203 (2016)
34. T.H. Lee, Measurement of ultrasonic nonlinearity and its application to nondestructive evaluation, Ph. D Dissertation, Hanyang University, 2010
35. L. Sun, S.S. Kulkarni, J.D. Achenbach, S. Krishnaswamy, Technique to minimize couplant-effect in acoustic nonlinearity measurements. J. Acoust. Soc. Am. **120**, 2500–2505 (2006)
36. I.-H. Choi, J.-I. Lee, G.-D. Kwon, K.-Y. Jhang, Effect of system dependent harmonics in the measurement of ultrasonic nonlinear parameter by using contact transducers. J. Korean Soc. Nondestr. Test. **28**, 358–363 (2008)
37. K.-J. Lee, Influence of transducer and signal processing on the measurement of ultrasonic nonlinear parameter, Master Thesis, Hanyang University, 2016
38. Y. Xiang, M. Deng, F.-Z. Xuan, Thermal degradation evaluation of HP40Nb alloy steel after long term service using a nonlinear ultrasonic technique. J. Nondestr. Eval. **33**, 279–287 (2014)
39. J.H. Cantrell, W.T. Yost, Determination of precipitate nucleation and growth rates from ultrasonic harmonic generation. Appl. Phys. Lett. **77**, 1952–1954 (2000)

40. A. Metya, M. Ghosh, N. Parida, S.P. Sagar, Higher harmonic analysis of ultrasonic signal for ageing behavior study of C-250 grade maraging steel. NDT E Int. **41**, 484–489 (2008)
41. W.F. Miao, D.E. Laughlin, Precipitation hardening in aluminum alloy 6022. Scr. Mater. V **40**, 873–878 (1999)
42. J. Kim, K.-Y. Jhang, Evaluation of ultrasonic nonlinear characteristics in heat-treated aluminum alloy (Al-Mg-Si-Cu). Adv. Mater. Sci. Eng. **407846**, 1–6 (2013)
43. J.H. Cantrell, X.G. Zhang, Nonlinear acoustic response from precipitate-matrix misfit in a dislocation network. J. Appl. Phys. **84**, 5469–5472 (1998)
44. M. Song, Modeling the hardness and yield strength evolutions of aluminum alloy with rod/needle-shaped precipitates. Mater. Sci. Eng., A **443**, 172–177 (2007)
45. R.S. Yassar, D.P. Field, H. Weiland, Transmission electron microscopy and differential scanning calorimetry studies on the precipitation sequence in an Al-Mg-Si alloy: AA6022. J. Mater. Res. **20**, 2705–2711 (2011)
46. S. Krishnan, M. O'Donnell, Transmit aperture processing for nonlinear contrast agent imaging. Ultrason. Imaging **18**, 77–105 (1996)
47. Y. Ohara, K. Kawashima, R. Yamada, H. Horio, Evaluation of amorphous diffusion bonding by nonlinear ultrasonic method. AIP Conf. Proc. **700**, 944–951 (2004)
48. A. Viswanath, B.P.C. Rao, S. Mahadevan, T. Jayakumar, B. Raj, Microstructural characterization of M250 grade maraging steel using nonlinear ultrasonic technique. J. Mater. Sci. **45**, 6719–6726 (2010)
49. F. Xie, Z. Guo, J. Zhang, Strategies for reliable second harmonic of nonlinear acoustic wave through cement-based materials. Nondestruct. Test. Eval. **29**, 183–194 (2014)
50. S. Choi, P. Lee, K.-Y. Jhang, A pulse inversion-based nonlinear ultrasonic technique using a single-cycle longitudinal wave for evaluating localized material degradation in plates. Int. J. Precis. Eng. Manuf. **20**, 549–558 (2019)
51. S. Zhang, X. Li, H. Jeong, S. Cho, H. Hu, Theoretical and experimental investigation of the pulse-echo nonlinearity acoustic sound fields of focused transducers. Appl. Acoust. **117**, 145–149 (2017)
52. H. Jeong, S. Cho, S. Zhang, X. Li, Acoustic nonlinearity parameter measurements in a pulse-echo setup with the stress-free reflection boundary. J. Acoust. Soc. Am. **143**, EL237–EL242 (2018)
53. C.-S. Kim, I.-K. Park, K.-Y. Jhang, N.-Y. Kim, Experimental characterization of cyclic deformation in copper using ultrasonic nonlinearity. J. Korean Soc. Nondestr. Test. **28**, 285–291 (2008)
54. S. Choi, H. Seo, K.-Y. Jhang, Noncontact evaluation of acoustic nonlinearity of a laser-generated surface wave in a plastically deformed aluminum alloy. Res. Nondestr. Eval. **26**, 13–22 (2013)
55. S. Choi, T. Nam, K.-Y. Jhang, C.S. Kim, Frequency response of narrowband surface waves generated by laser beams spatially modulated with a line-arrayed slit mask. J. Korean Phys. Soc. **60**, 26–30 (2012)
56. S. Kenderian, D. Cerniglia, B.B. Djordjevic, R.E. Green, Laser-generated acoustic signal interaction with surface flaws on rail wheels. Res. Nondestr. Eval. **16**, 195–207 (2005)
57. C.B. Scruby, L.E. Drain, *Laser Ultrasonics Techniques and Applications* (Adam Hilger, Bristol, 1990)
58. J. Li, L. Dong, C. Ni, Z. Shen, H. Zhang, Application of ultrasonic surface waves in the detection of microcracks using the scanning heating laser source technique. Chin. Opt. Lett. **10**, 111403–111406 (2012)
59. D. Dhital, J.R. Lee, A fully non-contact ultrasonic propagation imaging system for closed surface crack evaluation. Exp. Mech. **52**, 1111–1122 (2011)
60. C. Ni, L. Dong, Z. Shen, J. Lu, The experimental study of fatigue crack detection using scanning laser point source technique. Opt. Laser Technol. **43**, 1391–1397 (2011)
61. S.-K. Park, S.-H. Baik, H.-K. Cha, Y.-M. Cheong, Y.-J. Kang, Nondestructive inspection system using optical profiles and laser surface waves to detect a surface crack. J. Korean Phys. Soc. **56**, 333–337 (2010)

62. A. Moura, A.M. Lomonosov, P. Hess, Depth evaluation of surface-breaking cracks using laser-generated transmitted Rayleigh waves. J. Appl. Phys. **103**, 084911 (2008)
63. C.-S. Kim, K.-Y. Jhang, Acoustic nonlinearity of a laser-generated surface wave in a plastically deformed aluminum alloy. Chin. Phys. Lett. **29**, 120701 (2012)

Chapter 3
Measurement of Nonlinear Guided Waves

Cliff J. Lissenden and Mostafa Hasanian

Abstract Characteristics of nonlinear guided waves provide information about the current state of the material that comprises the waveguide. Notably, the information is related to the material's microstructure, which in turn influences the strength properties. However, the dispersive nature of guided waves makes their nonlinear characteristics more complicated than those of bulk waves. This chapter strives to describe nonlinear features of guided wave propagation including higher harmonic generation and wave mixing. It provides a methodology for selecting wave modes and frequencies that provide the best opportunities for measurements as well as measurement techniques and a glimpse of some recent results.

Nomenclature

I	second rank identity tensor
tr(.)	trace of the tensor
Re(.)	real part of the complex argument
u	displacement vector
U(Z)	displacement profile through thickness of plate (i.e., wavestructure)
H	displacement gradient tensor
E	Lagrange strain tensor
T	Cauchy stress tensor
P	First Piola-Kirchhoff stress tensor
S	Second Piola-Kirchhoff stress tensor
n	Outward normal unit vector
ρ	mass density

C. J. Lissenden (✉) · M. Hasanian
Department of Engineering Science and Mechanics, The Pennsylvania State University, University Park, USA
e-mail: Lissenden@psu.edu

M. Hasanian
e-mail: hasanian.mostafa@gmail.com

K.-Y. Jhang et al. (eds.), *Measurement of Nonlinear Ultrasonic Characteristics*, Springer Series in Measurement Science and Technology,
https://doi.org/10.1007/978-981-15-1461-6_3

λ_L, μ	Lame's constants
A, B, C	Landau-Lifshitz 3rd order elastic constants
$\beta' = A_2/A_1^2$	relative nonlinearity parameter, where A_1 and A_2 are the amplitudes of the primary and secondary wave fields respectively
W	strain energy function
h	half-thickness of plate
d	plate thickness
λ	wavelength
k	wavenumber
$\mathbf{K}$	wavevector
$\mathbf{p}$	position vector
M_{ab_r}	mixing power associated with the interaction of waves a and b, which generates the internally resonant mode r
θ	wave interaction angle between the primary waves
γ	direction of secondary wave vector with respect to wave a
ω	circular frequency
f	frequency in Hz
c_L, c_T	longitudinal wave speed, transverse wave speed
c_p, c_g	phase velocity (speed of an individual wave), group velocity (speed of a group of waves having similar frequencies, i.e., rate at which the envelope travels)
V_{pp}	peak-to-peak voltage

3.1 Introduction

This chapter describes propagation characteristics of weakly nonlinear elastodynamic guided waves with emphasis on measurement technology. It is limited to waveguides having finite dimensions. Most of the applications for these measurements are in the ultrasonic regime and are aimed at nondestructive characterization of material degradation. The wave nonlinearity is linked directly to the material nonlinearity, but there are complications that need to be addressed. Two key characteristics of weakly nonlinear guided waves are that the signal of interest occurs at a frequency other than that of the primary waves and its amplitude is small relative to the primary wave's amplitude. In fact, the models to date presume that the primary wave's amplitude remains constant, despite the fact that there must be power flux from the primary waves to the secondary waves. The change in frequency content occurs because the primary waves become distorted by self-interactions or mutual interactions between primary waves having different frequencies. The limitation that the nonlinearity is weak has two important consequences:

(1) It enables implementation of a perturbation solution to the nonlinear wave equation;

(2) It means that the signal of interest is quite small relative to the primary waves, demanding careful measurements and due consideration of measurement system nonlinearities.

There is a strong desire to improve the signal-to-noise ratio (SNR) as much as possible to enhance the reliability of the nonlinear measurements. Thus, it is common to use finite amplitude primary waves in an attempt to improve the SNR of the nonlinearity-generated secondary waves. However, finite amplitude elastodynamic waves may have amplitudes only on the order of tens of nanometers, so they are still quite small. The point is to use primary waves having the largest practical amplitude.

Even in their pristine state, metals display anharmonicity; i.e., their elastic oscillations are not at a single frequency. As material degradation occurs, an extrinsic nonlinearity alters the intrinsic anharmonicity. Although not yet realized, nonlinear guided waves have exciting potential for characterization of material degradation at an early state that could change the maintenance paradigm from schedule-based to condition-based. Detection of early-stage degradation or incipient damage enables tracking damage over the service lifetime and provides opportunity for decision-making and time for logistical planning. The early-stage degradation occurs at the microstructure level through mechanisms such as slip bands, dislocation substructures, precipitates, inclusions, and microcracks; and can be viewed for our purposes as continuum damage. Ultrasonic nondestructive techniques based on linearity are typically insensitive to this early-stage degradation. Not included in this chapter, although an active area of recent research, is the nonlinearity associated with a single discontinuity, such as a single macroscale crack. Interested readers can search the literature on contact acoustic nonlinearity.

The remainder of this chapter is divided into five sections. Section 3.2 provides a background on linear guided wave propagation and a brief history of the milestones in nonlinear guided wave propagation. Then the modeling necessary to select the primary waves that will in-turn generate strong secondary waves is provided in Sect. 3.3. Useful measurements of nonlinear guided waves are unlikely without intelligent selection of the primary waves. Thus, the formulation is included here for completeness even though it is given elsewhere, e.g., [1–3]. A list of nomenclature is included at the beginning of the chapter to provide clarity. Methods to actuate the primary waves and receive the secondary guided waves are described in Sect. 3.4, followed by a description of instrumentation and signal processing techniques in Sect. 3.5, and finally measurements of progressive degradation are provided and discussed in Sect. 3.6.

3.2 Background

Consider wave propagation in a waveguide having a traction-free lateral boundary at $Z = \pm h$, and that the waveguide is a lossless homogeneous isotropic material.

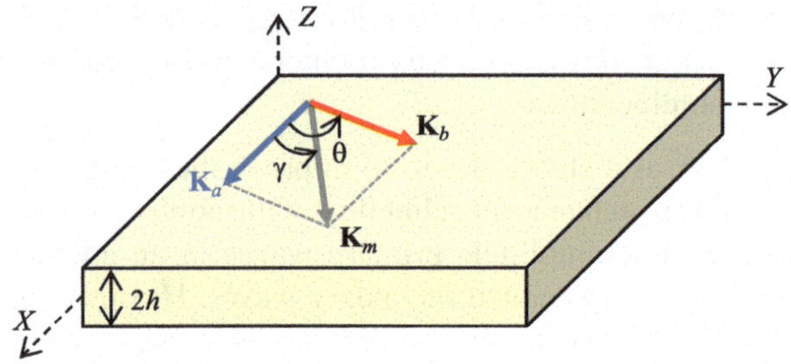

Fig. 3.1 Plate coordinate system and wave vectors related to the primary waves a and b whose interaction generates the secondary waves m. Wave vector $\mathbf{K}_a$ is in the X direction and θ is the interaction angle between waves a and b

For simplicity, this section is limited to planar waves in flat plates in a vacuum, for which the coordinate system is shown in Fig. 3.1.

3.2.1 Linear Guided Wave Propagation

For the problem just defined, at any frequency there are multiple wave modes that propagate, and they are in general dispersive, meaning that the velocity and wavenumber depend on the frequency. Dispersive waves exhibit pulse-spreading, which results in amplitude reduction as the waves travel. The relationships between the velocity, wavenumber, and frequency are found from the linear wave equation

$$[\lambda_L + \mu]\nabla[\nabla \cdot \mathbf{u}] + \mu\nabla^2\mathbf{u} = \rho\ddot{\mathbf{u}} \tag{3.1}$$

and the traction-free lateral surfaces

$$\mathbf{T} \cdot \mathbf{n} = 0 \text{ for } Z = \pm h. \tag{3.2}$$

Presuming wave propagation in the X-direction, plane strain conditions, and some tedious manipulations using the Helmholtz decomposition leads to the Rayleigh-Lamb dispersion relations [4]

$$\frac{\tan(qh)}{\tan(ph)} = -\left[\frac{4k^2pq}{\left[q^2 - k^2\right]^2}\right]^{\pm 1} \tag{3.3}$$

where the sign of the exponent indicates symmetric (+1) and antisymmetric (−1) modes respectively, and

$$\begin{aligned} p^2 &= \left[\tfrac{\omega}{c_L}\right]^2 - k^2 \\ q^2 &= \left[\tfrac{\omega}{c_T}\right]^2 - k^2. \end{aligned} \tag{3.4}$$

The transcendental Eq. 3.3 is solved numerically for the wavenumber, from which the phase velocity can be computed, $c_p = \omega/k$, and then the group velocity can be determined, $c_g = d\omega/dk$. Each propagating wave has a unique wavestructure (i.e., displacement profile) at each frequency. For these Lamb waves, the names symmetric modes and antisymmetric modes are based upon the displacement component in the *X*-direction. In summary, Lamb waves are multi-modal, dispersive, and each wave mode-frequency combination has a unique wavestructure. Thus, received wave signals are often difficult to analyze because there can be multiple modes propagating at different group velocities.

The solution to Eqs. 3.1 and 3.2 is much simpler for shear-horizontal (SH) wave modes, where the particle motion is solely in the *Y*-direction for a wave vector in the *X*-direction. The SH wave dispersion relations are simply

$$2qh = n\pi \tag{3.5}$$

where $n = 0, 1, 2, \ldots$ and symmetric modes occur for even values of n while antisymmetric modes occur for odd values of n.

The phase velocity dispersion curves and group velocity dispersion curves for both Lamb and SH waves are plotted in Fig. 3.2 for an aluminum plate. The modes are identified, except for those having higher cutoff frequencies.

3.2.2 Brief History of Nonlinear Guided Waves

While nonlinear acoustics in solid media has been studied for over 50 years, Deng [5–7] appears to be the first to study nonlinear ultrasonic guided waves in plates. Deng applied the partial waves method to analyze second harmonics (i.e., wave motion at twice the excitation frequency) of SH waves. He found that if the phase velocity of the dispersive SH primary wave was equal to the longitudinal wave speed then the second harmonic was a symmetric (Lamb) mode polarized perpendicular to the SH wave and that it increased linearly in amplitude. DeLima and Hamilton [1] formulated the problem in terms of the guided wave modes using the normal mode expansion, which simplifies the analysis, makes it more comprehensive, and leads to solution by successive approximations (see also Deng's work [8]). They show that cumulative second harmonics, whose amplitudes increase linearly with propagation distance, require that the ***internal resonance conditions*** be met; i.e.,

- the second harmonic is a propagating mode,
- the primary and secondary waves are phase-matched (i.e., synchronized), and
- non-zero power is transferred from the primary to secondary waves.

Phase-matching and non-zero power flux were studied by a number of authors. Srivastava and Lanza di Scalea [9] theoretically studied the symmetry characteristics of the primary and secondary waves (including higher harmonics above the second) and conducted a set of conclusive experiments. Likewise, Müller et al. [10] used parity

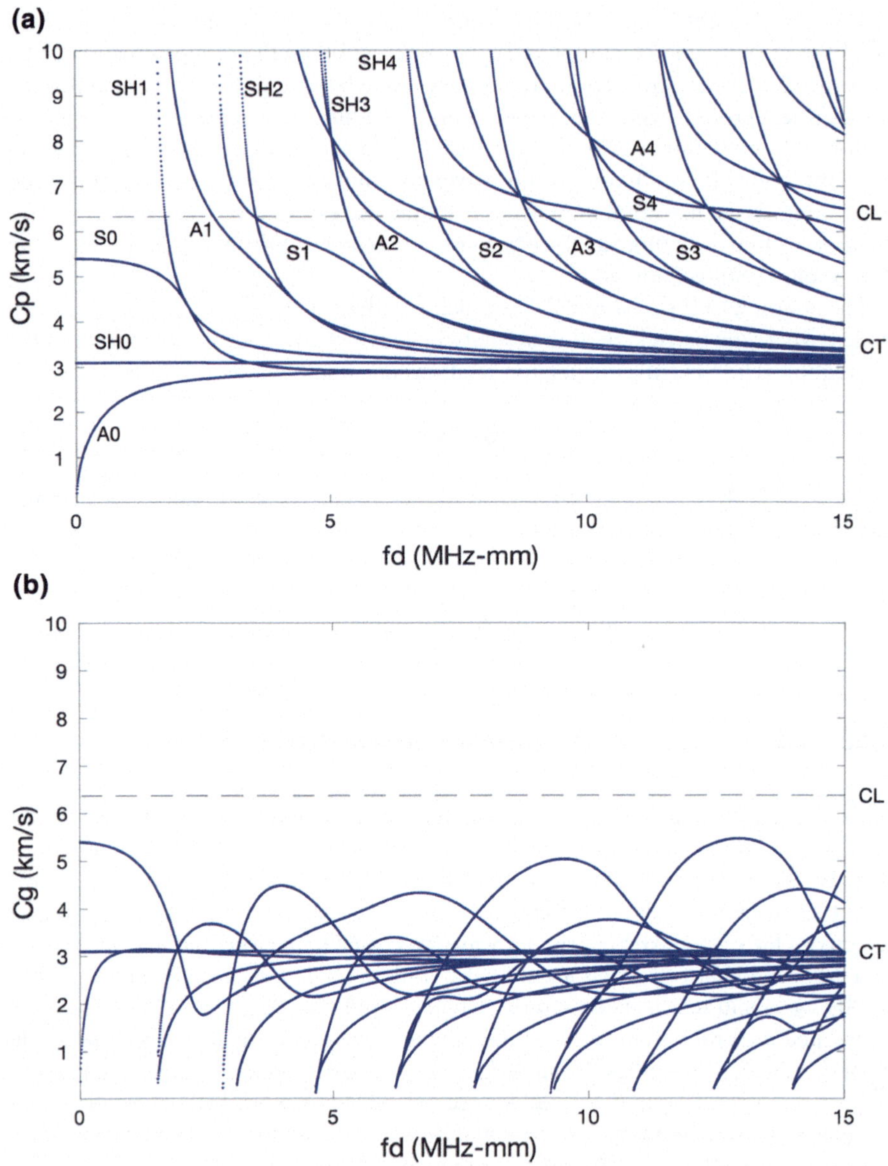

Fig. 3.2 Dispersion curves for an aluminum plate: **a** phase velocity and **b** group velocity. $\rho = 2700\ \text{kg/m}^3$, $c_L = 6300$ m/s, $c_T = 3100$ m/s ($\lambda_L = 55.27$ GPa, $\mu = 25.95$ GPa)

analysis based on the symmetry/antisymmetry of the displacement profile to show to which secondary modes there is no power flux, and also assessed group velocity matching. Matsuda and Biwa [11] identified all of the phase-matching points for Lamb waves in a plate, and also assessed group velocity matching. Then Chillara and Lissenden [12] formulated the problem in terms of self and mutual wave interactions, which enables setting up the various boundary value problems.

Now we highlight some of the early measurements of nonlinear guided waves, while more recent advances are discussed later in the chapter. Deng et al. [13] showed the cumulative nature of the second harmonic generated by Lamb waves at the S2/A2 mode-crossing point using angle-beam transducers on an aluminum plate (see Sect. 3.3.3.1). Then Bermes et al. [14] showed that S2 second harmonic Lamb waves generated by S1 primary Lamb waves are cumulative as well. They used an angle-beam transducer for generation and a laser interferometer for reception. Pruell et al. [15] employed similar measurements as Bermes et al. to show the sensitivity of the second harmonic to plastic deformation. Their results also show the challenges associated with generating a single dominant Lamb wave mode above the first cutoff frequency. In a subsequent study, Pruell et al. [16] used a similar test setup to assess the sensitivity of second harmonic generation of the S2 Lamb mode to low-cycle fatigue damage. They reported a monotonic increase in the normalized acoustic nonlinearity with fatigue life fraction. Realizing that Lamb wave modes with out-of-plane displacement components are more conducive for generation and reception, Lee et al. [17] actuated the A1 Lamb mode with an angle-beam transducer, which is both phase-matched and group velocity matched with the A2 Lamb mode. While perturbation analysis and confirming experiments have shown that second harmonics are symmetric modes [9], in contrast the nonlinearity measured by Lee et al. [17] increased with propagation distance.

Readers interested in other descriptions of nonlinear guided waves may find very useful information in Chap. 20 in [4], Chaps. 1, 6, 9 in [3], and [18–20]. Likewise, readers interested in nonlinear Rayleigh waves, which have many applications in nondestructive evaluation, are referred to Refs. [3] and [18]. To enhance focus, the scope of this chapter is limited to guided waves that propagate in finite cross-sections, and are dispersive, which is not the case for Rayleigh waves.

3.3 Primary Wave Selection for Secondary Wave Generation

This section provides a mathematical basis for understanding and modeling nonlinear guided wave propagation. We consider that planar primary waves are propagating in arbitrary directions within a waveguide, and that these waves interact to generate secondary waves at different frequencies. We will see that these could be self-interactions, which generate secondary waves at integer multiples of the primary waves' frequency. Alternatively, there could be mutual interactions between waves *a* and b having frequencies f_a and f_b, which generate secondary waves at the sum and difference frequencies $f_a \pm f_b$ (or higher order combinations thereof). It is the secondary waves that provide information about the material nonlinearity that we seek, thus the interrogation signals have frequencies other than the excitation frequencies of the primary waves.

3.3.1 Basic Principles

The combination of guided waves to propagate long distances and interrogate otherwise inaccessible material domains with the cumulative nature of internally resonant secondary waves has excellent potential for nondestructive material characterization. Thus, spurred by Deng's early findings [5–7], the hunt was on to determine which primary wave modes would generate the secondary waves most sensitive to a particular type of material degradation. In fact, as will be shown, only in very special cases does a propagating mode exist at the phase-matched velocity of the secondary waves. The basic principles for how to develop nonlinear guided waves into a viable method for nondestructive characterization of materials became apparent. First—determine which wave combinations satisfy the internal resonance conditions. Second—identify which primary guided wave modes provide strong power flux to the secondary modes. Third—discover which measurement setups are least sensitive to instrumentation nonlinearity. The remainder of Sect. 3.3 addresses these first two principles, while the third principle permeates Sects. 3.4–3.6.

3.3.2 Theoretical Formulation

Assume that the lossless material is homogeneous, isotropic, and can be described as hyperelastic with the cubic strain energy function

$$W = \frac{\lambda_L}{2}[\mathrm{tr}(\mathbf{E})]^2 + \mu \mathrm{tr}(\mathbf{E}^2) + \frac{A}{3}\mathrm{tr}(\mathbf{E}^3) + B\mathrm{tr}(\mathbf{E})\mathrm{tr}(\mathbf{E}^2) + \frac{C}{3}[\mathrm{tr}(\mathbf{E})]^3. \tag{3.6}$$

The second Piola-Kirchhoff stress is then determined from

$$\mathbf{T}_{RR} = \frac{\partial W}{\partial \mathbf{E}} \tag{3.7}$$

and is related to the first Piola-Kirchhoff stress by

$$\mathbf{S} = [\mathbf{I} + \mathbf{H}]\mathbf{T}_{RR}. \tag{3.8}$$

In addition, the strain-displacement relation is

$$\mathbf{E} = \frac{1}{2}\left[\mathbf{H} + \mathbf{H}^T + \mathbf{H}^T\mathbf{H}\right]. \tag{3.9}$$

Let us consider the interaction of two guided waves a and b, such that the displacement fields can be decomposed into primary and secondary components. Thus, the displacement gradient field can be written as

$$\mathbf{H} = \mathbf{H}_a + \mathbf{H}_b + \mathbf{H}_{aa} + \mathbf{H}_{bb} + \mathbf{H}_{ab} \tag{3.10}$$

for up to second order interactions, where the single subscripts a and b refer to the primary waves, the subscripts aa and bb refer to self-interactions, and the subscript ab denotes the mutual interaction between waves a and b.

It is now possible to decompose the second Piola-Kirchhoff stress tensor into linear and nonlinear parts based on the displacement gradient terms

$$\mathbf{S}(\mathbf{H}) = \mathbf{S}_L(\mathbf{H}_a) + \mathbf{S}_L(\mathbf{H}_b) + \mathbf{S}_L(\mathbf{H}_{aa}) + \mathbf{S}_L(\mathbf{H}_{bb}) + \mathbf{S}_L(\mathbf{H}_{ab}) + \mathbf{S}_{NL}(\mathbf{H}_a + \mathbf{H}_b) \tag{3.11}$$

and to write the nonlinear terms of order 2 in the displacement gradient due to interaction between waves a and b as

$$\mathbf{S}_{NL}(\mathbf{H}_a + \mathbf{H}_b) = \mathbf{S}_{NL}(\mathbf{H}_a, \mathbf{H}_a, 2) + \mathbf{S}_{NL}(\mathbf{H}_b, \mathbf{H}_b, 2) + \mathbf{S}_{NL}(\mathbf{H}_a, \mathbf{H}_b, 2) \tag{3.12}$$

where the first two terms on the right-hand-side are due to self-interaction and the third term is due to mutual interaction. The mutual interaction term is

$$\begin{aligned}
\mathbf{S}_{NL}(\mathbf{H}_a, \mathbf{H}_b, 2) = {} & \frac{\lambda_L}{2}\mathrm{tr}\left(\mathbf{H}_b + \mathbf{H}_b^T\right)\mathbf{H}_a + \mu\mathbf{H}_a\left[\mathbf{H}_b + \mathbf{H}_b^T\right] + \frac{\lambda_L}{2}\mathrm{tr}\left(\mathbf{H}_a + \mathbf{H}_a^T\right)\mathbf{H}_b \\
& + \mu\mathbf{H}_b\left[\mathbf{H}_a + \mathbf{H}_a^T\right] + \frac{\lambda_L}{2}\mathrm{tr}\left(\mathbf{H}_a^T\mathbf{H}_b + \mathbf{H}_b^T\mathbf{H}_a\right)\mathbf{I} + 2C\mathrm{tr}(\mathbf{H}_a)\mathrm{tr}(\mathbf{H}_b)\mathbf{I} \\
& + \mu\left[\mathbf{H}_a^T\mathbf{H}_b + \mathbf{H}_b^T\mathbf{H}_a\right] + B\mathrm{tr}(\mathbf{H}_a)\left[\mathbf{H}_b + \mathbf{H}_b^T\right] + B\mathrm{tr}(\mathbf{H}_b)\left[\mathbf{H}_a + \mathbf{H}_a^T\right] \\
& + \frac{B}{2}\mathrm{tr}\left(\mathbf{H}_a\mathbf{H}_b + \mathbf{H}_b\mathbf{H}_a + \mathbf{H}_a^T\mathbf{H}_b + \mathbf{H}_b^T\mathbf{H}_a\right)\mathbf{I} \\
& + \frac{A}{4}\left[\mathbf{H}_a\mathbf{H}_b + \mathbf{H}_b\mathbf{H}_a + \mathbf{H}_a^T\mathbf{H}_b^T + \mathbf{H}_b^T\mathbf{H}_a^T + \mathbf{H}_a^T\mathbf{H}_b + \mathbf{H}_b^T\mathbf{H}_a + \mathbf{H}_a\mathbf{H}_b^T + \mathbf{H}_b\mathbf{H}_a^T\right]
\end{aligned} \tag{3.13}$$

and the self-interaction terms in Eq. 3.12 are analogous.

The boundary value problem is given by the balance of linear momentum

$$\mathrm{Div}(\mathbf{S}(\mathbf{H})) = \rho\ddot{\mathbf{u}} \tag{3.14}$$

and the traction-free lateral boundaries

$$\mathbf{S}(\mathbf{H}) \cdot \mathbf{n} = 0 \text{ for } Z = \pm h. \tag{3.15}$$

Due to the weak nonlinearity ($\mathbf{u}_{aa}, \mathbf{u}_{bb}, \mathbf{u}_{ab} \ll \mathbf{u}_a, \mathbf{u}_b$), a perturbation method based on successive approximations can be employed to reorganize it into five separate boundary value problems:

$$\begin{aligned} \mathrm{Div}(\mathbf{S}_L(\mathbf{H}_a)) - \rho\ddot{\mathbf{u}}_a &= 0 \\ \mathbf{S}_L(\mathbf{H}_a)\cdot\ \mathbf{n} = 0 \text{ for } Z &= \pm h \end{aligned} \tag{3.16}$$

$$\begin{aligned} \mathrm{Div}(\mathbf{S}_L(\mathbf{H}_b)) - \rho\ddot{\mathbf{u}}_b &= 0 \\ \mathbf{S}_L(\mathbf{H}_b)\cdot\ \mathbf{n} = 0 \text{ for } Z &= \pm h \end{aligned} \tag{3.17}$$

$$\begin{aligned} \mathrm{Div}(\mathbf{S}_L(\mathbf{H}_{aa})) - \rho\ddot{\mathbf{u}}_{aa} &= -\mathrm{Div}(\mathbf{S}_{NL}(\mathbf{H}_a, \mathbf{H}_a, 2)) \\ \mathbf{S}_L(\mathbf{H}_{aa})\cdot\ \mathbf{n} &= -\mathbf{S}_{NL}(\mathbf{H}_a, \mathbf{H}_a, 2)\cdot\ \mathbf{n} \text{ for } Z = \pm h \end{aligned} \tag{3.18}$$

$$\begin{aligned} \mathrm{Div}(\mathbf{S}_L(\mathbf{H}_{bb})) - \rho\ddot{\mathbf{u}}_{bb} &= -\mathrm{Div}(\mathbf{S}_{NL}(\mathbf{H}_b, \mathbf{H}_b, 2)) \\ \mathbf{S}_L(\mathbf{H}_{bb})\cdot\ \mathbf{n} &= -\mathbf{S}_{NL}(\mathbf{H}_b, \mathbf{H}_b, 2)\cdot\ \mathbf{n} \text{ for } Z = \pm h \end{aligned} \tag{3.19}$$

$$\begin{aligned} \mathrm{Div}(\mathbf{S}_L(\mathbf{H}_{ab})) - \rho\ddot{\mathbf{u}}_{ab} &= -\mathrm{Div}(\mathbf{S}_{NL}(\mathbf{H}_a, \mathbf{H}_b, 2)) \\ \mathbf{S}_L(\mathbf{H}_{ab})\cdot\ \mathbf{n} &= -\mathbf{S}_{NL}(\mathbf{H}_a, \mathbf{H}_b, 2)\cdot\ \mathbf{n} \text{ for } Z = \pm h. \end{aligned} \tag{3.20}$$

The displacement fields associated with the primary waves can be written as

$$\begin{aligned} \mathbf{u}_a &= \mathrm{Re}\left(\mathbf{U}_a(Z)e^{i[\mathbf{K}_a\cdot\ \mathbf{p}(X,Y)-\omega_a t]}\right) \\ \mathbf{u}_b &= \mathrm{Re}\left(\mathbf{U}_b(Z)e^{i[\mathbf{K}_b\cdot\ \mathbf{p}(X,Y)-\omega_b t]}\right). \end{aligned} \tag{3.21}$$

Substituting the primary wave fields into the nonlinear stress components, Eq 3.12, leads to the possible secondary wave fields being expressible in terms of the exponential functions

$$e^{\pm i\left[[\mathbf{K}_a \pm \mathbf{K}_b]\cdot\ \mathbf{p}(X,Y) - [\omega_a \pm \omega_b]t\right]}$$

for mutual interactions when $\omega_a > \omega_b$. The subscript b can be replaced by a and vice versa to get the exponential functions associated with the self-interactions aa and bb respectively—giving the second harmonic and the quasi-static pulse in the special case of single frequency primary waves. As shown in Fig. 3.1, the interaction angle between the primary wave fields a and b is θ, with waves a being in the X-direction. Wave interactions are classified as co-directional ($\theta = 0$), counter-propagating ($\theta = 180°$), and non-collinear ($\theta \neq 0$ and $\theta \neq 180°$). The angle γ defines the direction of the secondary wave field with respect to waves a.

It remains to determine whether propagating waves exist at the wavenumber required for the secondary frequency. We follow the solution approach used by de Lima and Hamilton [1] for the mutual interaction problem, which is based on the normal mode expansion and the complex reciprocity theorem [21]. The self-interaction problem solution can be extrapolated from the mutual interaction solution.

$$\mathbf{S}_L(\mathbf{H}_{ab}) = \mathrm{Re}\left(\sum_{m=1}^{\infty} A_m(X, Y)\mathbf{S}_m e^{-i[\omega_a \pm \omega_b]t}\right) \quad (3.22)$$

$$\dot{\mathbf{u}}_{ab} = \mathrm{Re}\left(\sum_{m=1}^{\infty} A_m(X, Y)\mathbf{V}_m e^{-i[\omega_a \pm \omega_b]t}\right) \quad (3.23)$$

where the modal variables are connected in a linear elastic sense:

$$\begin{aligned} \mathbf{S}_m &= \tfrac{\lambda_L}{2}\mathrm{tr}\left(\mathbf{H}_m + \mathbf{H}_m^T\right)\mathbf{I} + \mu\left(\mathbf{H}_m + \mathbf{H}_m^T\right) \\ \mathbf{V}_m &= \dot{\mathbf{U}}_m, \ \mathbf{H}_m = \nabla\mathbf{U}_m. \end{aligned} \quad (3.24)$$

Applying Auld's complex reciprocity theorem [21] yields a partial differential equation in the plane of the plate:

$$\begin{aligned} 4\mathbf{P}'_{mn} \cdot \mathbf{n}_X\left[\frac{\partial}{\partial X} - i\mathbf{K}_n^* \cdot \mathbf{n}_X\right]A_m(X, Y) + 4\mathbf{P}'_{mn} \cdot \mathbf{n}_Y\left[\frac{\partial}{\partial Y} - i\mathbf{K}_n^* \cdot \mathbf{n}_Y\right]A_m(X, Y) \\ = \left[f_{ab\_n}^{surf} + f_{ab\_n}^{vol}\right]e^{i[\mathbf{K}_a \pm \mathbf{K}_b]\cdot \mathbf{p}(X,Y)} \end{aligned} \quad (3.25)$$

where

$$\mathbf{P}'_{mn} = -\frac{1}{4}\int_{-h}^{h} \frac{\mathbf{S}_m\mathbf{V}_n^* + \mathbf{S}_n^*\mathbf{V}_m}{4} dZ \quad (3.26)$$

$$P_{mn} = \mathbf{P}'_{mn} \cdot \mathbf{r}_m \quad (3.27)$$

$$f_{ab\_n}^{surf} = -\frac{1}{2}\mathbf{S}_{NL}(\mathbf{H}_a, \mathbf{H}_b, 2)\mathbf{V}_n^* \cdot \mathbf{n}_Z\Big|_{-h}^{h} \quad (3.28)$$

$$f_{ab\_n}^{vol} = -\frac{1}{2}\int_{-h}^{h} \mathrm{Div}(\mathbf{S}_{NL}(\mathbf{H}_a, \mathbf{H}_b, 2)) \cdot \mathbf{V}_n^* dZ. \quad (3.29)$$

$\mathbf{P}'_{mn}$ is the Poynting vector integrated through the thickness of the plate, while P_{mn} projects that vector in the direction of the secondary waves. Likewise, $f_{ab\_n}^{surf}$ and $f_{ab\_n}^{vol}$ can be thought of as nonlinear surface and volumetric (body) forces respectively. The solution for the modal amplitudes is:

$$A_m(X, Y) = \frac{f_{ab\_n}^{surf} + f_{ab\_n}^{vol}}{4P_{mn}} \frac{[\mathbf{K}_a \pm \mathbf{K}_b] \cdot \mathbf{p}(X, Y)}{\mathbf{K}_a \pm \mathbf{K}_b} e^{i[\mathbf{K}_a \pm \mathbf{K}_b]\cdot \mathbf{p}(X,Y)} \quad \text{if } \mathbf{K}_n^* = \mathbf{K}_a \pm \mathbf{K}_b \quad (3.30)$$

$$A_m(X, Y) = -\frac{i\left[f_{ab\_n}^{surf} + f_{ab\_n}^{vol}\right]}{4P_{mn}\left|\mathbf{K}_n^* - [\mathbf{K}_a \pm \mathbf{K}_b]\right|}\left[e^{i\mathbf{K}_n^*\cdot \mathbf{p}(X,Y)} - e^{i[\mathbf{K}_a \pm \mathbf{K}_b]\cdot \mathbf{p}(X,Y)}\right] \quad \text{if } \mathbf{K}_n^* \neq \mathbf{K}_a \pm \mathbf{K}_b \quad (3.31)$$

Equation 3.30 defines the internal resonance conditions for this general case of mutual wave interactions. The first fraction shows that power must be transferred to the secondary waves. The second fraction indicates that the amplitude of the secondary waves increases linearly with propagation distance. The condition 'if' indicates that Eq 3.30 is the solution when the secondary wave vector is the vector sum (or difference) of the primary wave vectors, which is a generalization of phase matching. See [1] for a discussion on the propagation distance upper limit for linear increase in amplitude. Equation 3.31, on the other hand, represents the bounded oscillation solution associated with the beat phenomenon. Some readers might be interested that Mazilu et al. [22] have shown that the summation of wave vectors does not apply for evanescent waves.

Orthogonality dictates that power transfer only occurs when $m = n$, which leads to internal resonance, thus we will denote the secondary wavefield with the subscript r,

$$\mathbf{u}_r = \mathrm{Re}\big(A_r(X, Y)\mathbf{U}_r(Z)e^{-i[\omega_a \pm \omega_b]t}\big) \tag{3.32}$$

which can be also written as

$$\mathbf{u}_r = \mathrm{Re}\Big(M_{ab\_r}\mathrm{Amp}(\mathbf{U}_a)\mathrm{Amp}(\mathbf{U}_b)\hat{\mathbf{U}}_r(Z)\big[\mathbf{r}_r \cdot \mathbf{p}(X, Y)\big]e^{i[\mathbf{K}_r \cdot \mathbf{p}(X,Y) - \omega_r t]}\Big) \tag{3.33}$$

where

$$M_{ab\_r} = \frac{f_{ab\_r}^{surf} + f_{ab\_r}^{vol}}{4P_{rr}} \frac{\mathrm{Amp}(\mathbf{U}_r)}{\mathrm{Amp}(\mathbf{U}_a)\mathrm{Amp}(\mathbf{U}_b)} \tag{3.34}$$

$$\hat{\mathbf{U}}_r(Z) = \frac{\mathbf{U}_r(Z)}{\mathrm{Amp}(\mathbf{U}_r(Z))} \tag{3.35}$$

$$\mathbf{r}_r = \frac{\mathbf{K}_a \pm \mathbf{K}_b}{|\mathbf{K}_a \pm \mathbf{K}_b|}. \tag{3.36}$$

Call M_{ab_r} the mixing power associated with the mutual wave interaction, although it will also be applied to the self-interaction case. It quantifies the power flux from primary waves to secondary waves and provides a convenient way to compare different types of wave interactions. It does not however consider the size of the interaction zone because the waves are considered to be planar continuous waves. $\hat{\mathbf{U}}_r(Z)$ is the normalized displacement profile through the thickness. Finally, $\mathbf{r}_r$ is the unit vector in the direction of the secondary wave field.

3.3.3 Self-interaction

Parity analysis based on the symmetry of the plate geometry shows that it may not be possible to transfer power from prescribed primary waves to a certain type of secondary waves [9, 10]. Second order self-interactions of any Lamb or SH primary waves transfer power only to symmetric Lamb waves [23], but third order interactions are more complicated as indicated in Table II of [24]. The phase matching points for primary waves that are Lamb waves and SH waves are shown in Fig. 3.3. These points will be discussed in the subsequent sections. Outlined below are some of the secondary waves generated by self-interaction that satisfy the internal resonance conditions. We emphasize that this is not a comprehensive list.

3.3.3.1 Primary Lamb Waves in Plate

Table 3.1 provides the specifics for the internal resonance points shown in Fig. 3.3a. These three points are the most commonly studied in the literature to date. The primary waves for points 1 and 2 are symmetric Lamb waves at the longitudinal wave speed (i.e., $c_p = c_L$), which are special points because the out-of-plane displacement at the plate surface is zero, meaning that ultrasonic energy will not leak into any fluid on the plate surface. All of the symmetric Lamb modes having $c_p = c_L$ are equally spaced along the *fd* axis,

$$(fd)_n = \frac{nc_T}{\sqrt{1-[c_T/c_L]^2}}.$$

Thus, a symmetric mode always exists at twice the excitation frequency. Furthermore, the group velocities at $c_p = c_L$ are all the same. While the dispersivity of the S1 and S2 modes at $c_p = c_L$ is unfortunately high, the S4 mode at $c_p = c_L$ has relatively low dispersivity, meaning that the S4 second harmonic will suffer little pulse spreading. The mixing power computed from Eq. 3.34 for self-interaction is also reported for each internal resonance point. It is clear that the most power flux occurs for S2 primary waves and S4 secondary waves. The higher frequencies are associated with higher power flux. It is helpful to use the mixing power computed for planar bulk longitudinal waves to secondary longitudinal waves as a reference. In this case, for a primary wave frequency of 1.00 MHz the mixing power is 1.89 mm^{-2}, while for 3.56 MHz the mixing power is 24.6 mm^{-2}, showing how the mixing power increases with frequency. Likewise, the mixing power of 24.6 mm^{-2} for a longitudinal wave at 3.56 MHz is comparable to point 1 in Table 3.1 having the mixing power 15.4 mm^{-2} for a 1 mm thick plate (i.e., $d = 1$ mm).

Internal resonance point 3 occurs at the crossing point of the S2 and A2 Lamb modes and generates the S4 Lamb mode. All of these modes are highly dispersive, and despite the relatively high frequencies the mixing power is the lowest in Table 3.1. In addition, the group velocities of the A2, S2, and S4 modes at point 3 are less than

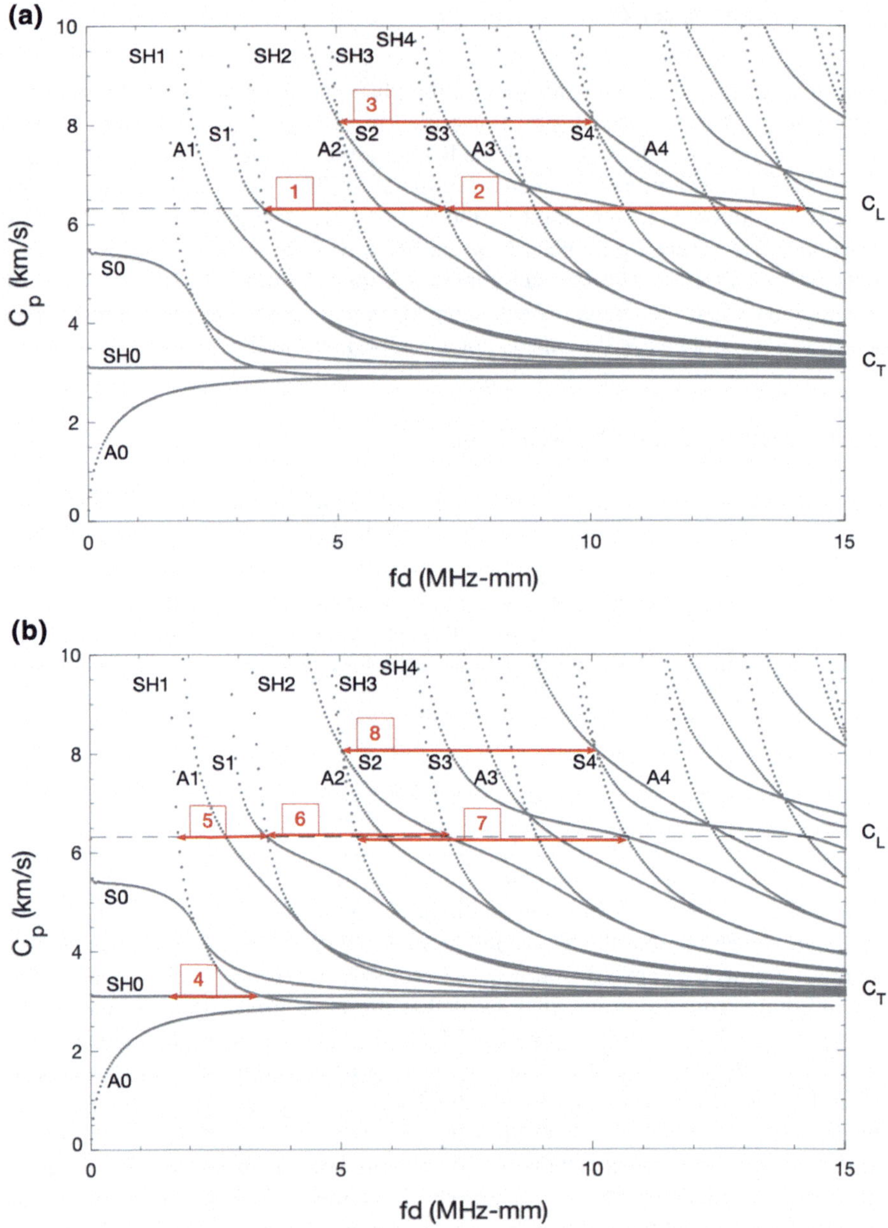

Fig. 3.3 Dispersion curves with internal resonance points marked for aluminum plate **a** primary Lamb waves and **b** primary SH waves

Table 3.1 Internal resonance points for Lamb primary waves in an aluminum plate (see Fig. 3.3a)

Point	Mode pair	*fd* (MHz mm)	c_p (mm/μs)	c_g (mm/μs)	M_{aa_r} (mm^{-2})
1	S1–S2	3.56	c_L	4.3–4.3	15.4
2	S2–S4	7.12	c_L	4.3–4.3	59.8
3	S2/A2–S4	~5.0	~8.1	~3.7–1.7	10.2

the group velocities of many other modes, which complicates the signal processing. However, Deng et al. [13] showed that the secondary waves are cumulative.

Müller et al. [10] analyzed the types of internal resonance points shown in Table 3.1 and Fig. 3.3a and Matlack et al. [25] conducted experiments using internal resonance points 1 and 2, where both primary and secondary waves have phase velocities of c_L. They used salol (phenyl salicylate) to bond an angle-beam actuator to the plate instead of using gel couplant in order to promote excitability of these modes that have no out-of-plane wave motion at the plate surface. They used a thin film of oil to couple the receiving angle-beam transducer to the plate because they found fluid couplant to provide less variability, although the amplitudes are lower. While the increase in the second harmonic amplitude with propagation distance was 4.23 times higher for point 2 than point 1, it is quite difficult to predominantly generate the S2 mode due to the multiplicity of modes at the higher frequency. Primary wave selection to generate cumulative second harmonics must consider the internal resonance conditions, as well as group velocity matching in order for the waves to interact as they propagate. Point 3, which occurs at a mode crossing, does not have matching group velocities. Transducer selection must also consider excitability of the primary waves and beam spreading.

Strictly speaking, it is not necessary to require phase matching, but rather nearly phase-matched waves can suffice [26–28]. Thus, the S0 mode at low frequencies can be useful because it is nearly nondispersive, which is explored in Sect. 3.6.3.1.

3.3.3.2 Primary Shear-Horizontal Waves in Plate

Deng's initial research on nonlinear guided waves analyzed SH primary waves [5–7]. The internal resonance points 4–8 given in Table 3.2 are shown in Fig. 3.3b. Liu et al.

Table 3.2 Internal resonance points for SH primary waves in an aluminum plate (see Fig. 3.3b)

Point	Mode pair	*fd* (MHz mm)	c_p (mm/μs)	c_g (mm/μs)	M_{aa_r} (mm^{-2})
4	SH0–S0	1.68	c_T	3.1–2.4	4.48
5	SH1–S1	1.78	c_L	1.5–4.3	6.32
6	SH2–S2	3.56	c_L	1.5–4.3	22.15
7	SH3–S3	5.34	c_L	1.5–4.3	56.29
8	SH3–S4	~5.0	~8.1	1.3–1.4	51.43

[23] demonstrated that the S3 secondary waves are cumulative at internal resonance point 8 using magnetostrictive transducers to both send SH3 waves and receive S4 waves.

The generation of secondary waves having a polarity completely different from that of the primary waves is a unique consequence of nonlinearity, which is concomitant with the coupling in the displacement gradient that comes from Eqs. 3.6 and 3.9. More discussion is provided in Refs. [23] and [29]. In this case SH waves generate symmetric Lamb waves. The difference in polarity must be accounted for when developing the reception system for making measurements, especially if there is a desire to receive both the primary waves and the secondary waves. A benefit of the secondary waves having a polarity different from the primary waves is that nonlinearities in the actuation system (e.g., synthesizer, amplifier, filters, transducer, couplant) are less likely to corrupt the material nonlinearity that is the goal of the measurement.

Group velocity matching is an important consideration for the generation of strong secondary waves from self-interaction when the primary waves are tone burst pulses, as is often the case. The internal resonance conditions implied by Eq. 3.30 were derived assuming that the primary waves are continuous waves. If the primary and secondary waves are finite wave packets having different group velocities, then they can separate and cease interacting. While the group velocities do not need to match exactly, the closer together they are, the longer they will interact. A simple analytical model was developed to understand what happens to secondary waves having a different group velocity than the primary waves [2]. The result for internal resonance point 4 in Table 3.2 is shown in Fig. 3.4. As the size of the interaction region grows the amplitude of the secondary waves increase. Then while the size of the interaction region remains constant the amplitude of the secondary waves remains constant, but the length of the wave packet increases. Furthermore, the amplitude of the secondary waves correlates directly with the primary pulse width. Analogous results were obtained by Xiang et al. [30] for internal resonance point 3 (i.e., the mode-crossing point) in Table 3.1 using finite element simulations and laboratory experiments. Additional discussion of group velocity can be found, for example, in Refs. [10, 11, 20, 31].

3.3.3.3 Axisymmetric Waves in Pipe

There is a close relationship between the dispersion curves for plates and axisymmetric waves in large-radius pipes with respect to second harmonic generation [32]. The dispersion curves for an aluminum pipe having an inner radius of 50 mm are shown in Fig. 3.5. The longitudinal modes L(m,n) are shown in Fig. 3.5a and the torsional modes T(m,n) are shown in Fig. 3.5b, where m and n represent the circumferential order and group order respectively. Axisymmetric longitudinal modes L(0,n) are comparable with Lamb waves, while torsional modes T(0,n) are comparable to SH waves. The modes are numbered in such a way that even values of n are quasi-symmetric, meaning the displacement profiles are close to, but not actually

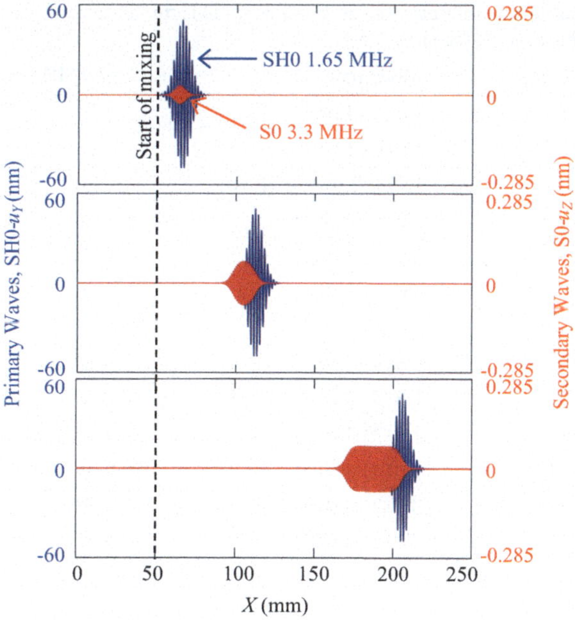

Fig. 3.4 Primary SH0 and secondary S0 wave packets at 5, 20, and 50 μs for internal resonance point 4 in a 1 mm thick aluminum plate. The primary pulse width is 27 mm and self-interaction begins at 50 mm. After Hasanian and Lissenden [2], with permission

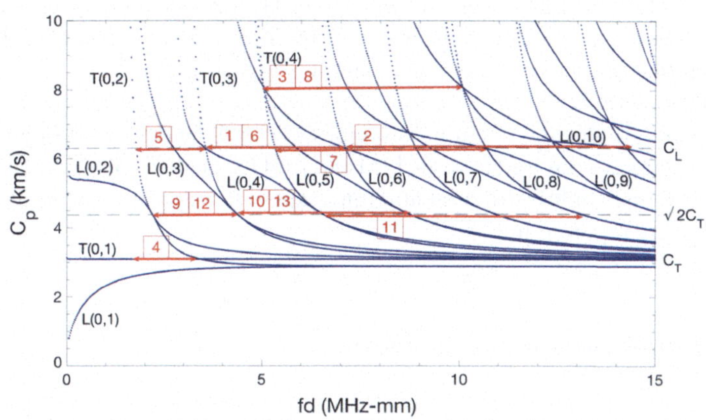

Fig. 3.5 Dispersion curves for aluminum pipes having a 50 mm inner radius

symmetric. The internal resonance points for axisymmetric waves in steel pipes were identified by Liu et al. [33]. Internal resonance points for aluminum pipes are identified in Fig. 3.5 as well as Table 3.3. Liu et al. [33] performed finite element analyses on internal resonance points 1, 6, and 12 in Table 3.3 to confirm their cumulative nature. Despite the lack of symmetry in the displacement profiles for L(0,n) and T(0,n) modes, power can only be transferred to L(0,n) modes due to the axisymmetry of the problem [33]. In addition to internal resonance points at the longitudinal

Table 3.3 Internal resonance points for axisymmetric waves in an aluminum pipes having a 50 mm inner radius (see Fig. 3.5)

Point	Mode pair	*fd* (MHz mm)	c_p (mm/μs)
1	L(0,4)–L(0,6)	3.56	c_L
2	L(0,6)–L(0,10)	7.12	c_L
3	L(0,5)/L(0,6)–L(0,10)	~5.0	~8.1
4	T(0,1)–L(0,2)	1.71	c_T
5	T(0,2)–L(0,4)	1.78	c_L
6	T(0,3)–L(0,6)	3.56	c_L
7	T(0,4)–L(0,8)	5.34	c_L
8	T(0,4)–L(0,10)	~5.0	~8.1
9	L(0,2)–L(0,3)	2.18	$\sqrt{2}c_T$
10	L(0,3)–L(0,5)	4.37	$\sqrt{2}c_T$
11	L(0,4)–L(0,8)	6.55	$\sqrt{2}c_T$
12	T(0,2)–L(0,3)	2.18	$\sqrt{2}c_T$
13	T(0,3)–L(0,5)	4.37	$\sqrt{2}c_T$

wave speed, the shear wave speed, and mode crossing points, there are also internal resonance points at the Lame wave speed, $\sqrt{2}c_T$. The phase matching at these points for a plate, where $c_p = \sqrt{2}c_T$, was analyzed by Matsuda and Biwa [11]. Liu et al. [33] showed that the power flux is quite small, suggesting that these are not points of practical concern.

Experiments conducted on aluminum pipes [34] will be described in Sect. 3.6.3. Interested readers are referred to Refs. [35] and [36] for analysis of nonlinear flexural waves. Finally, on a slightly different note, analyses of waveguides having an arbitrary cross-section are reported in [37–39].

3.3.4 Mutual Interaction in Plate

Consider the interaction of waves *a* and waves *b* that generates internally resonant waves *r*. Waves *a*, *b*, and *r* form what we will call a wave triplet and a vector analysis is use to satisfy the phase-matching condition [40]. A parametric analysis based on the symmetry features of Lamb-type and SH-type guided waves in plates was conducted to assess which primary waves are incapable of transferring power to certain secondary waves [2]. The mutual interactions that transfer nonzero power to specific types of secondary waves are shown in Table 3.4. It is interesting that non-collinear guided wave interactions are less restrictive than are collinear interactions.

The use of wave mixing to create mutual wave interactions is important for at least two reasons: (1) it provides a much broader range of opportunities to measure

Table 3.4 Types of guided wave interactions having nonzero power flux to secondary waves

Wave description (symmetry-type)	Primary waves types	Secondary waves types	
		Primary waves in any direction	Non-collinear primary waves
Like-same	S–S, A–A, SSH–SSH, ASH–ASH	S	S, SSH
Unlike-same	S–A, SSH–ASH	A	A, ASH
Like-mixed	S–SSH, A–ASH	SSH	SSH, S
Unlike-mixed	S–ASH, A–SSH	ASH	ASH, A

Legend: *S* sym Lamb, *A* antisym Lamb, *SSH* sym SH, *ASH* antisym SH

material-based nonlinearity and (2) it avoids making measurements at integer multiples of the excitation frequency, as is the case for second harmonic generation, where nonlinearities from the measurement system itself are common.

The important issue of group velocity matching for self-interactions is part of a broader topic for mutual interactions of tone burst-pulsed waves, which is about the finite size of the interaction (i.e., wave mixing) zone. Group velocity is just one of the variables that dictate the size of the interaction zone. Other variables include the interaction angle θ, the pulse durations, and dispersion. What size wave mixing zone is necessary to transfer sufficient power to the secondary waves is a question that must be answered on a case-by-case basis; and numerical simulations are an effective tool for this purpose. The subsequent three subsections discuss co-directional, counter-propagating, and non-collinear mutual interactions respectively.

3.3.4.1 Co-directional Waves, $\theta = 0°$

The main advantage of co-directional primary waves is that they can provide a large wave mixing zone, thus a lot of power can be transferred to the secondary waves if the group velocities are close together. However, this is not always the case, for example when the secondary waves propagate in the opposite direction to the primary waves ($\gamma = 180°$), which is known as one-way mixing. Such an arrangement could be valuable if it is beneficial for primary wave actuation and secondary wave reception to occur in the same vicinity, e.g., to interrogate an inaccessible region. Eight example sets of wave triplets are shown in Table 3.5 for co-directional wave interactions. Both waves *a* and *b* have $\theta = 0°$. The table provides the wave type and frequency for waves *a*, *b*, and *r*. It also provides the mixing power M_{ab_r}, which is a normalized measure of power flux to the secondary waves, the direction of the secondary waves, and whether the frequency combination is a sum or difference of the primary waves. Each wave triplet set satisfies the nonzero power flux criteria shown in Table 3.4. Experimental results for wave triplet set 1 are presented in Sect. 3.6.3.

Table 3.5 Co-directional wave interactions in 1 mm thick aluminum plate

Set	Waves a mode f (MHz)	Waves b mode f (MHz)	Waves r mode f (MHz)	M_{ab_r} $\times 10^6$ (m^{-2})	γ (°)	Combination
1	SH0 2.60	SH0 0.70	S0 3.30	2.81	0	Sum
2	A1 2.78	S0 1.10	A0 1.70	5.11	0	Difference
3	A0 1.46	S0 1.06	A1 2.52	4.63	0	Sum
4	A1 3.00	A0 0.42	S0 2.66	3.97	0	Difference
5	S1 3.92	A0 3.00	A0 0.92	0.91	180	Difference
6	S0 2.24	A0 1.90	A0 0.34	2.62	180	Difference
7	S1 3.98	S0 2.90	S0 1.08	1.51	180	Difference
8	S0 1.08	SH0 0.86	SH0 0.22	2.35	180	Difference

After satisfying the geometric constraint considerations for transducer placements, the mixing power is an important parameter because it characterizes power flux to the secondary waves. The larger the power flux, the more sensitive the setup should be to material nonlinearity.

A possible disadvantage of co-directional wave mixing can also a be disadvantage of self-interactions, i.e., the results are averaged over the entire distance between the actuator and the sensor. Thus, if the material degradation is localized to a region significantly smaller than the wave propagation distance, then sensitivity will be diminished. Sensitivity to localized material degradation is an advantage for counter-propagating wave interaction.

Li et al. [41] analyzed co-directional interactions between primary A1 and S0 modes in a 0.95 mm thick aluminum plate. They considered both second and third order interactions.

3.3.4.2 Counter-Propagating Waves, $\theta = 180°$

Examples of wave triplets based on counter-propagating primary waves are shown in Table 3.6, which is formatted like Table 3.5. In each of these cases the wave triplet

Table 3.6 Counter-propagating wave interactions in 1 mm thick aluminum plate

Set	Waves a mode f (MHz)	Waves b mode f (MHz)	Waves r mode f (MHz)	M_{ab_r} $\times 10^6$ (m^{-2})	γ (°)	Combination
1	A0 1.82	A0 1.02	S1 2.84	2.31	0	Sum
2	SH0 1.82	SH0 1.02	S1 2.84	1.45	0	Sum
3	SH0 1.72	SH0 0.34	S0 2.06	2.12	0	Sum
4	S0 1.16	A0 0.26	A0 0.90	2.56	0	Difference

is defined such that the secondary waves are in the same direction as waves *a*. Wave triplet sets 1 and 2 are unique in that the secondary waves are the S1 Lamb wave mode at 2.84 MHz, where the group velocity is zero (although the phase velocity is nonzero). Thus, the secondary waves are stationary, and we can conjecture that their amplitude will increase over time as the waves interact even though the secondary waves do not propagate. Experimental results for wave triplet set 3 are presented in Sect. 3.6.3. Since the primary waves are counter-propagating, the size of the wave mixing zone can be quite limited, unless long tone bursts are used. However, this disadvantage can be transformed into an advantage if the material nonlinearity is localized [42, 40]. The small wave mixing zone can provide good resolution for localized material degradation, and by phasing the primary waves the wave mixing zone can be moved, enabling a scan of the material between the two actuators.

3.3.4.3 Non-collinear Waves, $0° \neq \theta \neq 180°$

Non-collinear primary waves propagate at any interaction angle except 0° or 180°. Obviously, there are infinitely many possibilities. The wide space in which to design material degradation interrogation instrumentation is attractive. Table 3.7 shows some examples for 90° wave interactions. Ishii et al. [43] analyzed the full range of interaction angles. Ishii et al. [44] performed finite element simulations of non-collinear interaction of two A0 waves at various frequencies and describe the generation of symmetric Lamb and SH secondary waves at the sum frequency. They include the effects of tone bursts and finite width wave beams in the analysis.

3.4 Actuation of Primary Waves and Sensing of Secondary Waves

Selection of primary waves that are phase matched (or nearly so) and provide strong power flux to the secondary waves is just the first step in using nonlinear guided waves

Table 3.7 Non-collinear wave interactions in 1 mm thick aluminum plate with $\theta = 90°$

Set	Waves *a* mode *f* (MHz)	Waves *b* mode *f* (MHz)	Waves *r* mode *f* (MHz)	M_{ab_r} $\times 10^6$ (m^{-2})	γ (°)	Combination
1	A0 1.12	A0 0.62	SH0 1.74	1.36	33.2	Sum
2	A1 2.90	A0 0.90	SH0 2.00	1.45	−40.1	Difference
3	S0 2.40	S0 2.40	S1 4.80	3.60	45	Sum
4	S0 0.84	S0 0.32	SH0 0.52	0.89	−21.8	Difference
5	SH0 1.50	SH0 0.78	S0 2.28	1.15	27.5	Sum
6	SH0 1.10	SH0 1.10	S0 2.20	1.25	45	Sum

to detect material degradation. The next step is to choose transducers for actuation and reception in a through-transmission or pitch-catch modality. The wavestructure plays a large role in determining the excitability; i.e., how effectively a prescribed transducer can predominately actuate primary wave modes at a prescribed frequency. Likewise, the analog is true for reception of the secondary waves.

3.4.1 Actuation of Lamb Waves

The fact that there exist multiple propagating Lamb wave modes at all frequencies has always presented challenges, especially with regard to interpreting received signals. Thus, it is often preferable to use techniques that enable actuation of a single predominant mode. The use of angle beam transducers, comb transducers, and phased arrays to preferentially actuate prescribed Lamb wave modes at a certain frequency is well-known [4]. On the other hand, surface-bonded piezoelectric wafers provide very little control over which modes get generated other than the natural frequency tuning [45]. In all cases, the actuators are coupled to the plate at one surface, or in a few cases at both surfaces. Thus, it is difficult for an actuator to assimilate the desired displacement profile of the desired mode and frequency. The couplant between the transducer and the plate is usually a liquid gel or an adhesive. Variations in the gel (e.g., thickness, viscosity) are notorious for causing variations in the guided wave signals, especially the amplitude. The term excitability can be used to quantify how effectively a specific mode can be activated by an actuator at a given frequency.

Some key points regarding common actuators are highlighted below as a convenience for those readers having less experience with ultrasonic guided waves. Normally, an angle-beam transducer is a piston-like piezoelectric contact transducer mounted on an acrylic wedge as depicted in Fig. 3.6. The angle of the wedge necessary to activate a prescribed phase velocity can be determined from Snell's law

$$\frac{c_w}{\sin(\phi_w)} = \frac{c_p}{\sin(90^\circ)} \Rightarrow c_p = \frac{c_w}{\sin(\phi_w)}.$$

Thus, the activation line on a phase velocity dispersion curve for an angle-beam transducer is a horizontal line determined by the angle of oblique incidence and the longitudinal wave speed in the wedge. Knowing the central frequency of the transducer, one can read the phase velocity for the desired mode at that frequency and then use Snell's law to determine which angle wedge to use. The phase velocities must be larger than the longitudinal wave speed in the wedge. Also note that if liquid couplant is used between the wedge and the plate that only longitudinal waves will propagate through the couplant, which can limit the excitability of wave modes having a large in-plane displacement component at the surface of the plate (e.g., the symmetric modes at $c_p = c_L$).

Figure 3.6 may be deceiving because it suggests that an angle-beam transducer can be used to activate a single Lamb wave mode, when this is not typically the

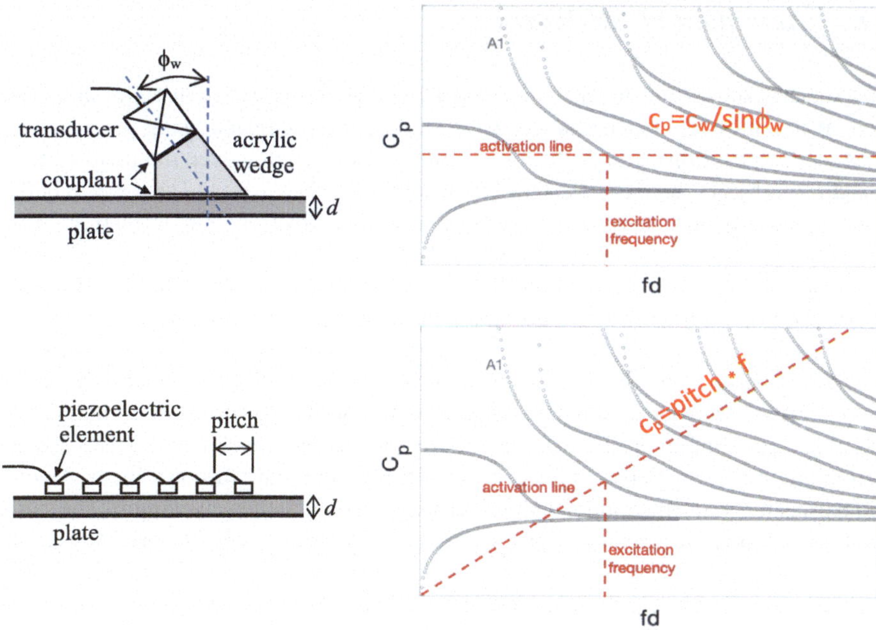

Fig. 3.6 Theory of operation of an angle-beam transducer (top) and a comb transducer (bottom) for Lamb wave actuation

case. In reality, a zone rather than a point is activated due to the finite size of the transducer, which creates a phase velocity range. In addition, the use of a tone burst pulse creates a frequency range. These effects are referred to as a source influence, and are discussed and quantified in Chap. 13 of Rose's book [4].

A typical comb transducer is comprised of a sequence of piezoelectric elements at a fixed center-to-center spacing, known as the pitch (see Fig. 3.6). One electrical signal is sent simultaneously to all elements and all elements have the same polarity, thus the pitch defines the wavelength activated. Interdigitated transducers have elements with alternating polarities, thus the wavelength they activate is twice the pitch. The fundamental equation for wave propagation, i.e., $c_p = \lambda f$, provides the activation line on dispersion curves for comb transducers. A comb transducer can be transformed into a phased array by providing distinct electrical inputs (i.e., phase delays and amplitudes) to each element. Phase delays can make the array function as though it has a different pitch. An excellent discussion of comb transducers and phased arrays for activation of guided waves is given in Chap. 19 of Rose's book [4].

Axisymmetric L(0,n) modes in pipe can be actuated by either an array of angle-beam transducers that wrap-around the pipe circumference or a comb transducer whose elements are rings. In all of the transducers described above, a voltage is sent to the transducer, which is converted by the piezoelectric elements into a mechanical disturbance.

3.4.2 Actuation of SH Waves

Piezoelectric shear transducers can be used to generate SH waves in plates and preferential actuation of a particular SH wave mode can be achieved using piezoelectric shear elements within an angle-beam or comb transducer. However, magnetostrictive transducers (MSTs) are very well suited for actuating SH waves and are easy to build. A mechanical disturbance can be generated in a magnetostrictive material (e.g., iron, remendur, galfenol) by subjecting it to an alternating magnetic field. A simple MST to actuate SH waves in a plate consists of a magnetostrictive layer having domains magnetically polarized in the transverse direction and a meandering electrical coil through which a tone burst signal is sent. The plate is coupled to the magnetostrictive layer (frequently with adhesive or through friction), with the electrical coil placed on top. Often, a rare earth magnet is placed on top of the electrical coil to magnetize and align the domains in the magnetostrictive layer. The mechanical waves generated in the magnetostrictive layer get transferred to the plate through shear. As in a comb transducer, the meanders in the electrical coil dictate the wavelength. The same MST arrangement can be used for reception. Furthermore, an MST used to actuate SH waves can also actuate Lamb waves simply by rotating the rare earth magnet 90° to change the magnetic polarization of the domains. An MST on an aluminum plate is shown in Fig. 3.7. MSTs can also generate the T(0,1) waves in a pipe by wrapping the MST around the pipe circumference.

Fig. 3.7 Magnetostrictive transducer on an aluminum plate; 0.06 mm thick remendur foil adhesively bonded to the plate and electric coil with 5.4 mm meander. The biasing rare earth magnet is not shown

3.4.3 Sensing

The transducers described above for actuation of Lamb and SH waves can also be used to receive them. Often, it is desirable to measure both the primary and secondary waves in order to compute the nonlinearity parameter, in which case broadband transducers are useful. In addition, noncontact air-coupled transducers and laser interferometers permit reception without concerns about couplant variability or interfering with the wavefield. Air-coupled transducers designed for the large acoustic impedance mismatch between metals and air can provide repeatable results if the liftoff and angle of incidence are carefully maintained. Second harmonic measurements with air-coupled transducers are reviewed by Matlack et al. [18] and have been used for nonlinear Rayleigh wave measurements [46].

Laser interferometers can be used to receive nonlinear ultrasound [47, 48]. Bermes et al. [14] used a heterodyne laser interferometer to receive secondary Lamb waves. Many researchers have used laser interferometers to receive Rayleigh waves. The distinguishing features of laser interferometer measurements are that they are broadband noncontact point measurements of the out-of-plane displacement (or velocity). All of the other sensors described above provide a signal averaged over the surface area of the active portion of the transducer.

Polyvinylidene difluoride (PVDF) films, which are electroactive polymers, make versatile sensors [49]. PVDF films are:

- extremely compliant and can be bonded to curved surfaces;
- very broadband (e.g., 0.2–3 MHz);
- readily formed into comb transducers and multi-element arrays having minimal crosstalk between the elements, which enables determination of wavenumber spectra;
- suitable for Lamb wave reception or simultaneous SH wave and Lamb wave reception [50];
- cost effective.

PVDF reception of nonlinear guided waves has been reported by Li and Cho [34], Cho et al. [51], and Zhu et al. [52].

Lastly, we mention that piezoelectric wafers adhesively bonded to the plate have been used for actuation and reception; e.g. by Hong et al. [53]. Note that the omnidirectionality of disc-like wafers used for actuation violates the assumption of plane waves in the theoretical formulation, Thus, the amplitude of the circular-crested primary waves decreases with propagation distance.

3.4.4 Effects of Diffraction

The diffraction, or spreading, of an ultrasonic beam generated by a transducer depends on the size of the transducer. The extreme cases are an infinite strip from which plane waves are generated, and a point source from which circular-crested waves emanate. Actuators of guided waves will operate somewhere within this range. With respect to nonlinear guided waves meeting the internal resonance conditions, the amplitude of the secondary wave field increases linearly with propagation distance (Eq. 3.30), presuming these are plane waves and the waveguide is lossless. Likewise, the relative nonlinearity parameter, β', increases with propagation distance as well. The problem is that this analysis is based on the amplitude of the primary waves remaining constant, which is not the case when diffraction occurs. Thus, in order to use the relative nonlinearity parameter it is important to: neglect diffraction when it is sufficiently small, account for it [25], or re-analyze the problem without the plane wave assumption.

3.5 Instrumentation and Signal Processing

As noted earlier, the weak nonlinearity makes the signals of interest quite small. Thus, measurement instrumentation, setup, and analysis techniques are crucial to obtaining meaningful results. This section highlights hardware and methodologies important for nonlinear guided waves, many of which are common to bulk wave measurements, but there are some important distinctions due to the multi-modal nature of guided waves and the use of primary and secondary waves having different polarity.

3.5.1 Instrumentation

The instrumentation and test setup are extremely important to successful use of nonlinear guided waves for nondestructive material characterization. The wave signals that interrogate the material are associated with weak nonlinearity from the interaction of finite amplitude waves and are at frequencies other than the excitation frequency. The weak nonlinearity results in the interrogation signals often being 40–60 dB smaller than the excitation signal, thus the signal-to-noise ratio (SNR) is crucial. If the amplitude of the excitation signal is increased in order to improve the SNR, care must be taken to not introduce distortion through the instrumentation. In addition, while finite amplitude waves are desired, they must remain elastic waves and not cause plastic deformation.

The components that comprise a nonlinear guided wave measurement system are described below. In many cases it is sufficient to use a subset of these components, and of course in research new or different instruments and methods are constantly being developed, so this is not intended to be an all-inclusive list. The order of the list follows the signal path from generation to storage.

- *High power signal generator*. An electrical tone burst signal is generated and amplified to high voltage. The ability to (i) generate long tone bursts having a narrow frequency bandwidth, (ii) linear amplification, and (iii) no leakage when the generator is turned off between pulses are key considerations. The SNAP system by Ritec (Warwick, Rhode Island, USA) can provide tone burst outputs as high as 5 kW root-mean-squared into a 50 Ω load. Nondestructive testing based on mutual wave interactions requires two excitation frequencies. If the mutual interactions are co-directional, the two frequencies can be sent to a single transducer, although the wave mixing in the active element of the transducer could be undesirable. Otherwise, two channels are necessary for two different actuators.

- *Low pass filter*. An analog high-power low-pass filter can be used to filter higher harmonics created by the signal generator.
- *Matching network*. An impedance matching network, typically consisting of inductors and capacitors, can be used to reduce reflections from the actuator. The components need to be compatible with the high-power signal. While matching networks are not common for piezoceramic transducers, they are very beneficial for magnetostrictive transducers.
- *Actuators*. The active element of the actuating transducer converts the electrical signal into a mechanical disturbance that propagates as an elastic wave. Conventional piezoceramic transducers in angle-beam or comb transducer configurations, magnetostrictive transducers, and piezoelectric wafers are discussed in Sect. 3.4. Obviously, transducer selection should take the breakdown voltage of piezoceramics into consideration.
- *Actuator couplant*. The actuated waves must be transmitted by couplant to the test object. Acoustic impedance matching between the actuator and the test object is the first consideration, but the nonlinearity of the couplant should also be considered. Additionally, the excitability of the primary wave mode at the excitation frequency requires consideration of the wavestructure, and the displacement components at the surface where the actuator is coupled in particular. Finally, the lack of repeatability of the couplant from test to test can result in irreproducible results.
- *Test object*. In laboratory tests the geometry of the specimens can be designed to optimally utilize the transducers, wave modes, frequencies, and tone burst durations. Typically, end-wall reflection interference with the primary or secondary waves need to be avoided. If multiple primary modes are excited, then it is helpful if their wave packets do not overlap at the point of reception. While for field testing, all of the test parameters must be selected based on the geometry of the test object.
- *Sensor couplant*. If couplant other than air is used, it needs to be consistent from test to test. In some setups it is necessary to scan the sensor along the surface to judge the cumulative nature of the secondary waves or to interrogate localized material degradation, doing so affects the choice of couplant. Moreover, the couplant should not distort the wave field that is being measured.
- *Sensor*. The sensor converts the mechanical disturbance into an electrical signal. Sensors are discussed in Sect. 3.4. Sensor selection should consider the size of the test object's area from which the signal is derived with respect to the wavelengths and localized material variability.

- *Matching network*. Analogous to the matching network for actuation, an impedance matching network can provide larger amplitude electrical signals.
- *Preamplifier*. The electrical signal can be amplified (e.g., 40 dB) with a low-noise, high-gain amplifier once it's linearity is assessed.
- *Receiver/oscilloscope*. Finally, the electrical signal is stored or further manipulated with a low-noise receiver or oscilloscope. The analog-to-digital conversion must have high resolution to record the signal distortion characteristic of weakly nonlinear guided waves.

3.5.2 Signal Processing

Signal processing is performed after actuating the primary waves and the signals are received. Signal averaging is the simplest method used to increase the signal-to-noise ratio (SNR). Anywhere from 32 to 1000 signals can be averaged together before storing the time-domain signal (A-scan) [25, 54].

The fast Fourier transform (FFT) is the most commonly used method to determine the frequency spectrum of the interrogation signal. However, the FFT can only be used if the guided wave modes are separated, otherwise contributions from the unwanted modes will influence the results. If the guided wave modes of interest are separated from other unwanted modes, then the type and size of the window need to be selected. Liu et al. [33] studied the effect that different types of windows has on the frequency spectrum and found that a Tukey window with the cosine-tapered ratio of 0.9 provides good resolution of the primary and secondary peaks relative to the other types of windows. Regardless of which type of window, it is imperative to maintain a fixed window size for each data set to provide uniformity. In addition, the window size must account for the different group velocities at which the primary and secondary waves travel, if indeed they are different. It may be necessary to zero pad the signal to ensure that a sufficient number of bins exist to provide an accurate FFT. It's well worth pointing out that the stored waveforms are recorded in units of volts, and the system calibration must be known if they are to be converted to units of displacement (say nm). Moreover, it is difficult to compare the results from one test setup to another due to differences in excitability of a type of transducer and amplification.

Some researchers split the signal received by the sensor into two channels [55]. One channel is stored to provide the primary wave signal. A high pass filter and a preamplifier are placed in the other channel to filter out the primary wave signal and retain the amplified secondary wave signal, which is then stored separately. The two wave forms can be processed separately in either the frequency domain or the time domain.

The multimodal nature of guided waves or primary and secondary waves having different group velocities may make it impossible to separate the modes of interest from the unwanted ones, in which case the short time Fourier transform (STFT) can be applied [15, 25].

The phase inversion technique is valuable for second harmonic generation tests [56]. After one signal is sent, received, and stored, the phase is inverted and the new signal is sent, received, and stored. The two recorded signals are simply summed, which removes the odd harmonics (including the primary waves) leaving only the even harmonics. The original signal and the summed signal can be processed in either the frequency domain [56] or the time domain [57].

Finally, chaotic oscillators have been investigated as a way to extract small nonlinear signals from noisy signals [58, 59]. We leave it to the reader to explore this type of signal processing through the references cited. We close this section with a simple proposition that bridges between instrumentation/signal processing and measurement considerations. Repetitive tests involving disassembly of at least a portion of the test setup should be routinely performed in order to quantify the repeatability of the nonlinearity measurements.

3.6 Measurement Considerations

The goal of this chapter is to describe how nonlinear guided waves can be used for nondestructive characterization of material nonlinearity associated with material degradation. As mentioned in Sect. 3.1, the weak nonlinearity means that the signal-to-noise ratio (SNR) for the secondary waves is relatively small, which means that the measurements must be made carefully, taking into account all other potential sources of nonlinearity. Numerical simulations provide a means to control which nonlinearities are included in the analysis. The finite element analyses of nonlinear guided waves reported in the literature typically include material nonlinearity and/or geometric nonlinearity, see [26] for example. Nonlinearities associated with amplifiers, transducers, couplant, filters, etc. are specifically excluded from these analyses. Nonetheless, the utility of nonlinear guided waves ultimately comes down to making measurements in the lab and the field.

3.6.1 Measurement System Nonlinearities

When using nonlinear guided waves for nondestructive material characterization, diligent investigators are wary about the source of the nonlinearity that they are measuring. They must find evidence that the nonlinearity is from the material and not from other sources—i.e., the measurement system, or at least they might identify changes in the nonlinearity specifically associated with the material. One way to assess the measurement system nonlinearity is to put the actuator and sensor back-to-back and record signals. For example, Deng et al. [13] replaced the wedges for their angle-beam transducers with a glass block known to have low nonlinearity and recorded signals.

Changes in material nonlinearity can be isolated from system nonlinearities by comparing the relative nonlinearity parameter $\beta' = A_2/A_1^2$ computed for one material state with another material state. For each material state a sequence of measurements are made where the excitation voltage is incremented and β' is computed from the slope of the plot of A_2 as a function of A_1^2. Presuming that the measurement system is identical for each measurement, the system nonlinearities will be the same. Thus, changes in β' must be attributed to the material. System nonlinearities are discussed in more detail by Matlack et al. [18].

For the case of the secondary waves having a polarity different from that of the primary waves (i.e., SH primary waves generating S0 secondary waves), Shan et al. [60] were able to dissipate transducer nonlinearity with gel because the S0 Lamb mode at the secondary frequency has a large out-of-plane displacement on the surface that causes the wave energy to leak into the gel. Thus, the received signal originates in the material beyond the gel and is attributed to material nonlinearity.

Avoiding measurement system nonlinearities is one of the arguments for using mutual interactions of primary waves to generate sum and difference frequencies that are away from integer multiples of the excitation frequency, where system nonlinearities tend to exist. Finally, an advantage that guided waves (including Rayleigh waves) have over bulk waves is that they propagate longer distances, making it easier to quantify the cumulative nature of the secondary waves. The cumulative effect is a consequence of the internal resonance conditions (implied by Eq. 3.30) being met. While the linear increase of the secondary wave's amplitude is limited to planar waves continuously interacting in a lossless waveguide, system nonlinearities are independent of propagation distance. Scanning the sensor to receive waveforms that have traveled different distances is a powerful technique for measuring material nonlinearity.

3.6.2 Material Nonlinearities

To this point, the question of what constitutes material nonlinearity has not been addressed other than that the Landau-Lifshitz strain energy function (Eq. 3.6) represents the nonlinearity through the third order elastic constants *A*, *B*, and *C*. The implicit assumption is that changes in the waveguide material's microstructure changes the values of *A*, *B*, and *C* while having little effect on the linear elastic properties. The diagnostic capabilities of nonlinear guided waves are based on detecting changes in the nonlinearity parameter or related features. But the prognostic capabilities of nonlinear guided waves rely on being able to correlate the current material state (e.g., dislocation substructures, persistent slip bands, precipitates) to the ultrasonic nonlinearity. Matlack et al. [18] review some of the research along this track. The initial modeling work was done by Hikata and co-workers [61], here we simply site some of the influential and more recent work [62–67].

3.6.3 *Measuring Progressive Degradation*

This section provides a number of examples that show the sensitivity of nonlinear guided waves to material nonlinearity. Our intent is to show the diverse range of methods and types of guided waves that have been investigated. We start with self-interaction of Lamb waves, which includes internal points 1 (S1–S2 mode pair) and 3 (S2/A2–S4 mode pair) and the S0–S0 mode pair below the first cutoff frequency, which is not exactly phase matched. Then we describe a set of experiments on small-diameter aluminum pipes. Just a few sets of experiments based on mutual interaction have been conducted to date. We summarize results from counter-propagating and co-directional SH waves that generate the S0 Lamb mode as secondary waves. Co-directional mixing of S0 Lamb waves to localized creep degradation is described as well.

3.6.3.1 Self-interaction: Lamb Waves

Pruell et al. [16] used the S1–S2 mode pair to assess low-cycle fatigue damage in a 1.6 mm thick 1100-H14 aluminum plate. The ultrasonic nonlinearity increases at a decreasing rate with fatigue cycles. The cumulative generation of S2 Lamb waves from self-interaction of S1 Lamb waves (internal resonance point 1 in Table 3.1) was shown by Matlack et al. [25]. They used angle-beam transducers on an undamaged 1.6 mm thick 6061-T6 aluminum plate. The narrowband 2.25 MHz piezoelectric actuator was given a 35 cycle tone burst of ~660 V_{pp}. Reception was through a narrowband 5 MHz transducer. We note that the bandwidth of the sensor can be important because if it is too narrow then the primary wave amplitudes can be reduced or filtered, resulting in an artificially high relative nonlinearity parameter $\beta' = A_2/A_1^2$ values. Moreover, if only changes in β' are of interest, say with respect to propagation distance or changing material state, then this is somewhat irrelevant. Linear increase in the second harmonic was measured for propagation distances up to 420 mm.

Metya et al. [68] used internal resonance point 1 in Table 3.1 to assess the effect of tempering temperatures on 9Cr–1Mo steel. They used Plexiglas wedge transducers oil-coupled to a 2.5 mm thick steel plate. A 5-cycle tone burst with 2 MHz central frequency was actuated by a 2.25 MHz narrowband transducer, while reception was by a 5 MHz broadband transducer. A typical A-scan and the STFT results are shown in Fig. 3.8. The amplitudes in Fig. 3.8b are A_1 and A_2 respectively and are used to compute β'. The authors plotted β' for input voltages from 600–1200 V_{pp}. This plot should be linear for material nonlinearity, thus input voltages where it is nonlinear should be avoided. They also show that β' increases linearly with propagation distance for the range they wanted to use, 50–80 mm. After a microstructural examination of the samples and discussion of the relationships between precipitate pinning of dislocations and changes in β', the authors compare the material hardness parameter and β' over a range of tempering temperatures. As shown in Fig. 3.9, hardness and β' follow the same general trend, which is not monotonic.

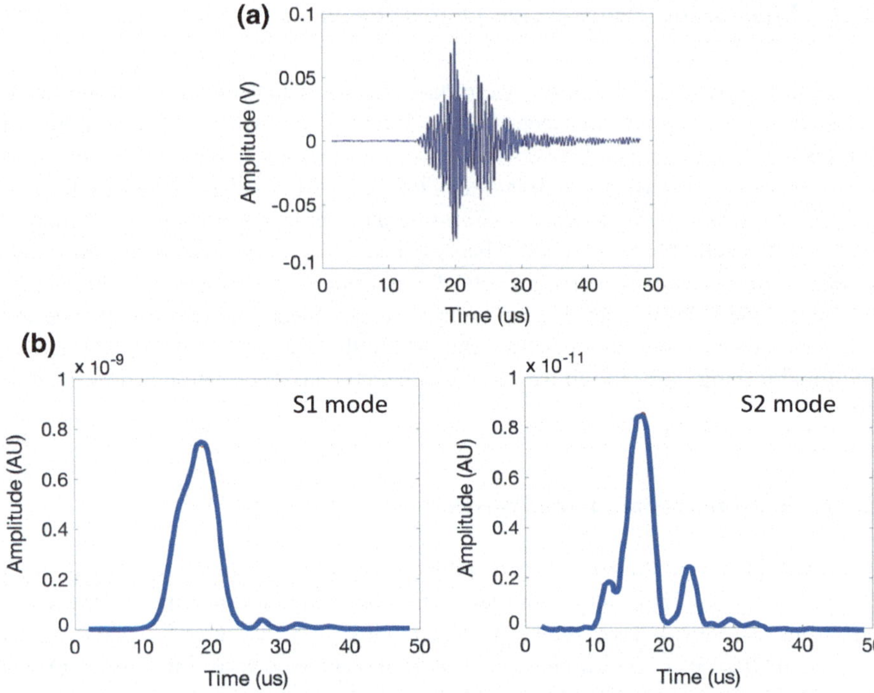

Fig. 3.8 Sample results from the S1–S2 mode pair, **a** A-scan and **b** results of STFT for the primary S1 wave mode at 2 MHz and the secondary S2 wave mode at 4 MHz. After Metya et al. [68] with permission

The sensitivity of nonlinear guided waves to material nonlinearity associated with incipient, or early-stage, material degradation makes nonlinear guided wave techniques extremely well-suited for structural health monitoring applications. However, to be successful the measurement methods need to incorporate stay-in-place transducers that are tolerant of the in-service operational and environmental conditions. Cho [69] found that PVDF comb transducers functioned well in this capacity for internal resonance point 1 in Table 3.1. PVDF is discussed in Sect. 3.4.3 for reception, but not in Sect. 3.4.2 for actuation because it is generally thought of as a weak actuator and finite amplitudes are strongly preferred for nonlinear guided waves. However, PVDF has quite a large breakdown voltage, so reasonable amplitudes can be actuated if the voltages are high enough. A caveat is that the melting temperature of PVDF is around 170 °C. The transducers are very simple to make; cyanoacrylate is used to bond a PVDF film to the plate, silver electrodes are deposited through a mask onto the PVDF film, and then a copper lead wire is attached. For actuation, the five electrodes are 41 mm by 2.7 mm and there is a 2.7 mm gap between them. For reception, there is a single electrode 30 mm by 1 mm.

Cho installed the PVDF transducers 203 mm apart on a 3.2 mm thick dog-boned 2024-T3 aluminum plate and subjected the plate to tensile cycling with a fatigue

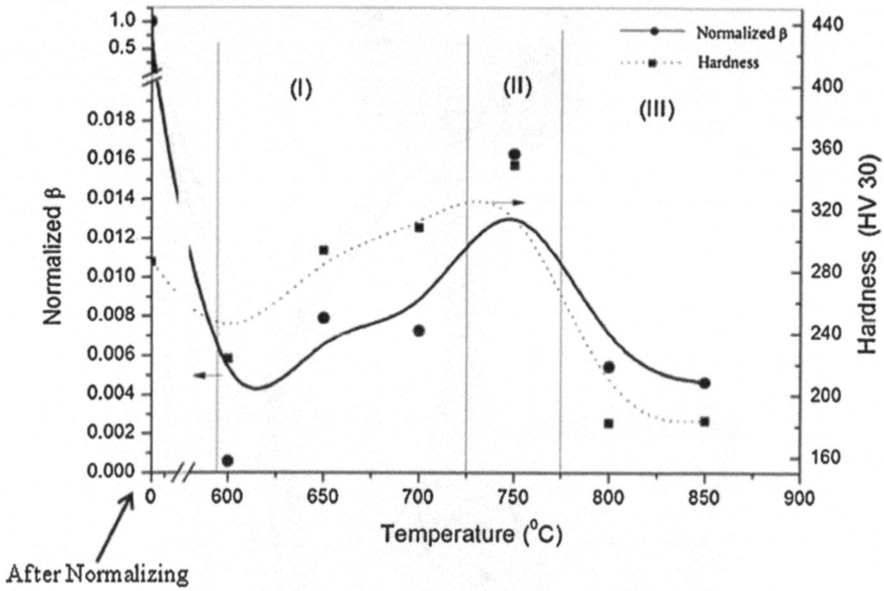

Fig. 3.9 Dependence of the nonlinearity parameter and hardness on the tempering temperature of 9Cr–1Mo steel. After Metya et al. [68] with permission

ratio of 0.011. The maximum tensile stress of 308 MPa is slightly higher than the listed yield strength of 283 MPa. The compliance of the PVDF film enables it to endure the cyclic loading with no obvious degradation to itself or the bond. The fatigue test was interrupted every 2000 cycles for ultrasonic interrogations, which were conducted at a constant stress of 3 MPa. As a side note, nonlinear guided wave results obtained during dynamic mechanical loading have been reported elsewhere [70]. The ultrasonic interrogations involve sending a 5-cycle 1440 V_{pp} tone burst with a central frequency of 1.1 MHz to the PVDF actuator. The received signal is acquired at a sampling frequency of 1 GHz and 512 signals are averaged together. An aluminum plate sample in the mechanical test frame is shown in Fig. 3.10, where the PVDF transducers above and below the dog-boned portion of the sample are also visible.

A sample A-scan taken after 40,000 cycles is shown in Fig. 3.11 along the frequency spectrum obtained by FFT of the windowed portion of the A-scan. Since the S1 Lamb wave mode has the fastest group velocity, and the group velocity of the S2 secondary waves matches the S1 mode, an FFT is sufficient. There was no visual evidence of fatigue damage until the sample failed after 62,530 cycles. The normalized primary and secondary wave amplitudes A_1 and A_2, as well as the nonlinearity parameter are plotted as a function of the number of mechanical cycles in Fig. 3.12. In summary, the results illustrate that the primary wave amplitude remains relatively constant (with the exception of the final data point), while the secondary wave amplitude increases with a decreasing rate.

Fig. 3.10 Aluminum alloy instrumented with PVDF transducers in the mechanical test rig. After Cho [69] with permission

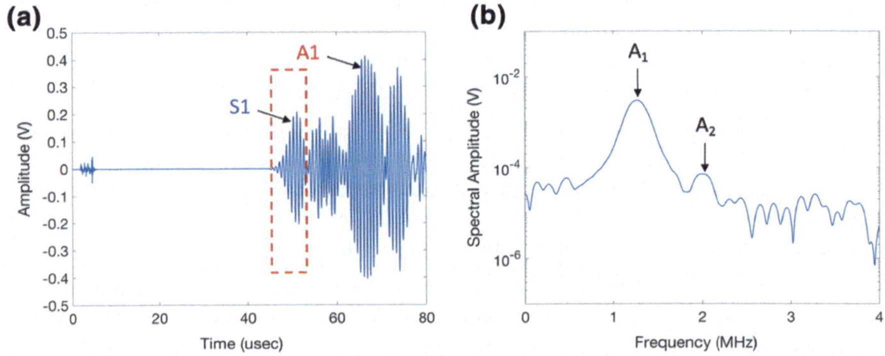

Fig. 3.11 Sample **a** A-scan and **b** frequency spectrum obtained by FFT after 40,000 cycles. After Cho [69] with permission

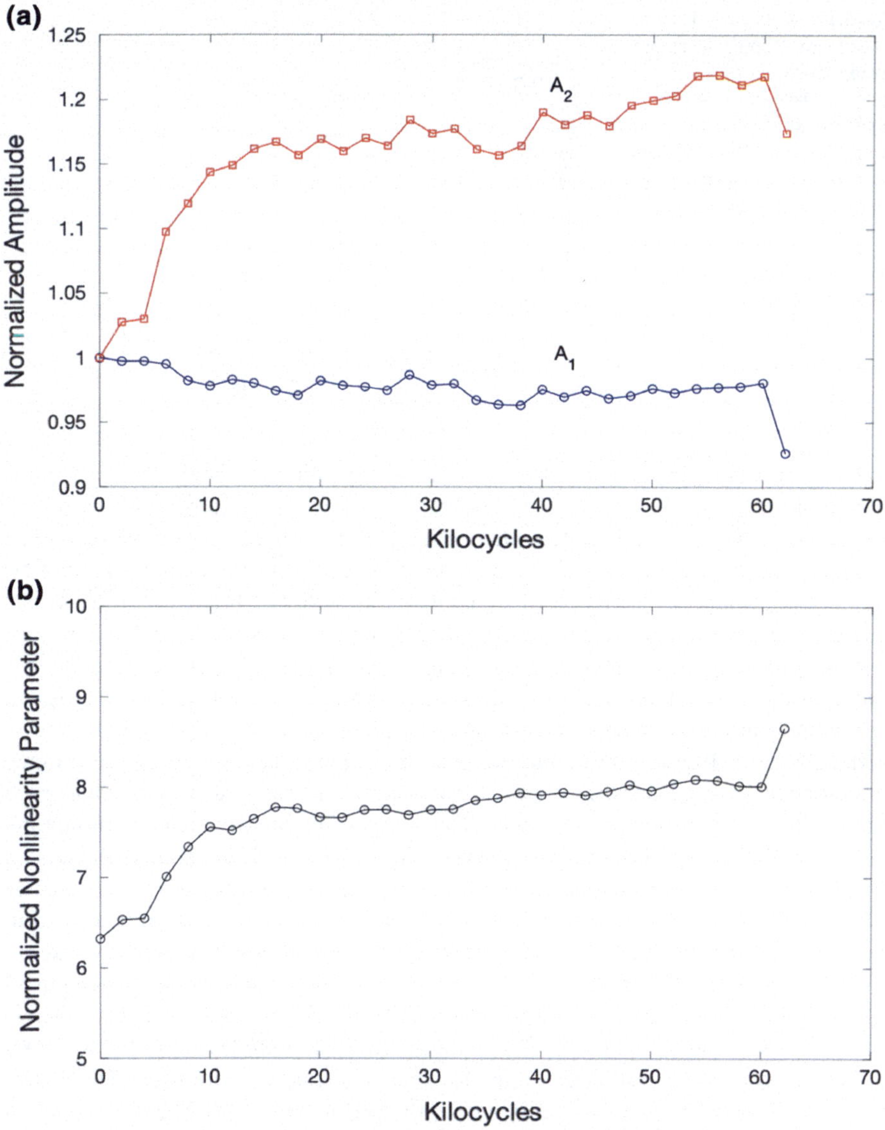

Fig. 3.12 Evolution of the primary and secondary wave amplitudes as a function of fatigue cycles: **a** S1 mode amplitude A_1 and S2 mode amplitude A_2 normalized with respect the pristine material condition and **b** the normalized nonlinearity parameter β'. After Cho [69] with permission

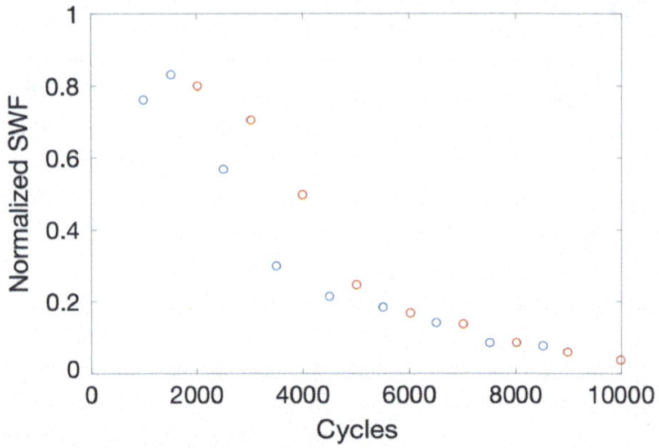

Fig. 3.13 Normalized stress wave factor (SWF) decreases during tensile fatigue cycling. Data from two experiments are shown. After Deng and Pei [55] with data replotted under the Creative Commons license

Deng and Pei [55] actuated the mode crossing point (internal resonance point 3 in Table 5.1) in 1.85 mm thick aluminum plates during interruptions of tensile fatigue tests. As previously noted, Deng et al. [13] measured cumulative secondary wave generation at this point. Deng and Pei [55] used fluid-filled wedges and 2 cm diameter transducers. The received signal was split with one branch sent to a sensor (to measure the primary waves) and the other branch was passed through a high-pass filter and then amplified 60 dB before being sent to a sensor (to measure the secondary waves). Signal processing entailed transforming the secondary waves' signals to the frequency domain and plotting the spectral amplitude squared as a function of frequency, and also integrating the spectral amplitude squared between the primary and secondary frequencies. The latter result is called the stress wave factor and gives one value for the frequency spectrum. The stress wave factor decreases monotonically with fatigue life fraction as shown in Fig. 3.13, which is anomalous because metal fatigue processes include the formation of dislocation substructures, persistent slip bands, and microcracks that generally increase the material nonlinearity, thus mechanical fatigue damage typically increases the generation of secondary waves.

As noted at the end of Sect. 3.3.3.1, phase-matching is not necessarily required, near-matching of phases can be sufficient. Zhu et al. [71] studied 300 kHz S0 waves in eleven 2 mm thick heat treated 7075 aluminum plate samples subjected to 3000–25,000 fatigue cycles. Oil-coupled angle-beam transducers were used. The results from S0–S0 and S1–S2 mode pairs are duplicated in Fig. 3.14. The S0–S0 pair is observed to be less sensitive to fatigue damage than is the S1–S2 pair.

3.6.3.2 Self-interaction: Guided Waves in Pipe

Thermal fatigue damage (5 and 10 cycles from room temperature to 240 °C) in aluminum pipes (10 mm outer radius and 3 mm wall thickness) was correlated with the nonlinearity parameter by Li and Cho [34]. The authors used wrap-around PVDF comb transducers to send axisymmetric L(0,6) primary waves, which in turn generate

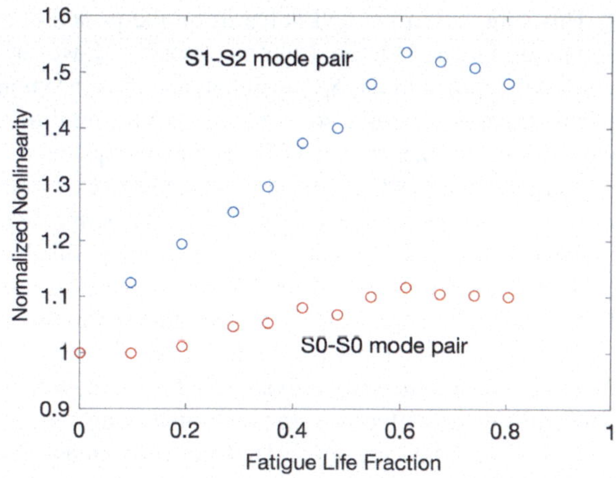

Fig. 3.14 Comparison of normalized nonlinearity parameter for the S0–S0 mode pair (300 kHz excitation) and S1–S2 mode pair (1.81 MHz excitation). After Zhu et al. [71] with data replotted under the Creative Commons license

L(0,10) secondary waves as shown in Table 3.3 (although the radius is different). The secondary waves are shown to be cumulative up to a propagation distance of 170 mm. While the description of experimental details is limited, the slope of the plot of β' versus propagation distance increases with thermal cycling, implying that the second harmonic generation of the L(0,10) wave mode is sensitive to increasing material nonlinearity from thermal damage.

Shear waves in pipes have also been shown to correlate with material degradation. One study used the T(0,1) mode generated at 0.83 MHz by an MST to generate third harmonic T(0,1) waves by self-interaction [72]. In this case the material degradation due to combined constant-tension and cyclic-shear loading at 850 °C in Inconel 617 pipes having a 9 mm inner diameter and 1.5 mm wall thickness was investigated. The normalized nonlinearity parameter for third harmonics, A_3/A_1^3, increased monotonically with the damage fraction. These third harmonic tests in pipe were a natural extension of third harmonic tests in plates using the nondispersive fundamental SH0 mode. The generation of third harmonics in aluminum plates have been correlated with plastic deformation [54] and fatigue [73].

3.6.3.3 Mutual Interaction: Counter-Propagating SH0 Waves

The counter-propagating SH0 waves detailed as set 3 in Table 3.6 were employed to indicate thermal damage using angle-beam reception [40] and fatigue damage using PVDF [51] on 7075-O aluminum plates. In both sets of experiments a subtraction method was implemented to isolate as best possible the nonlinearity from mutual interaction. The subtraction method used is based upon receiving three consecutive signals: waves *a*, waves *b*, and then waves *a* and *b* together. Then, the difference signal is computed as

$$S_{Diff} = S_{[a+b]} - S_{[a]} - S_{[b]}.$$

This subtraction method is based on the principle of linear superposition; if the signals are linear, then the difference is zero. However, if there is mutual interaction it will only appear in the $S_{[a+b]}$ signal, and thus the subtraction operation will isolate it. This method is similar to one proposed for nonlinear phased array imaging [74], and we note that Bruno et al. [75] have also proposed a scaled subtraction method. As an example, two MSTs are used to actuate waves a (traveling to the right) and waves b (traveling to the left) at 1.7 and 0.31 MHz respectively. A 2.25 MHz angle-beam transducer oriented to receive the out-of-plane displacement component of the S0 mode (traveling to the right) at 2 MHz receives the signals shown in Fig. 3.15. The difference signal is reasonably large compared to the noise.

The effect of local thermal damage (due to a temperature excursion to 327 °C) on the ultrasonic nonlinearity is evident in Fig. 3.16, which shows a schematic of the test setup, the ratio of difference signals from mixing zones 1 and 2, and sample difference signals. Time delays are applied to force the primary SH0 waves to mix either in zone 1 or zone 2. The secondary waves generated in mixing zone 1 must travel 162 mm to be received by the angle-beam transducer, while the secondary waves generated in mixing zone 2 only travel 41 mm to the angle-beam transducer. Therefore, if the

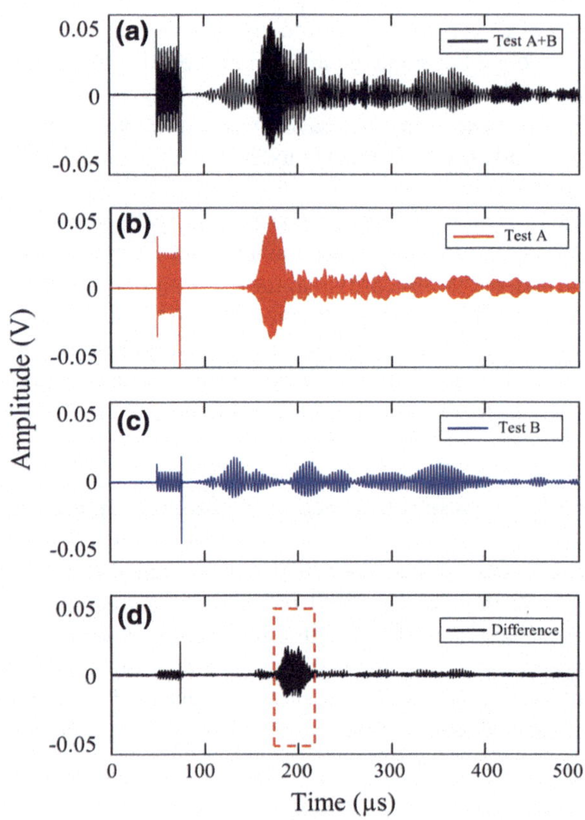

Fig. 3.15 Signals received or computed from an angle-beam transducer. MSTs actuated SH0 waves: **a** test [A + B], **b** test [A], **c** test [B], and **d** the difference signal. After Hasanian and Lissenden [40] with permission

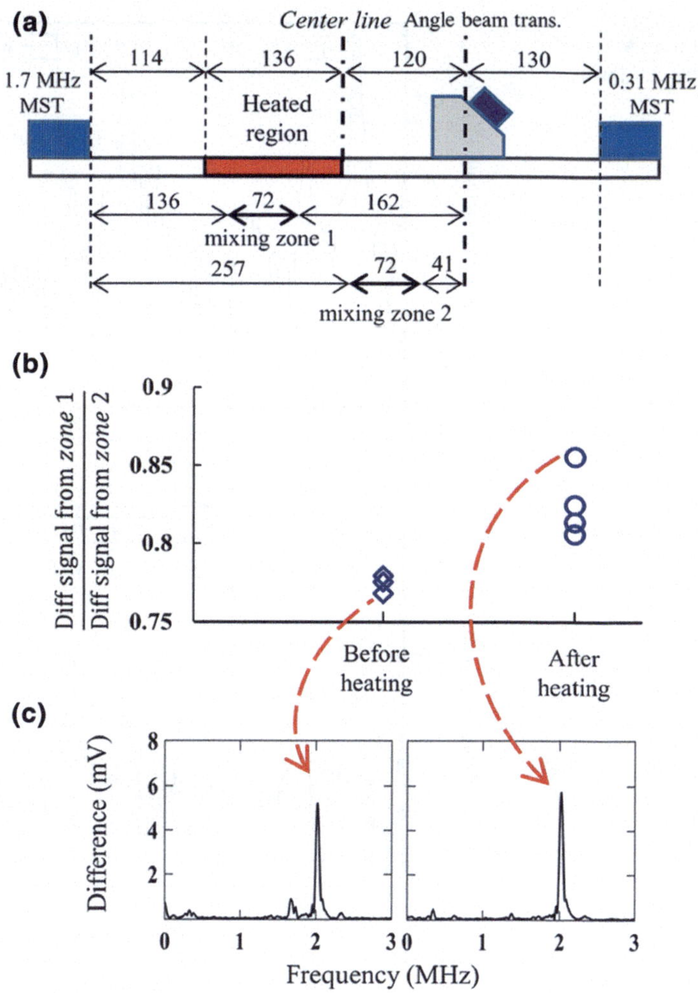

Fig. 3.16 **a** Schematic of test setup showing MSTs, angle-beam transducer, wave mixing zones, and heated region. Dimensions are in mm. **b** Ratios of difference signals in zones 1 and 2 before and after heating, and **c** sample 'before and after' difference signals. After Hasanian and Lissenden [40] with permission

material nonlinearity is uniform, then the difference signal received from zone 1 is expected to be smaller than the one received from zone 2 due to attenuation. The increase in the difference signal due to thermal damage is appreciable.

Cho et al. [51] used anisotropic PVDF film oriented at 45° in order to receive both primary SH0 waves and secondary S0 waves, for which a frequency spectrum from test [A + B] is shown in Fig. 3.17. Similarly, the frequency spectrum from the difference signal is plotted in Fig. 3.18. Notched 1 mm thick plates were cycled in tension to 60% of the fatigue life. Fatigue degradation is expected to be localized at

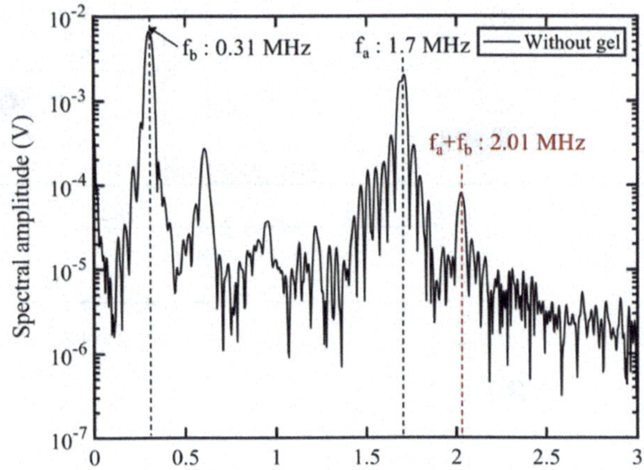

Fig. 3.17 Frequency spectrum for test [A + B]. After Cho et al. [51] with permission

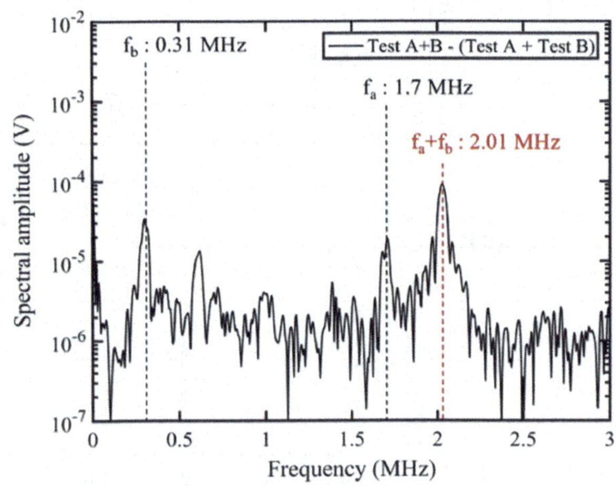

Fig. 3.18 Frequency spectrum for difference signal. After Cho et al. [51] with permission

the root of the notch. Transducers were then installed as shown in Fig. 3.19. To switch the MSTs, it was only necessary to interchange the electrical coils. The secondary S0 waves propagate in the same direction as the waves a. The PVDF enables computation of the nonlinearity parameter, $\frac{A_{ab}}{A_a A_b}$, based on the difference signal and the signals from waves a and waves b. Time delays were applied to the tone bursts sent to the MSTs in order to move the location of the wave mixing zone. The values of the nonlinearity parameter at different locations are shown in Fig. 3.20. Prior to cyclic loading the nonlinearity parameter values are reasonably uniform, but after cyclic loading its values are elevated in the notched region where fatigue damage occurred.

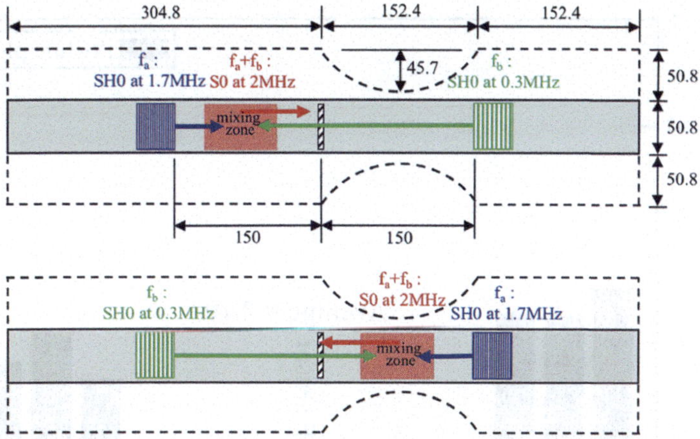

Fig. 3.19 Schematic of counter-propagating wave interaction experiments on aluminum plates. Dimensions are in mm. After Cho et al. [51] with permission

3.6.3.4 Mutual Interaction: Co-directional Lamb Waves

Metya et al. [76] studied co-directional Lamb wave mixing to assess creep degradation in 9Cr–1Mo steel. The S0 mode is actuated at 0.43 and 0.71 MHz in a 2 mm thick plate using a dual-frequency input signal to a broadband angle-beam transducer. This is not a phase-matched wave triplet. The difference in group velocities enable interrogation of localized material degradation. Time delays are imposed to create four different finite-sized wave mixing zones along the length of the plate. The amplitude of the frequency spectrum at the sum frequency is plotted for each wave mixing zone as a function of creep duration/strain in Fig. 3.21. In these experiments the nonlinearity increased substantially after roughly 40% of the time to rupture and was highest in the region where the rupture eventually occurred.

3.6.3.5 Mutual Interaction: Co-directional SH0 Waves

Set 1 of Table 3.5 indicates that the mutual interaction of two co-directional SH0 waves will generate the S0 mode at the sum frequency-thickness product of 3.34 MHz mm. In fact, any combination such that $[f_a + f_b]d = 3.34$ MHz mm will be phase-matched. Shan et al. [60] showed that the closer the values of f_a and f_b, the larger the amplitude of the secondary waves. Changes to the material state described by the Landau-Lifshitz constants can be correlated with the slope of a plot of nonlinearity parameter versus propagation distance. A set of 3.125 mm thick 2024-T3 aluminum plates were cycled in tension to 25, 50, and 75% of the fatigue life. Two MSTs generated SH0 waves at 0.75 and 0.32 MHz, while an air-coupled transducer received the secondary S0 waves at the sum frequency. An electromagnetic acoustic transducer was used to receive the primary waves. S0 waves are easily received over

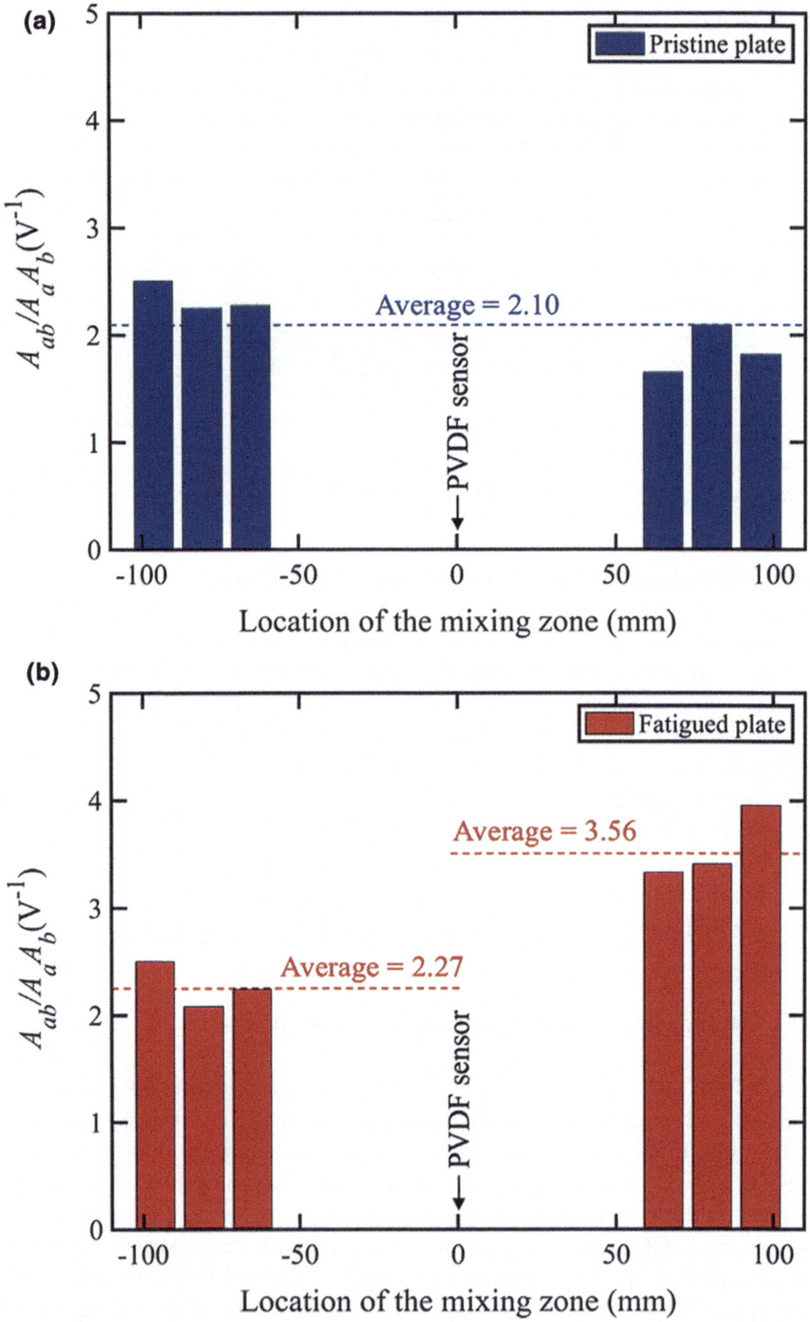

Fig. 3.20 Nonlinearity parameter values as a function of location in the plate sample for **a** the pristine plate and **b** the fatigued plate. After Cho et al. [51] with permission

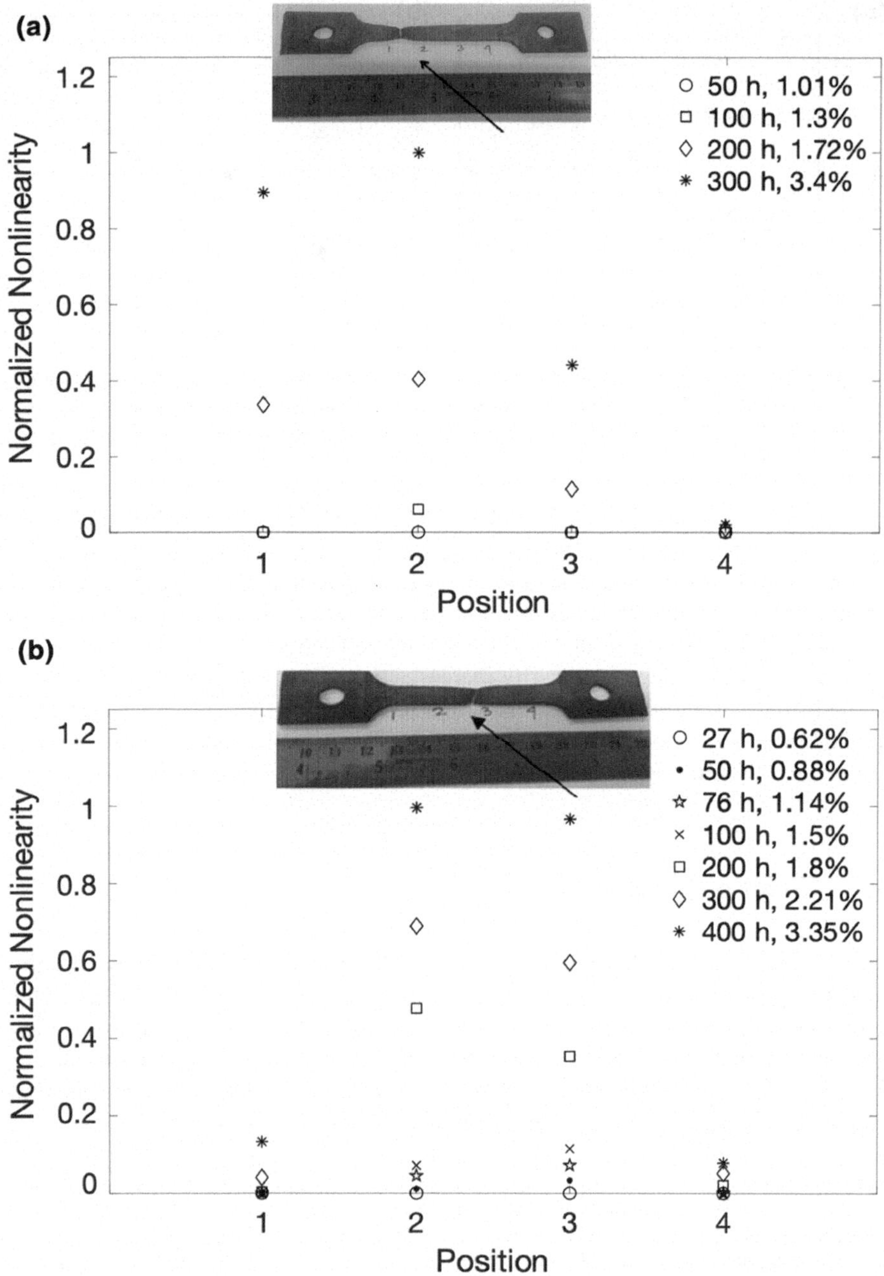

Fig. 3.21 Spectral amplitude at the sum frequency at four positions along creep samples at interrupted times. Legend provides the creep time and strain. **a** Rupture occurred between positions 1 and 2 after 396 h. **b** Rupture occurred between positions 2 and 3 after 440.5 h. After Metya et al. [76] with permission

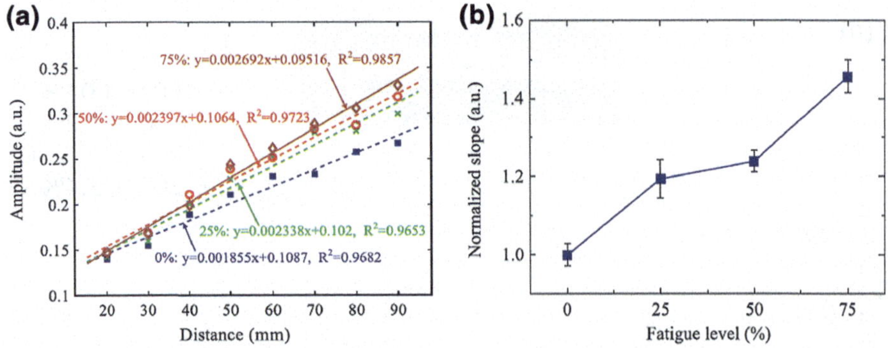

Fig. 3.22 Results from co-directional wave mixing in aluminum plate: **a** difference signal of secondary S0 waves increase linearly with propagation distance, and the slope **b** increases monotonically with fatigue damage. After Shan et al. [60] with permission

a range of propagation distances because the air-coupled transducer is noncontact. The subtraction method is applied to the signals from the air-coupled transducer. The difference signal is plotted as a function of propagation distance in Fig. 3.22a and linear regression is used to determine the slope for each sample. The slopes are then normalized with respect to the pristine sample and plotted as a function of fatigue level in Fig. 3.22b. The nonlinearity increases monotonically with fatigue damage. The sensitivity to fatigue damage in the first 25% of the fatigue life suggests that the method has a strong potential for application. These results were dependent on the use of a gel filter to reduce nonlinearity from the actuator.

3.7 Closing Comments

This chapter indicates that much progress has been made to understand and utilize nonlinear guided waves for nondestructive material characterization testing, whether it be in the laboratory or the field. However, there is a long way to go to make the measurements practical and reliable enough for industrial applications. We are optimistic that this will happen in the near future as the foundation is strong and the number of contributing researchers continues to increase.

Acknowledgements The first author acknowledges that our research in nonlinear guided waves was partially supported by projects funded by the Nuclear Energy Universities Program in the U.S. Department of Energy (awards 102,946 and 120,237) and the U.S. National Science Foundation (awards 1,300,562 and 1,727,292). In addition, he wants to acknowledge graduate and postdoc students Yang Liu, Vamshi Chillara, Gloria Choi, Hwanjeong Cho, Baiyang Ren, and Chung Seok Kim for their weighty contributions, as well as encouragement from his colleague Joseph Rose.

References

1. W.J.N. De Lima, M.F. Hamilton, Finite-amplitude waves in isotropic elastic plates. J. Sound Vib. **265**(4), 819–839 (2003)
2. M. Hasanian, C.J. Lissenden, Second order ultrasonic guided wave mutual interactions in plate: arbitrary angles, internal resonance, and finite interaction region. J. Appl. Phys. **124**(16) (2018)
3. T. Kundu, *Nonlinear Ultrasonic and Vibro-Acoustical Techniques for Nondestructive Evaluation* (2018)
4. J. L. Rose, *Ultrasonic Guided Waves in Solid Media*. (Cambridge University Press, Cambridge, 2014)
5. M. Deng, Second-harmonic properties of horizontally polarized shear modes in an isotropic plate. Jpn. J. Appl. Physics, Part 1 Regul. Pap. Short Notes Rev. Pap. **35**(7), 4004–4010 (1996)
6. M. Deng, Cumulative second-harmonic generation accompanying nonlinear shear horizontal mode propagation in a solid plate. J. Appl. Phys. **84**(7), 3500–3505 (1998)
7. M. Deng, Cumulative second-harmonic generation of Lamb-mode propagation in a solid plate. J. Appl. Phys. **85**(6), 3051–3058 (1999)
8. M. Deng, Analysis of second-harmonic generation of Lamb modes using a modal analysis approach. J. Appl. Phys. **94**(6), 4152–4159 (2003)
9. A. Srivastava, F. Lanza di Scalea, On the existence of antisymmetric or symmetric Lamb waves at nonlinear higher harmonics. J. Sound Vib. **323**(3–5), 932–943 (2009)
10. M.F. Müller, J.-Y. Kim, J. Qu, L.J. Jacobs, Characteristics of second harmonic generation of Lamb waves in nonlinear elastic plates. J. Acoust. Soc. Am. **127**(4), 2141–2152 (2010)
11. N. Matsuda, S. Biwa, Phase and group velocity matching for cumulative harmonic generation in Lamb waves phase and group velocity matching for cumulative harmonic generation in Lamb waves. J. Appl. Phys. **109**, 094903 (2011)
12. V. K. Chillara, C. J. Lissenden, Interaction of guided wave modes in isotropic weakly nonlinear elastic plates: higher harmonic generation. J. Appl. Phys. **111**(12) (2012)
13. M. Deng, P. Wang, X. Lv, Experimental verification of cumulative growth effect of second harmonics of Lamb wave propagation in an elastic plate. Appl. Phys. Lett. **86**(12), 1–3 (2005)
14. C. Bermes, J.Y. Kim, J. Qu, L.J. Jacobs, Experimental characterization of material nonlinearity using Lamb waves. Appl. Phys. Lett. **90**(2), 1–4 (2007)
15. C. Pruell, J.Y. Kim, J. Qu, L.J. Jacobs, Evaluation of plasticity driven material damage using Lamb waves. Appl. Phys. Lett. **91**(23) (2007)
16. C. Pruell, J.Y. Kim, J. Qu, L.J. Jacobs, Evaluation of fatigue damage using nonlinear guided waves. Smart Mater. Struct. **18**, 035033 (2009)
17. T.-H. Lee, I.-H. Choi, K.-Y. Jhang, The nonlinearity of guided wave in an elastic plate. Mod. Phys. Lett. B **22**(11), 1135–1140 (2008)
18. K.H. Matlack, J.Y. Kim, L.J. Jacobs, J. Qu, Review of second harmonic generation measurement techniques for material state determination in metals. J. Nondestruct. Eval. **34**(1) (2015)
19. V.K. Chillara, C.J. Lissenden, Review of nonlinear ultrasonic guided wave nondestructive evaluation: theory, numerics, and experiments. Opt. Eng. **55**(1) (2016)
20. W. Bin Li, M.X. Deng, Y.X. Xiang, Review on second-harmonic generation of ultrasonic guided waves in solid media (I): theoretical analyses. Chin. Phys. B **26**(11) (2017)
21. B.A. Auld, *Acoustic Fields and Waves in Solids*, vol. II (Wiley, 1973)
22. M. Mazilu, A. Demčenko, R. Wilson, J. Reboud, J.M. Cooper, Breaking the symmetry of momentum conservation using evanescent acoustic fields. Phys. Rev. Lett. **121**(24), 244301 (2018)
23. Y. Liu, V.K. Chillara, C.J. Lissenden, On selection of primary modes for generation of strong internally resonant second harmonics in plate. J. Sound Vib. **332**(19), 4517–4528 (2013)
24. Y. Liu, V.K. Chillara, C.J. Lissenden, J.L. Rose, Third harmonic shear horizontal and Rayleigh Lamb waves in weakly nonlinear plates. J. Appl. Phys. **114**(11) (2013)
25. K.H. Matlack, J.Y. Kim, L.J. Jacobs, J. Qu, Experimental characterization of efficient second harmonic generation of Lamb wave modes in a nonlinear elastic isotropic plate. J. Appl. Phys. **109**(1), 1–5 (2011)

26. V.K. Chillara, C.J. Lissenden, Nonlinear guided waves in plates: a numerical perspective. Ultrasonics **54**(6), 1553–1558 (2014)
27. N. Matsuda, S. Biwa, Frequency dependence of second-harmonic generation in Lamb waves. J. Nondestruct. Eval. **33**(2), 169–177 (2014)
28. P. Zuo, Y. Zhou, Z. Fan, Numerical and experimental investigation of nonlinear ultrasonic Lamb waves at low frequency. Appl. Phys. Lett. **109**(2) (2016)
29. V.K. Chillara, C.J. Lissenden, On some aspects of material behavior relating microstructure and ultrasonic higher harmonic generation. Int. J. Eng. Sci. **94**, 59–70 (2015)
30. Y. Xiang, W. Zhu, M. Deng, F.Z. Xuan, C.J. Liu, Generation of cumulative second-harmonic ultrasonic guided waves with group velocity mismatching: numerical analysis and experimental validation. Epl **116**(3) (2016)
31. W. Zhu, Y. Xiang, C.J. Liu, M. Deng, F.Z. Xuan, A feasibility study on fatigue damage evaluation using nonlinear Lamb waves with group-velocity mismatching. Ultrasonics **90**(June), 18–22 (2018)
32. V.K. Chillara, C.J. Lissenden, Analysis of second harmonic guided waves in pipes using a large-radius asymptotic approximation for axis-symmetric longitudinal modes. Ultrasonics **53**(4), 862–869 (2013)
33. Y. Liu, E. Khajeh, C.J. Lissenden, J.L. Rose, Interaction of torsional and longitudinal guided waves in weakly nonlinear circular cylinders. J. Acoust. Soc. Am. **133**(5), 2541–2553 (2013)
34. W. Li, Y. Cho, Thermal fatigue damage assessment in an isotropic pipe using nonlinear ultrasonic guided waves. Exp. Mech. **54**(8), 1309–1318 (2014)
35. Y. Liu, C.J. Lissenden, J.L. Rose, Higher order interaction of elastic waves in weakly nonlinear hollow circular cylinders. I. Analytical foundation. J. Appl. Phys. **115**(21) (2014)
36. Y. Liu, E. Khajeh, C.J. Lissenden, J.L. Rose, Higher order interaction of elastic waves in weakly nonlinear hollow circular cylinders. II. Physical interpretation and numerical results. J. Appl. Phys. **115**(21) (2014)
37. W.J.N. de Lima, M.F. Hamilton, Finite amplitude waves in isotropic elastic waveguides with arbitrary constant cross-sectional area. Wave Motion **41**(1), 1–11 (2005)
38. A. Srivastava, I. Bartoli, S. Salamone, F. Lanza di Scalea, Higher harmonic generation in nonlinear waveguides of arbitrary cross-section. J. Acoust. Soc. Am. **127**(5), 2790–2796 (2010)
39. C. Nucera, F. Lanza di Scalea, Higher-harmonic generation analysis in complex waveguides via a nonlinear semianalytical finite element algorithm. Math. Probl. Eng. **2012**, 1–16 (2012)
40. M. Hasanian, C.J. Lissenden, Second order harmonic guided wave mutual interactions in plate: vector analysis, numerical simulation, and experimental results. J. Appl. Phys. **122**(8) (2017)
41. W. Li, M. Deng, N. Hu, Y. Xiang, Theoretical analysis and experimental observation of frequency mixing response of ultrasonic Lamb waves. J. Appl. Phys. **124**(4) (2018)
42. G. Tang, M. Liu, L.J. Jacobs, J. Qu, Detecting localized plastic strain by a scanning collinear wave mixing method. J. Nondestruct. Eval. **33**(2), 196–204 (2014)
43. Y. Ishii, S. Biwa, T. Adachi, Non-collinear interaction of guided elastic waves in an isotropic plate. J. Sound Vib. **419**, 390–404 (2018)
44. Y. Ishii, K. Hiraoka, T. Adachi, Finite-element analysis of non-collinear mixing of two lowest-order antisymmetric Rayleigh-Lamb waves. J. Acoust. Soc. Am. **144**(1), 53–68 (2018)
45. V. Giurgiutiu, *Structural Health Monitoring with piezoelectric wafer active sensors*. (Elsevier, 2008)
46. S. Thiele, J.Y. Kim, J. Qu, L.J. Jacobs, Air-coupled detection of nonlinear Rayleigh surface waves to assess material nonlinearity. Ultrasonics **54**(6), 1470–1475 (2014)
47. A. Moreau, Detection of acoustic second harmonics in solids using a heterodyne laser interferometer. J. Acoust. Soc. Am. **98**(5), 2745–2752 (1995)
48. D.C. Hurley, C.M. Fortunko, Determination of the nonlinear ultrasonic parameter β using a Michelson interferometer. Meas. Sci. Technol. **8**(6), 634–642 (1997)
49. B. Ren, C.J. Lissenden, PVDF multielement lamb wave sensor for structural health monitoring. IEEE Trans. Ultrason. Ferroelectr. Freq. Control **63**(1), 178–185 (2016)
50. B. Ren, H. Cho, C.J. Lissenden, A guided wave sensor enabling simultaneous wavenumber-frequency analysis for both lamb and shear-horizontal waves. Sensors (Switzerland) **17**(3) (2017)

51. H. Cho, M. Hasanian, S. Shan, C.J. Lissenden, Nonlinear guided wave technique for localized damage detection in plates with surface-bonded sensors to receive Lamb waves generated by shear-horizontal wave mixing. NDT E Int. **102**, 35–46 (2019)
52. Y. Zhu, X. Zeng, M. Deng, K. Han, D. Gao, Detection of nonlinear Lamb wave using a PVDF comb transducer. NDT E Int. **93**, 110–116 (2018)
53. M. Hong, Z. Su, Q. Wang, L. Cheng, X. Qing, Modeling nonlinearities of ultrasonic waves for fatigue damage characterization: theory, simulation, and experimental validation. Ultrasonics **54**(3), 770–778 (2014)
54. C.J. Lissenden, Y. Liu, G.W. Choi, X. Yao, Effect of localized microstructure evolution on higher harmonic generation of guided waves. J. Nondestruct. Eval. **33**(2), 178–186 (2014)
55. M. Deng, J. Pei, Assessment of accumulated fatigue damage in solid plates using nonlinear Lamb wave approach. Appl. Phys. Lett. **90**(12), 1–4 (2007)
56. J.-Y. Kim, L.J. Jacobs, J. Qu, J.W. Littles, Experimental characterization of fatigue damage in a nickel-base superalloy using nonlinear ultrasonic waves. J. Acoust. Soc. Am. **120**(3), 1266–1273 (2006)
57. S. Shan, L. Cheng, P. Li, Adhesive nonlinearity in Lamb-wave-based structural health monitoring systems. Smart Mater. Struct. **26**(2), 1–17 (2017)
58. X. Liu et al., Detection of micro-cracks using nonlinear lamb waves based on the Duffing-Holmes system. J. Sound Vib. **405**, 175–186 (2017)
59. X. Liu et al., Locating and imaging contact delamination based on chaotic detection of nonlinear Lamb waves. Mech. Syst. Signal Process. **109**, 58–73 (2018)
60. S. Shan, M. Hasanian, H. Cho, C.J. Lissenden, L. Cheng, New nonlinear ultrasonic method for material characterization: codirectional shear horizontal guided wave mixing in plate. Ultrasonics **96**, 64–74 (2019)
61. A. Hikata, B.B. Chick, C. Elbaum, Dislocation contribution to the second harmonic generation of ultrasonic waves. J. Appl. Phys. **36**(1), 229–236 (1965)
62. J.H. Cantrell, Fundamentals and applications of nonlinear ultrasonic nondestructive evaluation," in *Ultrasonic Nondestructive Evaluation*, ed. by T. Kundu (Boca Raton: CRC Press, 2004), pp. 363–434
63. W.D. Cash, W. Cai, Dislocation contribution to acoustic nonlinearity: the effect of orientation-dependent line energy. J. Appl. Phys. **109**(1) (2011)
64. W.D. Cash, W. Cai, Contribution of dislocation dipole structures to the acoustic nonlinearity. J. Appl. Phys. **111**(7) (2012)
65. J. Zhang, F.Z. Xuan, A general model for dislocation contribution to acoustic nonlinearity. Europhys. Lett. **105**, 54005 (2014)
66. X. Gao, J. Qu, Acoustic nonlinearity parameter induced by extended dislocations. J. Appl. Phys. **124**, 125102 (2018)
67. X. Gao, J. Qu, Contribution of dislocation pileups to acoustic nonlinearity parameter. J. Appl. Phys. **125**, 215104 (2019)
68. A.K. Metya, M. Ghosh, N. Parida, K. Balasubramaniam, Effect of tempering temperatures on nonlinear Lamb wave signal of modified 9Cr-1Mo steel. Mater. Charact. **107**, 14–22 (2015)
69. H. Cho, *Toward Robust SHM and NDE of plate-like structures using nonlinear guided wave features*. The Pennsylvania State University, 2017
70. V. Chillara, H. Cho, M. Hasanian, C. Lissenden, Effect of load and temperature changes on nonlinear ultrasonic measurements: implications for SHM. Struct. Health Monit. **2015**, 783–790 (2015)
71. W. Zhu, Y. Xiang, C. Liu, M. Deng, C. Ma, F. Xuan, Fatigue damage evaluation using nonlinear Lamb waves with quasi phase-velocity matching at low frequency. Mater. (Basel) **11**(10), 1920 (2018)
72. G. Choi, Y. Liu, C.J. Lissenden, Nonlinear guided waves for monitoring microstructural changes in metal structures," in *Proceedings of the ASME 2015 Pressure Vessels and Piping Conference*, 2015, pp. PVP2015–45292
73. C.J. Lissenden, Y. Liu, J.L. Rose, Use of non-linear ultrasonic guided waves for early damage detection. Insight Non-Destructive Test. Cond. Monit. **57**(4), 206–211 (2015)

74. J.N. Potter, A.J. Croxford, P.D. Wilcox, Nonlinear ultrasonic phased array imaging. Phys. Rev. Lett. **113**(14), 1–5 (2014)
75. C.L.E. Bruno, A.S. Gliozzi, M. Scalerandi, P. Antonaci, Analysis of elastic nonlinearity using the scaling subtraction method. Phys. Rev. B Condens. Matter Mater. Phys. **79**(6), 1–13 (2009)
76. A.K. Metya, S. Tarafder, K. Balasubramaniam, Nonlinear Lamb wave mixing for assessing localized deformation during creep. NDT E Int. **98**(April), 89–94 (2018)

Part II
Measurements of Nonlinear Ultrasonic Characteristics Related with Contact Acoustic Nonlinearity

Chapter 4
Nonlinear Acoustic Measurements for NDE Applications: Waves Versus Vibrations

Igor Solodov

Abstract Majority of acoustic instruments widely used in industry and technology for non-destructive evaluation (NDE) make use of a linear elastic response of materials. The nonlinear approach to ultrasonic NDE is concerned with nonlinear material response, which is inherently related to the frequency changes of the input signal, and is a new technology for monitoring of deterioration in material properties and diagnostics of damage. The application field now includes both the nonlinear wave and nonlinear vibration modes. The former is based on the assumption and is applicable to the case studies of the distributed material nonlinearity. It profits from accumulation of the wave nonlinear response along the propagation distance and relies on the higher harmonic signals. A strong nonlinear response of non-bonded interfaces in planar defects introduces the nonlinearity localized in the defect area where the vibration nonlinearity steps up. The concept of local defect resonance (LDR) combined with its nonlinearity identifies a nonlinear inclusion as a nonlinear oscillator and brings about different dynamic and frequency scenarios in vibration nonlinear phenomena. The LDR-induced trapping of the nonlinearity generates a defect-selective nonlinearity and conditions for efficient and even noncontact nonlinear diagnostic imaging of damage.

4.1 Introduction

Majority of acoustic instruments widely used in industry and technology for non-destructive evaluation (NDE) and quality assessment make use of a linear elastic response of materials that generally results in the amplitude and phase variations of the input signal. The nonlinear approach to ultrasonic NDE is concerned with nonlinear material response, which is inherently related to the frequency changes of the input signal. Classical nonlinear acoustics of solids, systematically established in

I. Solodov (✉)
Institute for Polymer Technology, University of Stuttgart, Stuttgart, Germany
e-mail: igor.solodov@ikt.uni-stuttgart.de

K.-Y. Jhang et al. (eds.), *Measurement of Nonlinear Ultrasonic Characteristics*, Springer Series in Measurement Science and Technology,
https://doi.org/10.1007/978-981-15-1461-6_4

1960s–1970s [1, 2], dealt with homogeneous (quasi-flawless) materials, whose nonlinearity is associated with lattice anharmonicity and reveals nonlinear behaviour of inter-molecular forces. At the macro-scale, the dynamic stiffness of a nonlinear material is a function of strain that results in a local variation of the wave velocity, which leads to a waveform distortion and higher harmonic (HH) generation. However, the measurements showed that in basically all homogeneous and free from defects materials even for high acoustic strains ~10^{-4} the stiffness variation due to nonlinearity is below 10^{-3}. As a result, noticeable nonlinear effects are developed only due to accumulation of the nonlinear wave response along the propagation distance (*wave distributed nonlinearity*), and, in practical terms, solely the second and the third (less frequently) harmonic signals can be used for material characterization and NDE.

However, even in the first experimental studies a substantial increase in nonlinearity was noticed in materials with imperfections: a substantial enhancement of the second harmonic signal was measured in a high-purity Al single crystal in which a dislocation pattern was induced by mechanical stress applied [3]. Further investigations confirmed an important role of internal boundaries in acoustic nonlinearity enhancement for dislocations in fatigued materials [4] and matrix-precipitate interfaces in alloys [5].

A number of studies were implemented then to identify the mechanisms and manifestations of the interface nonlinearity. The experiments revealed a drastic increase of nonlinearity in non-bonded contacts for surface and bulk waves [6] due to specific contact acoustic nonlinearity (CAN) [7, 8]. CAN is an example of a *localized vibration nonlinearity* caused by constraints of interface vibrations (or fragments of cracked defects) for both normal ("clapping") and tangential (micro-slip) vibrations. The intact material outside CAN area can be considered as a "linear carrier" of an acoustic wave while the nonlinearity manifests in local defect vibrations only.

A further step in the development of the localized nonlinearity is concerned with the concept of Local Defect Resonance (LDR) [9–11], which enhances substantially the efficiency of acoustic activation of damage and has generated much interest in development and applications of the resonance techniques for ultrasonic testing and imaging of defects [12–19]. It occurs for a frequency match of a driving wave to a defect natural frequency and results in a resonant amplification of a standing wave vibration selectively in the damage area on the background of relatively small amplitude excitation in the rest of the specimen. The benefit of the LDR approach is, therefore, primarily usable in nonlinear acoustic methodology: the high amplitude vibrations developed locally in the damage area manifest pronounced nonlinearity even at moderate acoustic excitation level. The CAN combined with LDR also identifies a nonlinear inclusion as a nonlinear oscillator and brings about qualitatively different dynamic and frequency scenarios in nonlinear phenomena [20].

Unlike conventional nonlinear inspection concerned with the nonlinear wave propagation the LDR-CAN-based methodology shifts to nonlinear vibration effects in particular isolated areas of a nonlinearly inhomogeneous imperfect material. The both nonlinear approaches (nonlinear waves and vibrations) can operate with (or triggered by) various types of acoustic waves: bulk and/or surface waves in volumetric samples and guided waves in plate-like specimens. In this Chapter, the fundamentals and manifestations of the both approaches are introduced with particular emphasis on the experimental methods used for characterization and nonlinear NDE applications of various type acoustic waves.

4.2 Fundamental Effect in Nonlinear Acoustics: Higher Harmonic Generation for Longitudinal Acoustic Waves

The equation of motion for a finite amplitude longitudinal wave in an isotropic solid is [21]:

$$\rho \frac{\partial^2 u}{\partial t^2} = \frac{\partial \sigma}{\partial x}, \tag{4.1}$$

where u is the displacement component along the propagation direction x, ρ is the density of undeformed material, and the (engineering) stress σ is found as a derivative on the linear strain ($\varepsilon = \partial u/\partial x$) of the internal elastic energy of a unit volume U:

$$\sigma = \frac{\partial U}{\partial \varepsilon}. \tag{4.2}$$

By keeping the higher-order terms in the $U(\varepsilon)$ expansion (up to the fourth-order on ε) one obtains the nonlinear stress-strain relation in the form:

$$\sigma = \beta_1(\varepsilon - \frac{\beta_2}{2}\varepsilon^2 - \frac{\beta_3}{3}\varepsilon^3), \tag{4.3}$$

where $\beta_1 = \lambda + 2\mu$ is the linear elasticity for longitudinal wave (λ, μ are Lame constants), $\beta_2 = -\left(3 + \frac{2A+6B+2C}{\beta_1}\right)$ contains the third-order elastic constants, and $\beta_3 = -\left(\frac{3}{2} + \frac{6A+18B+6C+12(D+G+H+J)}{\beta_1}\right)$ also includes the fourth-order constants.

By using (4.3) in (4.1) the equation of motion for nonlinear longitudinal wave takes the form:

$$\frac{\partial^2 u}{\partial t^2} = c_0^2(1 - \beta_2\frac{\partial u}{\partial x} - \beta_3(\frac{\partial u}{\partial x})^2)\frac{\partial^2 u}{\partial x^2}, \tag{4.4}$$

where $c_0^2 = \beta_1/\rho$ is the linear wave velocity squared.

Equations (4.3) and (4.4) show that in nonlinear materials the stiffness depends on the wave amplitude and the depth of the stiffness modulation $\sim \sum\limits_n \beta_n \varepsilon^n$ is an overall characteristic of the nonlinear material properties.

The solution to nonlinear Eq. (4.4) is sought by using the perturbation theory: $u = u_\omega + u_{2\omega} + u_{3\omega}$, where the amplitudes of the higher harmonics are small compared to that of the fundamental wave $u_\omega \sin(\omega t - kx)$. The inhomogeneous equations of the consequent approximations yield the well-known relations for the amplitudes of the second and third harmonics (see also e.g. [22–24]):

$$u_{2\omega} = \frac{\beta_2}{8} k^2 u_\omega^2 x. \tag{4.5}$$

$$u_{3\omega} = \frac{\beta_2^2 k^4 u_\omega^3 x^2}{32} (1 + \frac{16}{9k^2 x^2} (1 - \frac{\beta_3}{\beta_2^2})^2)^{1/2}. \tag{4.6}$$

Equations (4.5) and (4.6) clearly illustrate the power-law dynamics of the both higher harmonics $u_{2\omega} \sim u_\omega^2$ and $u_{3\omega} \sim u_\omega^3$. The second harmonic amplitude increases linearly with distance x and is also proportional to the nonlinearity parameter β_2, which depends on the second and third-order elastic constants of the material. The latter enables to determine a certain combination of the third-order elastic constants [25]. It is instructive noting, that here (as well as generally in nonlinear acoustic of solids) β_2 is actually introduced by the ratio of the quadratic to linear terms in equation of motion (4.4). An alternative approach [26] comes from nonlinear acoustics of gases and liquids, (see, e.g. [27]) and uses the corresponding ratio in nonlinear stress-strain relation (4.3) that yields β_2 twice smaller and $\frac{\beta_2}{4}$ factor in Eq. (4.5).

The interpretation of the third harmonic response (4.6), which generally depends on both β_2 and β_3, is not as straightforward as that for the second harmonic [28]. To this end, we first write the radicand in (4.6) in the form $1 + \eta$. If $\eta \ll 1$, one can neglect the second term in brackets and Eq. (4.6) simplifies:

$$u_{3\omega} = \frac{\beta_2^2 k^4 u_\omega^3 x^2}{32}. \tag{4.7}$$

In this case, the third harmonic depends on β_2 only and is a quadratic function of the propagation distance. From (4.6), it may happen at large distances from the excitation source ($kx >> 1$). At smaller distances and when $\beta_3 > \beta_2$, the second term in (4.6) dominates and the third harmonic is fully determined by β_3 value and grows linearly with distance:

$$u_{3\omega} = \frac{\beta_3 k^3 u_\omega^3 x}{24}. \tag{4.8}$$

The difference between (4.7) and (4.8) illustrates the two possible mechanisms of the third harmonic generation: as the distance x increases, the four-wave interaction $\omega + \omega + \omega \to 3\omega$ changes for two subsequent three-wave interactions $\omega + \omega \to 2\omega$ and $2\omega + \omega \to 3\omega$.

As the distance x increases, $u_{3\omega}(x)$ deviates from (4.8) and changes for (4.7). The transition distance may be assumed to correspond to the approximate condition $\eta \approx 1$ which yields:

$$16(1 - \beta_3/\beta_2^2)^2 \approx 9k^2 x_t^2. \tag{4.9}$$

Equation (4.9) enables to estimate β_3/β_2 by measuring the transition distance x_t and will be used in the experimental study in the next Section.

4.2.1 Experimental Validation of Higher Harmonic Characteristics

Enormous number of the higher harmonic studies has been implemented for more than 50 years since the first experiments on the second harmonic generation in 1963 [3, 29]. The majority of the studies and applications are concerned with the second harmonic and are summarised in an extensive recent review [30]. Far less consideration has been given to the third harmonic experimental characteristics mainly studied in the context of the nonlinear frequency shift [31]. Here, some results on experimental validation of the higher harmonic characteristics will be shown to affirm the relations obtained above and illustrate the opportunities for NDE of the material nonlinearity parameters.

Growing spatial characteristics of the higher harmonics are of the most interest, however, their measurements offer some experimental problems. In transparent materials, the optical diffraction can be applied whereas the surface acoustic waves (SAW) (also known as Rayleigh waves in isotropic materials) are used in other cases [32]. The major advantages of the SAW for nonlinear applications are the lack of the velocity dispersion (which allows nonlinearity to accumulate with distance) and the high power density provided by inhomogeneous energy distribution in the near-surface area. They also permit the surface access to the harmonic fields that facilitates the experimental observations.

In the experiment [33], the SAW of the fundamental frequency 11 MHz are generated in the XY-cut of piezoelectric quartz specimen ($160 \times 60 \times 5$ mm^3) by using a Plexiglas wedge transducer. A burst signal up to 200 V amplitude was applied to the X-cut quartz plate attached to the wedge. To detect and monitor the second harmonic signal along the SAW propagation path, the interdigital transducer (Al array of $\approx$ 140 μm period, the aperture 8 mm) was deposited on the bottom face of the glass plate which was put in contact and can move freely on the substrate surface. The SAW in the XY-cut of quartz is known to manifest the beam steering effect so that the mobile receiving transducer moved along the group velocity path at $\approx 10°$ to the Y-axis.

Figure 4.1 shows the measured output signal of the second harmonic transducer (22 MHz) as a function of distance x for different input voltages. The amplitude increases linearly (accumulation of nonlinearity) for $x \leq 90$ mm and saturates then due to the impact of attenuation. This unique feature of nonlinearity is supported by the dynamics of the second harmonic in Fig. 4.2: its amplitude is a quadratic function of the fundamental wave amplitude in full accord with (4.5).

To evaluate the nonlinearity parameter in (4.5), the absolute values of SAW displacements are required. They can be derived from the measurements of the

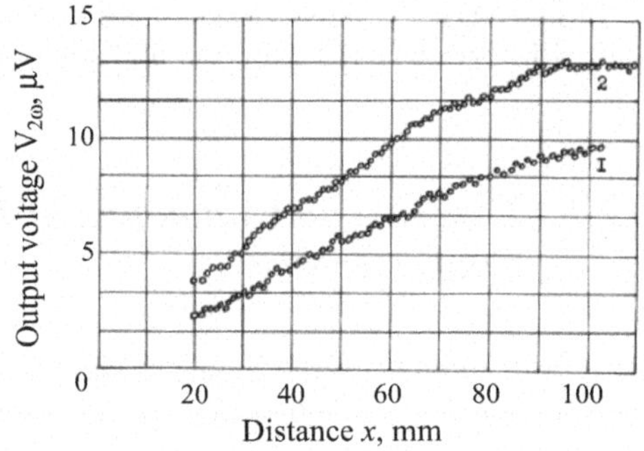

Fig. 4.1 Second harmonic output voltage as a function of distance in XY-quartz: 1—input voltage 140 V, 2—170 V

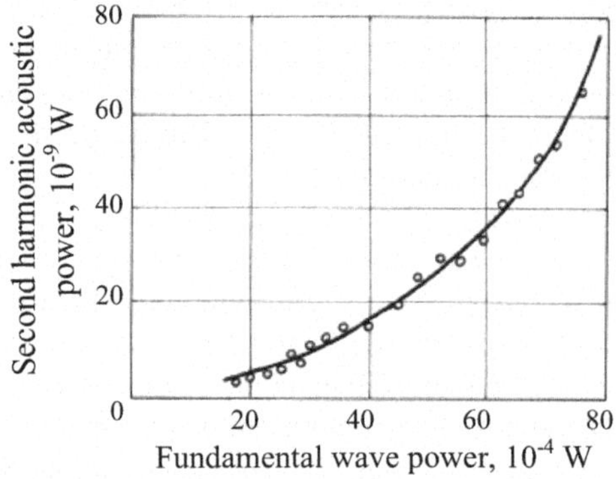

Fig. 4.2 Dynamics of the second harmonic signal in XY-quartz

input-output electrical signals by calculating the insertion loss factor for converting electrical power into acoustics and back in the line of two identical transducers: $L_{out}^{in} = 10\lg(P_{in}/P_{out})$. Due to reciprocity, a half of this value is the acoustic loss factor: $L_{ac}^{in} = 10\lg(P_{in}/P_{ac})$ which enables to estimate the power flow of the fundamental acoustic wave P_{ac}. In a similar way, a half of the insertion loss at the higher harmonic frequency is used for evaluating its acoustic power from the electric output: $L_{ac}^{out} = 10\lg(P_{ac}^{n\omega}/P_{out}^{n\omega})$. The SAW longitudinal displacements (for any frequency) can be found then by using the following relation [34]: $P_{ac}^{n\omega} = K(n\omega)WU_{Ln\omega}^2$, where W is the aperture of the SAW beam, K is the SAW power flow constant and used in (4.5), (4.7), (4.8) to estimate the nonlinearity parameters.

This methodology was applied to deriving the absolute values of the longitudinal displacements in the fundamental SAW and its second harmonic in SiO_2. For acoustic power of the SAW ≈ 2 mW, the displacement of the fundamental wave was found to be $\approx 10^{-9}$ m accompanied by the second harmonic of $\approx 5 \times 10^{-12}$ m amplitude at 10 cm distance in Fig. 4.1. With these data from Eq. (4.5) one obtains: $\beta_2 \approx 0.8$ which is in general agreement with the values for the bulk waves [35]. It is instructive noting that the SAW second harmonic of measurable amplitude ($u_{2\omega}/u_\omega \approx 0.5\%$) is generated in the crystalline material in mW power range when the fundamental strain is as low as $\sim 10^{-5}$. The SAW power density is nonetheless in quite a nonlinear

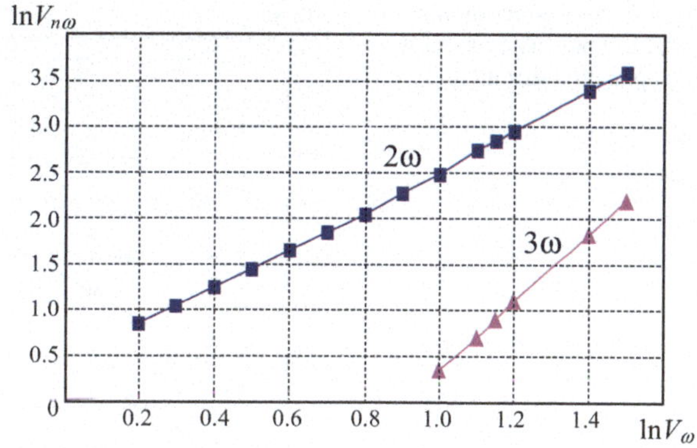

Fig. 4.3 Second and third harmonics as functions of fundamental frequency input voltage

range $\sim 5 \times 10^3$ W/m^2 due to the "skin-effect" of the near-surface localization for a high-frequency SAW.

The SAW higher harmonics (including the third harmonic) in a stronger piezoelectric material were studied in [28]. The interdigital transducers deposited on the substrate (YZ-$LiNbO_3$) surface were used for generation of the SAW bursts with fundamental frequency 15 MHz and detection of its higher harmonics. The second and third harmonic signals as functions of the fundamental frequency input are shown in Fig. 4.3. The logarithmic scale demonstrates the difference in the higher harmonic nonlinear dynamics: in accord with (4.5) and (4.6), the second harmonic is close to quadratic while the third harmonic to a cubic function of the fundamental SAW amplitude. The methodology of using the insertion loss factors described above was then applied to calculate the absolute values of the wave displacements and evaluate the nonlinearity parameter according to (4.5). For the YZ-$LiNbO_3$, this parameter was found to be higher than that for quartz: $\beta_2 \approx 4 \pm 1$.

To receive the signals of the higher harmonics and probe the distance characteristics in a piezoelectric substrate one can also use a non-contact metallic probe which detects the electric fields induced by the SAW. In the experiment, a thin steel plate (the edge thickness $d \approx 5$ µm, aperture 1 cm) was positioned in a close proximity to the substrate surface parallel to the wave front. Since the probe thickness $d \ll \lambda$ it is a wideband receiver and can be used to trace the higher harmonic distribution over the specimen length.

These plots obtained for the second and third harmonics are shown in Figs. 4.4 and 4.5 in logarithmic scale [28]. The slope of the second harmonic plot in Fig. 4.4 is evaluated as ≈0.9 while it changes from about this value at small distances (up to ≈ 10 mm) for ≈2.0 at the larger propagation path.

The transition distance in Fig. 4.5 $x_t \approx 10$ mm is used then in (4.9) to evaluate the higher-order material nonlinearity to obtain: $\beta_3 \approx 200\ \beta_2^2 \approx 650$.

The experimental estimates of the nonlinearity parameters β_2 and β_3 in $LiNbO_3$ enable to evaluate the combinations of the third- and fourth-order elastic constants of the material. By using these values in (4.3), for the third order term one obtains $|A + 3B + C| \approx 1.3 \times 10^{11}$ N/m^2 and $|D + G + H + J| \approx 5.5 \times 10^{12}$ N/m^2 for the

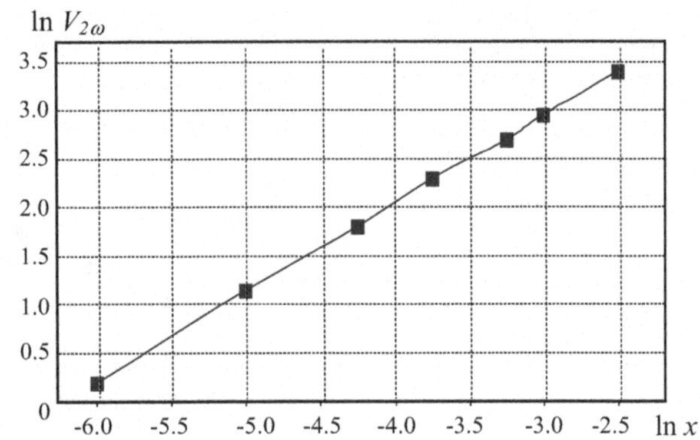

Fig. 4.4 Second harmonic in YZ-$LiNbO_3$ as a function of propagation distance

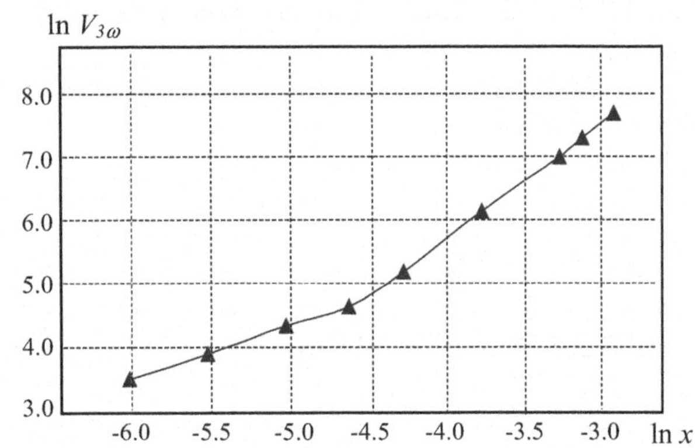

Fig. 4.5 Third harmonic distribution along the propagation path in YZ-$LiNbO_3$

fourth-order constants. Thus, the nonlinear elastic constants are generally higher than conventional (linear) material elasticity (for $LiNbO_3$ $\beta_1 = \rho c_0^2 \approx 5 \times 10^{10}$ N/m^2). The difference is within an order of magnitude for the third-order nonlinearity and about 2 orders of magnitude for the fourth-order constants. These findings agree quite well with the literature data for other materials [36, 37].

The results of similar second harmonic experiments for various crystalline materials are summarized in Table 4.1. The data confirm the fact (also mentioned in

Table 4.1 Nonlinearity parameters for some crystalline materials

Material	Cut, direction	Frequency (MHz)	β_2
SiO_2	XY	11	0.8 ± 0.4
$LiNbO_3$	YZ	44 15	5 ± 2 4 ± 1
$Bi_{12}GeO_{20}$	(001), [110]	10	6 ± 2
$\alpha-$ HJO_3	(100), [011]	20	8 ± 3

[35]) that the nonlinearity is higher in "soft" and "slow" materials: β_2 maximizes in $Bi_{12}GeO_{20}$ (SAW velocity $\approx 1.7 \times 10^3$ m/s) and α-HJO_3 (longitudinal wave velocity $\approx 2.5 \times 10^3$ m/s).

The linear distance dependence for the second harmonic amplitude is characteristic of the low-dissipative materials in the so-called fixed-field approximation for the fundamental wave (total acoustic power in the mW-range, acoustic intensity ~10^{-1} W/cm^2). For the higher wave intensities, the second harmonic generation can be so efficient that affects noticeably the fundamental wave. To study the nonlinearity under these conditions, the high-intensity high-frequency ($\omega/2\pi = 128$ MHz) SAW was excited in the YZ-$LiNbO_3$ by the interdigital transducers with extremely low acoustic loss factor: $L_{ac}^{out} \approx 4$ dB [38]. For a total acoustic power $P_{ac}^{\omega} \sim 1$ W, the acoustic intensity in the surface layer increases up to ~10^3 W/cm^2.

The acousto-optical Raman-Nath diffraction in the reflection mode was applied to probe the second harmonic field (Fig. 4.6). The relative intensity of the higher-order diffraction maxima is given by [39]:

$$I_m/I_0 \approx \frac{1}{m!}(DfP_{ac}^{\omega})^m, \tag{4.10}$$

where D is the constant of the SAW energy quality, which for the YZ-$LiNbO_3$ is $D = 1.5 \times 10^{-11}$ c/W.

For the experimental conditions used, a contribution of the fundamental frequency wave to the second-order diffraction from (4.10) is expected to be: $I_2/I_1 \approx 10^{-3}$. Figure 4.7 shows the distributions of the first and the second-order maxima along the SAW propagation distance. The ratio I_2/I_1 measured is about 2 orders of magnitude higher, i.e. it is fully determined by the second harmonic wave. From (4.10), $P_{ac}^{2\omega}/P_{ac}^{\omega} = 2I_{2\omega}/I_{\omega}$, i.e. according to Fig. 4.7, $P_{ac}^{2\omega} \approx 0.4P_{ac}^{\omega}$ at $x = 7$ mm from the source. At this distance, the fundamental wave displays a noticeable deviation from the fixed-field condition by about 25% loss of the power radiated. From Fig. 4.7,

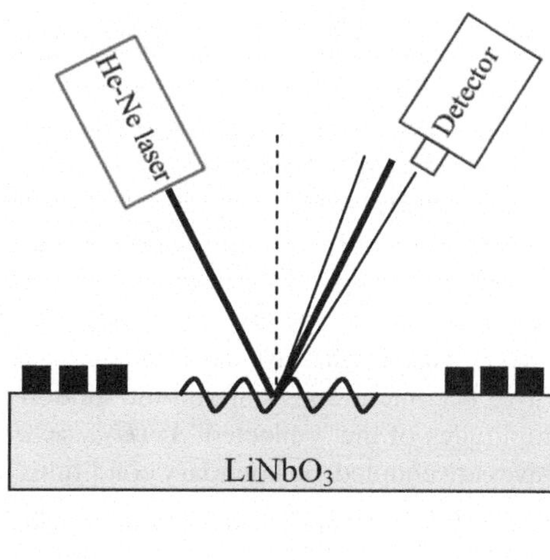

Fig. 4.6 Experimental setup for SAW higher harmonics optical probing

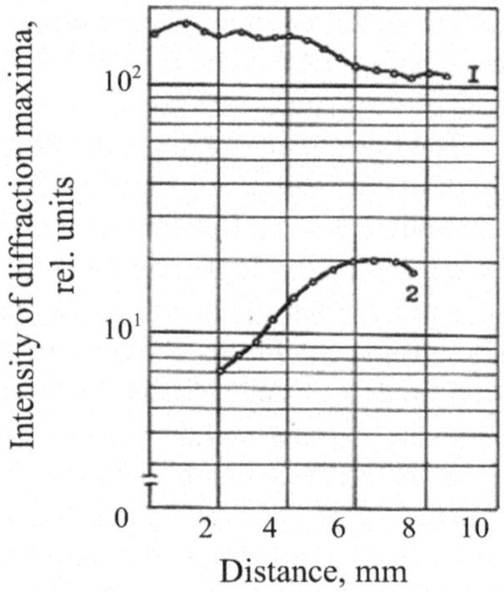

Fig. 4.7 Distribution of the first—(1) and second-order (2) diffraction maxima along the SAW propagation distance

about a half of this "loss" is pumped into the second harmonic. Such a nonlinear "depletion" of the fundamental wave feeds back to the second harmonic whose amplitude decays with the distance.

4.3 Nonlinear SAW: Theory

The advantages of the SAW for nonlinear applications demonstrated in the previous section are concerned with convenient and efficient excitation, lack of dispersion and the high power density that enables to attain the nonlinear regime at lower input power. However, an intricate structure of the SAW, which is a combination of the longitudinal and shear inhomogeneous waves, apparently obscures a fairy clear concept of the longitudinal wave nonlinearity given above. Here, the physical picture and an explicit analytic description of the SAW nonlinearity are highlighted [34] with particular emphasis on the aforementioned point. For the sake of clarity, only the second approximation is considered, the material is isotropic; both the dissipation and the energy exchange between the fundamental wave and the higher harmonics are not under consideration.

An elastic wave (saggital (xz)-polarization) freely propagating along the boundary can be considered as a combination of reflected (or transmitted) waves (longitudinal (L) and transversal (T) in the isotropic case) existing without an incident wave. If the phase velocity of this composition is lower than v_L and v_T velocities in the material (Rayleigh wave case $v_R < v_{L,T}$), the reflection angles are imaginary, the waves become evanescent and the motion amplitude decays from the surface. The amplitudes of the "reflected" L-($U_{L\omega} = div\bar{U}_R$) and T-($U_{T\omega} = (rot\bar{U}_R)_y$) partial waves are coupled via boundary conditions on a free surface to form the fundamental (linear) Rayleigh wave displacement pattern in the form:

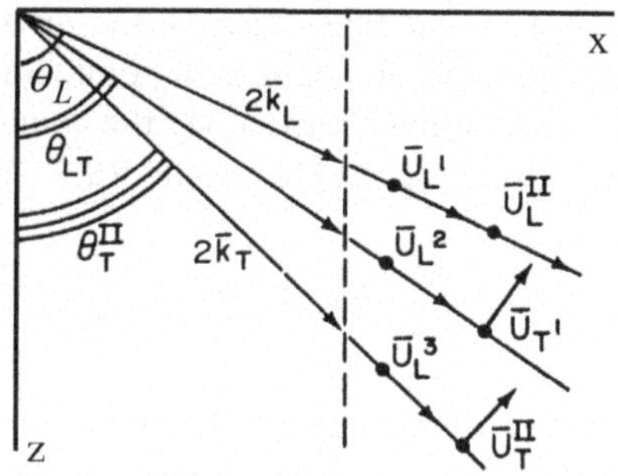

Fig. 4.8 Reflection model of the second harmonic SAW wave field

$$\bar{U}_R = U_{L\omega}\bar{R}(\omega)\exp[i(\omega t - kx)], \tag{4.11}$$

where the components of the complex vector $\bar{R}(\omega)$ characterize the depth-dependent wave pattern defined in terms of the amplitude of the L-partial wave [34].

In the perturbation approach, the L- and T-partial waves of the second-order wave field are the solutions to the inhomogeneous equations of motion:

$$\rho\frac{\partial^2}{\partial t^2}(div\bar{U}_{L2\omega}) - (\lambda + 2\mu)\Delta(div\bar{U}_{L2\omega}) = div\bar{F} \tag{4.12}$$

$$\rho\frac{\partial^2}{\partial t^2}(rot\bar{U}_{T2\omega})_y - \mu\Delta(rot\bar{U}_{T2\omega})_y = (rot\bar{F})_y, \tag{4.13}$$

where $\bar{F}$ is a quadratic function, which takes into account the self- and cross-interactions of the fundamental partial waves $U_{L\omega}$, and $U_{T\omega}$. From (4.12) and (4.13), the second harmonic field includes both the free partial waves (solutions to the homogeneous equations) and the driven waves produced by the right-hand side terms. The calculations [34, 40] show that the driven wave field contains three L-waves produced by the $L_\omega - L_\omega$ interaction (along $2\bar{k}_L$ direction), $L_\omega - T_\omega(\bar{k}_L + \bar{k}_T)$, and $T_\omega - T_\omega(2\bar{k}_T)$ shown in Fig. 4.8, as well as one T-wave generated by $L_\omega - T_\omega$ along $(\bar{k}_L + \bar{k}_T)$ direction. The sum of the free and driven waves in Fig. 4.8, however, reveals the only "resonant" term caused by the superposition of the phase-matched (equal velocities and propagation directions) free and driven waves. The other partial waves result in the oscillating pattern without accumulation of nonlinearity and do not contribute substantially to the second harmonic field.

Similar to the bulk L-wave case considered above, this "resonant" general solution provides the growth with distance for the second harmonic of L-partial wave. According to Fig. 4.8, for the growth along the surface we therefore obtain:

$$U_{L2\omega} = (\beta_2/8)x\sin\vartheta_L k_L^2 U_{L\omega}^2, \tag{4.14}$$

where θ_L is the angle of "reflection" for the "resonant" partial wave (Fig. 4.8) phase matched with SAW along the boundary so that $\sin\theta_L = v_L/v_R$.

Like in the linear case above, on the boundary the deformation in L-partial wave can not exist alone: in each point along the surface it activates T-partial second harmonic component to satisfy the boundary conditions. Both, therefore, are combined to form the second harmonic Rayleigh wave displacement pattern:

$$\bar{U}_{R2\omega} = U_{L2\omega}\bar{R}(2\omega)\exp[2i(\omega t - kx)] = (\beta_2/8)(v_L/v_R)k_L^2 U_{L\omega}^2 x\bar{R}(2\omega)\exp[2i(\omega t - kx)]. \tag{4.15}$$

According to (4.14, 4.15), the second harmonic is activated by the longitudinal partial wave of SAW; complementary T-partial wave arises because of the elastic coupling of nonlinear deformations on the boundary (coupled boundary nonlinearity) and completes the structure of $\bar{R}(2\omega)$. Overall nonlinearity develops analogously to the homogeneous (bulk) wave case: the SAW amplitude increases linearly with distance, and is a quadratic form of the fundamental partial wave amplitude and the wave number. However, the slowness of SAW results in additional accumulation of the nonlinearity that increases the second harmonic amplitude (v_L/v_R) times.

Harmonic accumulation due to nonlinear wave propagation leads to the distortion of SAW and formation of a shock-like wave. The solution obtained above is relevant to the initial stage of the wave distortion only, nevertheless, enables to preview the trend of the steady-state nonlinear waveform development. Unlike the bulk wave, the SAW nonlinear distortion is two-dimensional, so that one has to use a two-component representation and separate the distortions of vertical and horizontal displacements. To this end, the higher harmonic (4.15) and linear (4.11) solutions are written in the unified component presentations:

$$\begin{aligned} U^x_{Rn\omega} &= U_{Ln\omega}R^x_{n\omega}(n\omega)\exp[in(\omega t - kx)] \\ U^z_{Rn\omega} &= U_{Ln\omega}R^z_{n\omega}(n\omega)\exp[in(\omega t - kx)]. \end{aligned} \tag{4.16}$$

The nonlinear inhomogeneous wave-field pattern can then be obtained as a sum of the corresponding components (4.16) for $n = 1, 2$. For $n = 1$, one uses Eqs. (4.11) and (4.15) for $n = 2$ with all linear and nonlinear parameters determined from experimental conditions. The illustration of the SAW waveform distortion is given in Fig. 4.9, where the second harmonic amplitude $U_{L2\omega}$ is assumed to be 30% of that for the fundamental wave and about 10% of the third harmonic is added to cancel out the unwanted oscillations of the nonlinear SAW profile. The resultant nonlinear SAW field reveals that a weak periodic shock wave is being produced with saw-tooth vertical displacement profile.

4.3.1 Nonlinear Surface Waves: Experimental

The SAW higher harmonics studied in the previous section is an implicit proof for nonlinear wave distortion. However, firstly, their amplitudes are usually too small

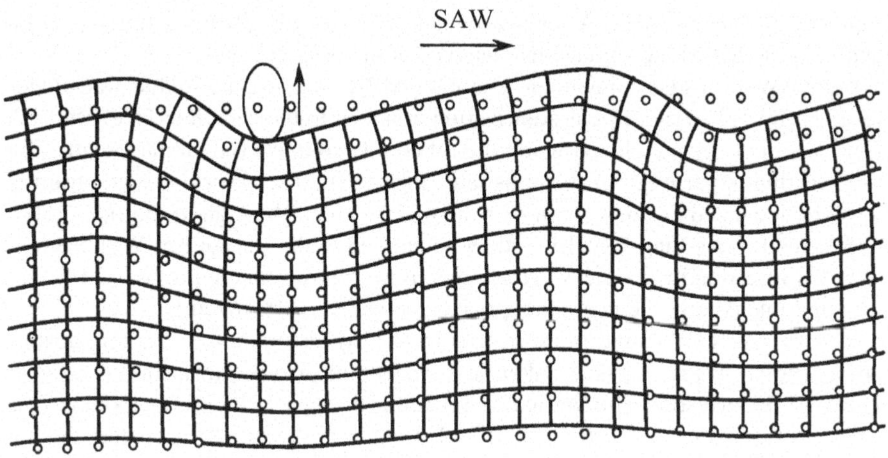

Fig. 4.9 Nonlinear SAW wave field calculated according to (4.16)

to cause noticeable distortion. Secondly, precise phase measurements of the higher harmonic are required to compose the distorted nonlinear field from the spectral measurements.

For a direct experimental observation of the nonlinear SAW waveform distortion [41], we proposed to use the so-called electro-dynamical transducer: a narrow metallic electrode deposited on the surface of the substrate placed in magnetic field $\bar{B}$ (Fig. 4.10). The vibrations of the electrode induced by a SAW cause the output voltage proportional to the vibration velocity of the transducer:

$$V = \left[\bar{v} \cdot \bar{B}\right]l, \tag{4.17}$$

where $\bar{v}$ is the vibration velocity and l is the length of the electrode.

When the width of the electrode $W \ll \lambda$, it is a wideband transducer applicable to the observation of the nonlinear vibrations, which contain the higher harmonics. By variation of the magnetic field orientation different components of the vibration velocity can be singled out. In the experiment, $W = 50\,\mu\text{m}$, $\lambda = 600\,\mu\text{m}$, the SAW

Fig. 4.10 Setup for observation of the waveform distortion of nonlinear SAW

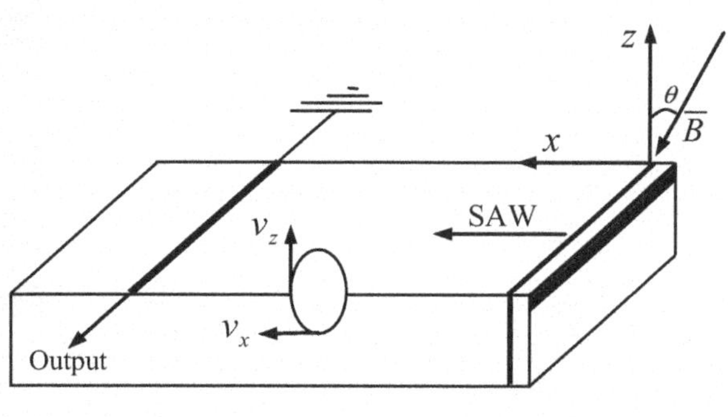

frequency $1/T = 5$ MHz. After a wideband receiver (20 dB amplification in the bandwidth 50 MHz) the signal was observed at the oscilloscope.

The SAW in a glass specimen was excited by the so-called edge transducer: piezo-ceramic plate of vertical polarization attached to the substrate face. One side of the plate is fully metallised while the other side contains only a narrow metallic strip near the upper edge of the substrate (Fig. 4.10). The piezo-induced strain is, therefore, localized exclusively in its near-surface area. The high-amplitude (150 V at 75 Ω load) 3 μs-long 5 MH-bursts were applied to the transducer. With account for 10 dB insertion losses the SAW acoustic power was estimated to be ≈14 W and the sound intensity in the near-surface area as high as ≈300 W/cm^2.

A harmonic SAW at the transducer (Fig. 4.11a) distorts along the propagation path as shown in Fig. 4.11b, c taken at $x = 80$ mm distance from the transducer. Figure 4.11b corresponds to vertical $\bar{B}$ position (in-plane velocity component v_x is active) while in Fig. 4.11c $\theta = \pi/2$ and it exhibits the vibration pattern for the out-of-plane velocity component v_z. Positive phases of the scope traces match to the velocity directions v_x, $v_z > 0$ indicated in Fig. 4.10. The results in Fig. 4.11b, c show that the nonlinear distortion leads to a saw-tooth-like wave on v_x (asymmetry between the trailing and leading edges $\Delta t \approx 0.12T$) and the "inverse bell" profile on v_z(difference in duration between positive and negative phases $\approx -0.18T$). The latter indicates that the harmonic vertical motion changes for short "splashes" above the surface characteristic of the saw-tooth nonlinear distortion shown in Fig. 4.9.

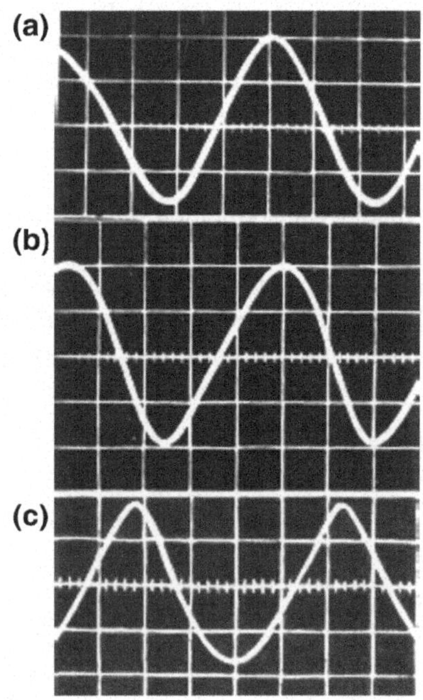

Fig. 4.11 Oscilloscope traces for nonlinear SAW at different distances x from the source and magnetic field orientations: **a** $x = 10$ mm, $\theta = 0°$; **b** $x = 80$ mm, $\theta = 0°$; **c** $x = 80$ mm, $\theta = 90°$

The distortion of the in-plane component (Fig. 4.11b) is similar to the nonlinear wave distortion in liquids so to quantify the effect we use the relation for the shift of the maximum in v_x determined in [1]:

$$\Delta x = \frac{\beta_2}{4} M x, \tag{4.18}$$

where $M = v/c_0$ is Mach number.

In the experiment, the output voltage $V = 1$ mV for $B = 8 \times 10^{-2}$ T so that from (4.17): $v_x \approx 1$ m/s and $M \approx 3 \times 10^{-4}$. According to Fig. 4.11b, $\Delta x/\lambda \approx 3.6 \times 10^{-2}$ and from (4.17) one obtains: $\beta_2 \approx 3.6$, i.e. a somewhat higher value than that for a similar crystalline material (SiO_2, Table 4.1).

4.4 Localized Nonlinearity of a Non-bonded Interface

The result obtained above is quite general: a violation of a highly ordered microscopic structure of a material by introducing the lattice defects of various order (dislocations, grain boundaries, micro-cracks, etc.) results in its higher nonlinearity parameter. The increase in nonlinearity becomes clear because the depth of the stiffness modulation caused by acoustic vibrations enhances in the presence of "soft" defects. For NDE applications, particularly important are the planar defects, like cracks, delaminations, debondings, impact and fatigue damages, etc. The vibration of the weakly-bonded interfacial area is accompanied by a strong local stiffness variation: the stiffness of the interface is substantially higher for compression than that for the tensile stress. The depth of a local stiffness modulation can, therefore, exceed considerably the values for the classical nonlinearity ~ $\beta_2\varepsilon$, i.e. 10^{-4} to 10^{-3} at the maximum strain. This makes CAN an efficient vibrational nonlinearity localized exclusively in the defect area (defect-selective) as opposed to the distributed wave nonlinearity considered above.

Experimental evidence of the CAN characteristics is demonstrated in [8, 42] for low-frequency vibrations ($\cong 300 Hz$) of the interface between two optically polished metal samples squeeze together with an ambient pressure. The correlation between the interface "clapping" and the threshold distortion of the contact vibrations is proved and shown to be accompanied by multiple higher harmonic generation. A distinctive feature of CAN is concerned with non-monotonic (oscillating) spectra: the higher harmonic amplitude modulation similar to the *sinc*-function was observed (Fig. 4.12a). Physically, this is attributed to a "pulse" modulation of the interface elasticity: the stiffness of the contact changes discontinuously when it alters from "open" to "close" recurrently by the driving vibration. As the driving amplitude increased, the vibration becomes unstable: a cascade process of multiple subharmonic generation was observed which ended up with chaotic vibration pattern (Fig. 4.12b, c).

The locality of the contact nonlinearity is also manifested in the nonlinear reflection effects for acoustic waves encounter with the CAN area. In the experiment [43],

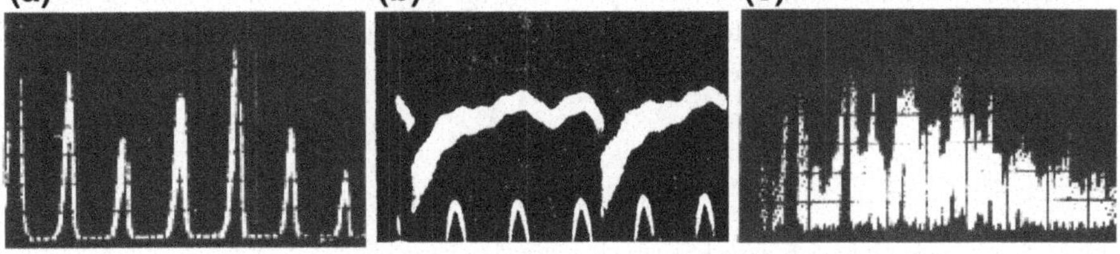

Fig. 4.12 *Sinc*-modulated higher harmonic spectrum (**a**), subharmonic (3d order) (**b**) and chaotic vibrations in simulated CAN experiment (**c**)

a 20 MHz SV-wave in a glass specimen is incident on the contact (glass-glass) interface at 45° (Fig. 4.13) and the spectrum of the reflected field is analyzed by using the wideband electro-dynamical transducer described above. Figure 4.14 shows that as the contact pressure increases, the fundamental wave reflection diminishes (due

Fig. 4.13 Experimental setup for observation of CAN in SV-wave reflection

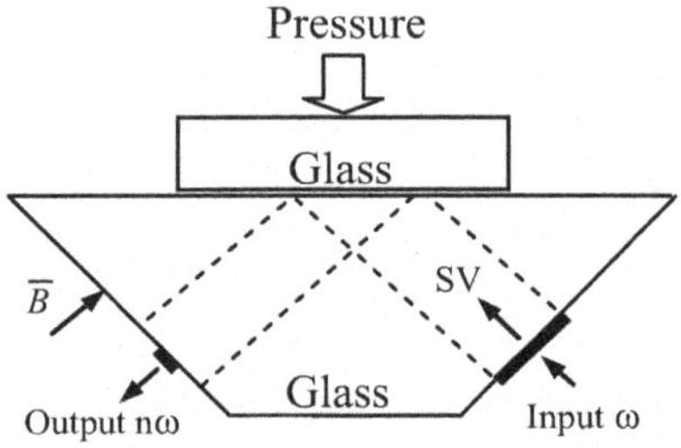

Fig. 4.14 Higher harmonic amplitudes ($\times 10^{-10}$ m) as functions of contact pressure at glass-glass interface: 1—fundamental wave, 2, 3, 4—second, third and fourth harmonics, respectively

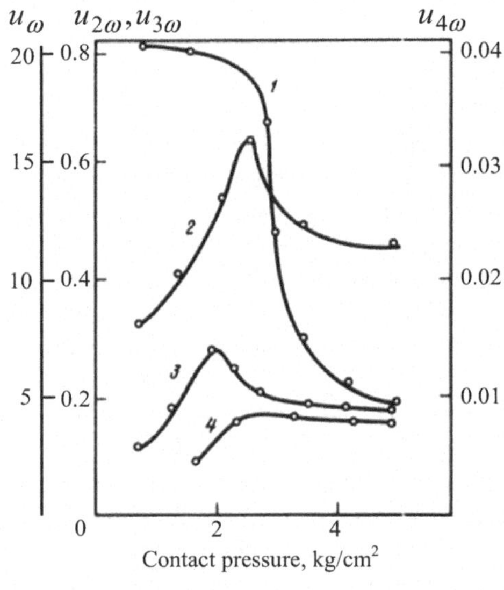

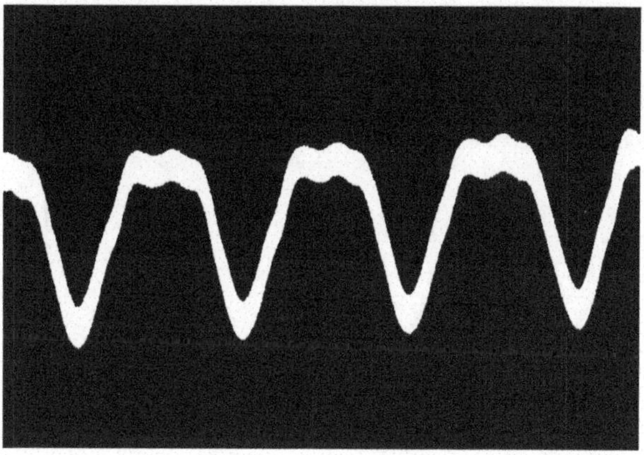

Fig. 4.15 Nonlinear waveform distortion for longitudinal wave transmitted through a contact interface

to increase in transmission) while the higher harmonics reach maxima in the transition pressure region. Their amplitudes are only within one-two orders of magnitude lower than the fundamental vibration, i.e. are much higher than those observed in the distributed case.

Such a strong localized nonlinearity is also manifested in an evident waveform distortion even without accumulation of the nonlinearity with distance. In the experiment [44], a 1.5 MHz-burst of intense longitudinal wave is generated in the fused silica sample ($5 \times 5 \times 47$ mm^3) with a PZT-ceramic transducer. The nonlinear contact is formed between optically polished surfaces of the sample and the buffer made of the same material. To observe the spectrum of the acoustic wave transmitted through the contact, a 20 MHz-output transducer was used with a wideband (flat) frequency response up to $\approx$15 MHz. At a moderate contact pressure, an anomalously strong nonlinear distortion of the "rectified sine" type was observed (Fig. 4.15), which is in a full compliance with the theoretical expectations of the "diode model" developed in [7].

Similarly strong nonlinear effects are also observed for propagating SAW in [45] and also used to assess the CAN efficiency in classical terms of the nonlinearity parameters. The SAW higher harmonic amplitudes were measured for a 15 MHz-fundamental SAW transmitted through a $\approx$5 mm CAN area formed by pressing an optically polished glass sample against $LiNbO_3$-substrate. At the optimal contact pressure ($\approx$1 MPa) the amplitudes of the second and third harmonics were measured to be $u_{2\omega} \cong 1.2$ Å and $u_{3\omega} \cong 0.32$ Å for the fundamental wave amplitude $u_{\omega} \cong 16$ Å. The nonlinearity parameter values calculated by using Eqs. (4.5) and (4.8) for the higher harmonic amplitudes are found to be: $\beta_2 \cong 100$; $\beta_3 \cong 1.5 \times 10^6$. Therefore, CAN exhibits an abnormally high level of both the quadratic and, particularly, cubic nonlinearity.

Since loading of the surface changes the wave velocity by only a few percent, one expects an insignificant linear SAW reflection from the contact area while CAN provides a strong generation of the backward propagating ("reflected") higher harmonics [8]. These results are shown in Fig. 4.16 for a 15 MHz SAW reflected from

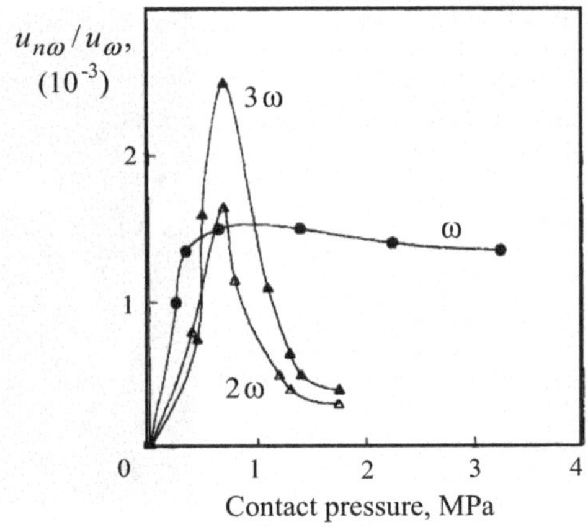

Fig. 4.16 SAW higher harmonic reflection from a contact interface as a function of contact pressure

≈4 mm area of $LiNbO_3$-glass contact. Similar to the bulk wave case, the SAW harmonics peak at a moderate contact pressure, but in such an effective way that the nonlinear spectrum is inverted in reflection: $u_{3\omega} > u_{2\omega} > u_{\omega}$ (Fig. 4.16). These data illustrate a possibility of a highly sensitive and selective nonlinear probing of fractured defects: it is especially useful in the cases when a linear acoustic contrast (impedance mismatch) is low whereas the nonlinear response could be extremely high due to CAN [46].

Some other "non-classical" features are demonstrated in the experiment which combines CAN and the resonance of the sample [44]. The experimental set-up (Fig. 4.17) uses the input CW-voltage with the amplitude up to ≅ 20 V in the frequency range ≅200–800 kHz to generate a standing-wave pattern at one of the higher natural frequencies of a steel acoustic resonator. The nonlinearity is introduced by the nonlinear contact with a glass buffer attached to the resonator. To adjust a contact pressure (and nonlinearity) a DC-bias voltage V_B was applied to an electromagnetic adjustment coil. For moderate contact pressure ($V_B \cong 5$ V), similar to the experiments discussed above, an efficient higher harmonic generation was observed accompanying by the depletion of the fundamental mode (Fig. 4.18a).

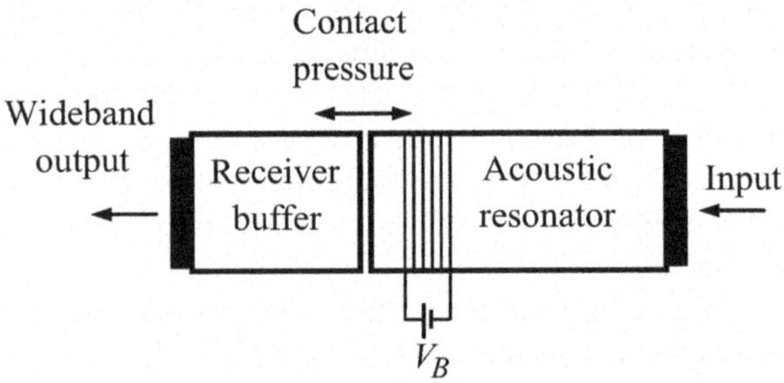

Fig. 4.17 Setup for observation of CAN effects in an acoustic resonator

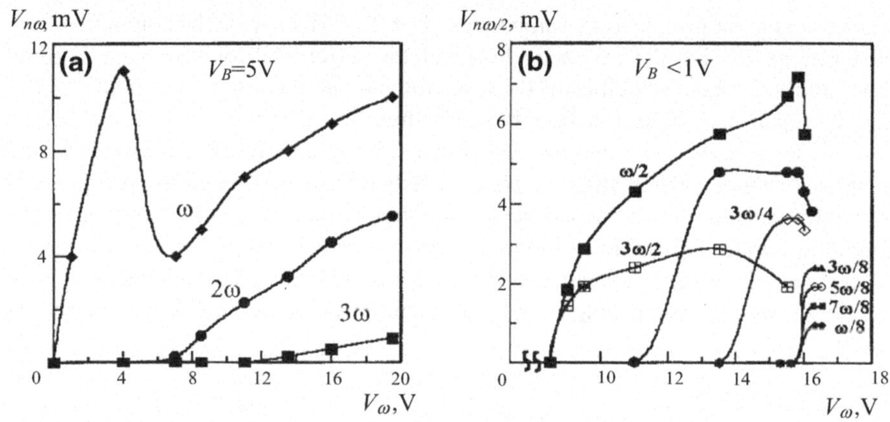

Fig. 4.18 Higher harmonic (**a**) and subharmonic (**b**) amplitudes as functions of the input voltage in acoustic resonator with nonlinear contact interface for different contact pressures (V_B)

For weakly pressed contacts ($V_B < 1$ V), the vibrations became unstable: a threshold cascade process of multiple subharmonic generation was observed as the input voltage increases (Fig. 4.18b). The successive bifurcations to periodic motions with periods n-fold to that of the driving vibration is a primary scenario for transition to chaotic vibrations in nonlinear systems, which was also observed in the simulation experiment [8] and in realistic fractured defects in composites [47].

4.5 Localized Nonlinearity for Monitoring of Bonding Quality

A strong nonlinearity of a non-bonded interface raises the question of feasibility of the nonlinear approach to NDE of imperfectly-bonded joints. In this case, the joining materials are bonded by means of an adhesive layer which forms two perfect interfaces between the materials and the adhesive. The joint can now be represented as a nonlinear spring with a stress-strain response dependent on elastic properties of the adhesive itself as well as thin boundary layers (~μm size) between the adhesive and the adherends [48]. Its nonlinear response is expected to increase with decline of adhesion quality and maximise in the extreme case of the lack of bonding.

The role of technology for joining and consolidating is of primary importance in manufacturing and maintenance of high performance composite components in automotive and aviation industries. Even minor deviations in the production and/or maintenance/repair processes are likely to degrade the adhesion strength and apparently to increase the joint nonlinear response. A conventional way to monitor the distributed nonlinearity in composite laminates is based on the second harmonic measurements for propagating Lamb modes [49–51]. To provide the cumulative second harmonic of a Lamb wave mode, however, the two crucial conditions are required to be satisfied: phase velocities of the fundamental and the second harmonic modes must be equal and the power flow between them must be non-zero. Due to Lamb wave dispersion, the first condition leads to a strict selection of the frequencies and the types of modes which are to be phase matched (see, e.g. [51]). The precise conditions for selection of the experimental parameters make the nonlinear plate wave technique barely applicable in industrial environment (various lay-ups, thickness, curved specimens, etc.). Instead, the approach based on the material localized vibration nonlinear response enables to recognise and evaluate the difference in bonding quality [52].

4.5.1 Methodology

The technique is based on the local generation of high amplitude vibrations and detecting the higher harmonics in the excitation area. An option for the higher amplitude excitation was found by using piezo-actuators manufactured by isi-sys GmbH (Germany) with a frequency response extended from low kHz into high kHz range (above 100 kHz). The actuators are vacuum attached to the specimens and can be used for on-site measurements of large aviation components. They are driven by a continuous wave voltage generated by the HP 33120A arbitrary waveform generator. The generator is combined with HVA-B100 amplifier to result in 10–40 V input amplitude for the piezo-actuator. The actuator is vacuum attached to one of the sides of the specimen and nonlinear vibrations produced locally in the area of excitation are measured on the opposite side of the specimen (Fig. 4.19).

Fig. 4.19 Excitation and detection of vibrations in CFRP specimen

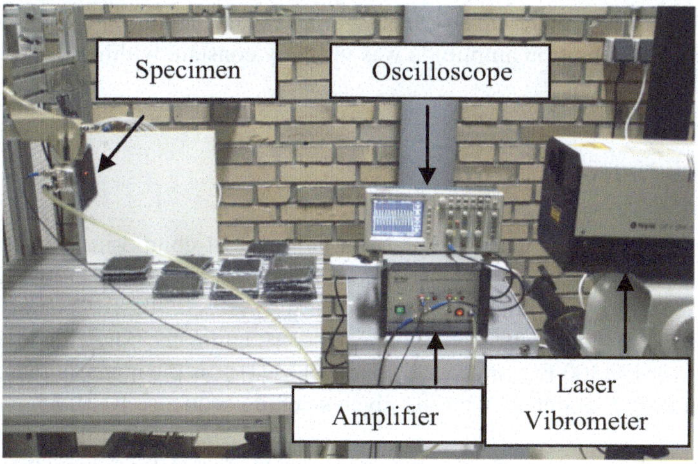

Fig. 4.20 Experimental setup used for excitation/detection of vibrations in CFRP specimen

To receive and analyze the frequency content of the vibrations generated locally in the excitation area a scanning laser vibrometer (SLV, Polytec 300) operating in the vibration velocity mode with maximum frequency bandwidth of 1.5 MHz was used (Fig. 4.20). The dynamic range of the SLV measurements (100–120 dB) is well beyond the level of nonlinear frequency components. To avoid an impact of reflections on the local vibration in the excitation area, the edges of specimens were covered with dissipative material and rather high vibration frequency was chosen (49 kHz). The spectrum and the temporal pattern of the vibrations are measured in the central (source) area (a circle with ≈5 mm radius) where no plate wave propagation is involved. For input voltage of 20 V, the vibration velocity amplitude measured at 49 kHz was in the range of 130–150 mm/s. The displacement amplitude is therefore in the range of (4–5) × 10^{-8} m so that a local strain developed in the excitation area is ~10^{-5} to 10^{-4}. This deformation is sufficient for manifestation of noticeable local nonlinearity in composite materials.

Each value of the fundamental frequency vibration velocity (v_0) and the higher harmonic components (v_n) measured in the probing area was used for evaluation of the nonlinear ratio $N_i = \sum_n v_n^2/v_0^2$. An average value of the nonlinear ratio in the probing area was then calculated $N = \sum_{i=1}^{m} N_i/m$ (where m is the number of measurements) and the standard deviation of the results ΔN was estimated. The measurements were repeated in various points over the central part of the specimen to provide the relative error $\Delta N/N \leq 10 - 20\%$.

The nonlinear ratio N_i is a part of the vibration energy converted into the higher harmonics so that it clearly quantifies material nonlinearity. To avoid the influence of the amplitude-dependent effects the input voltage of the transducer was kept constant (20 V) over the course of measurements. As a result, in all the specimens the fundamental vibration amplitude was virtually constant within ~10% deviation.

4.5.2 *Specimens*

The measurements were carried out for a large set of composite specimens with different bonding conditions. The samples (10×10 cm^2) manufactured according to Airbus Approval for composite manufacturing (AIPS 03-02-019) and comprised two 8-ply carbon fiber-reinforced polymer (CFRP) laminates bonded with an epoxy adhesive layer FM300 K from Cytec®. Hexcel M21E® is the material that was used for the composite laminates with the given stacking sequence [0, 0, 45, −45, −45, 45, 0, 0].

Two typical stages during the life of a structural part for which the adhesive properties of a bonding joint could be degraded were considered: the production process and the maintenance/repair scenario. For production scenario, three kinds of contaminants were studied: Release agent (RA), fingerprint (FP) and moisture (MO). For repair scenario, only one kind of contaminant was applied: De-icing fluid (DI). The other two processes that could result in a loss of adhesion due to external influences or errors during bonding process were also studied: Thermal degradation (TD) and Faulty Curing (FC). For further details on the specimen preparation, see [52].

For each type of contamination in both scenarios, a set of three specimens of every level of contamination was prepared for reliable statistics. Along with other three (non-contaminated) reference specimens (REFR and REFP) for each of the scenarios a total number of the specimens to be tested amounted to 60.

4.5.3 *Measurement Results*

The results of measurements are plotted in Figs. 4.21, 4.22, 4.23, 4.24, 4.25, 4.26, 4.27 and 4.28 in the way to trace the impact of the contamination level (the number right after the contamination notation (1, 2, 3)) for a particular specimen number (−1, −2, −3). According to Figs. 4.21 and 4.22, the reference specimens (without contamination of the adhesion layer) reveal the minimal values of $N \approx 3$ (here and further in 10^{-3} units). For each type of contamination (Figs. 4.23, 4.24, 4.25, 4.26, 4.27 and 4.28), the values of N change noticeably with variation of the contamination level that indicates the sensitivity to the changes in the thin boundary layer between the adhesive and the adherends.

The maximum nonlinear ratio N was obtained in TD 3-2 specimen ($N = 50 \pm 6$, outside the scale in Fig. 4.23). In this specimen, the N_i values were found to depend on the position of the measurement point. This fact, along with anomalously high value of N indicates the presence of delamination induced by TD.

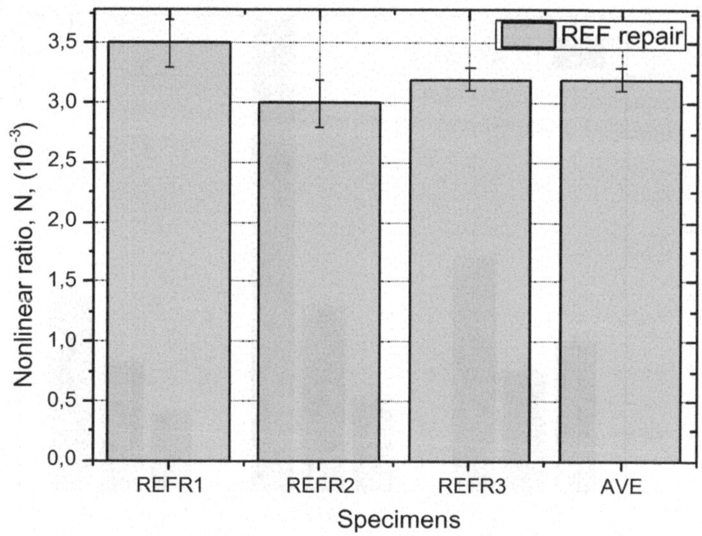

Fig. 4.21 Nonlinear ratios for reference specimens of repair scenario

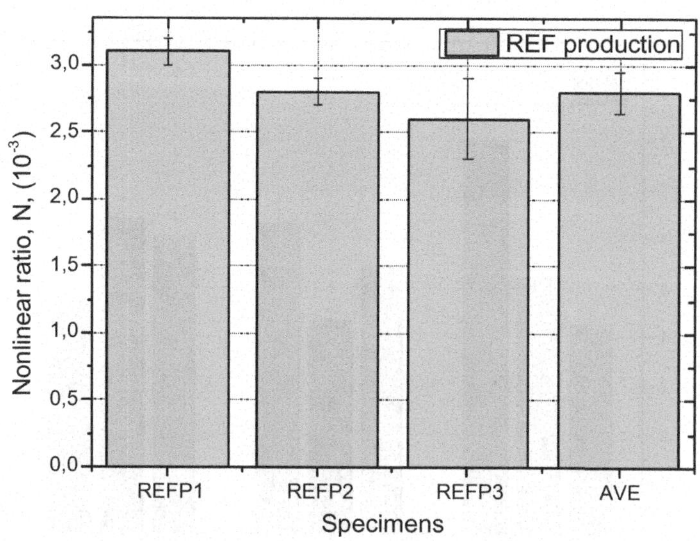

Fig. 4.22 Nonlinear ratios for reference specimens of production scenario

For the contaminations types TD, RA, FP, DI, and FC material nonlinearity increases with the increase of the level of contamination (Figs. 4.23, 4.24, 4.25, 4.26 and 4.27). Since nonlinearity enhances due to weakening of the material, the increase in N is an indication of degrading of the adhesive bonding. From this viewpoint, the strongest decrease in bonding strength is caused by level 3 TD-contamination. The

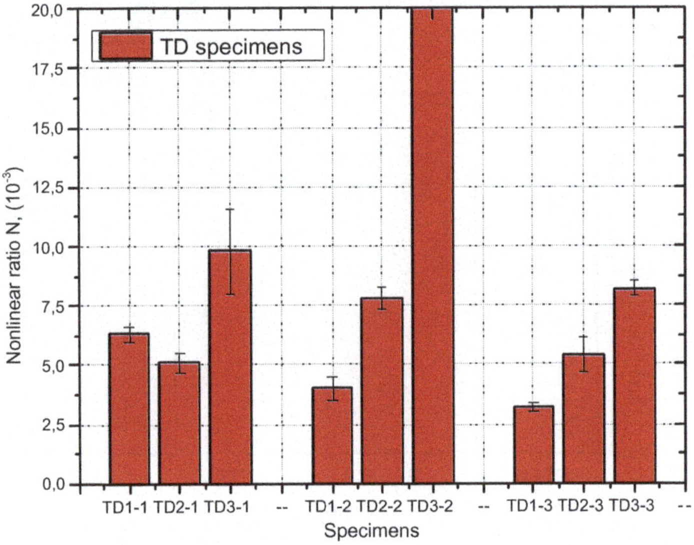

Fig. 4.23 Nonlinearity of each of the 3 specimens (−1, −2, −3) as a function of TD level of contamination

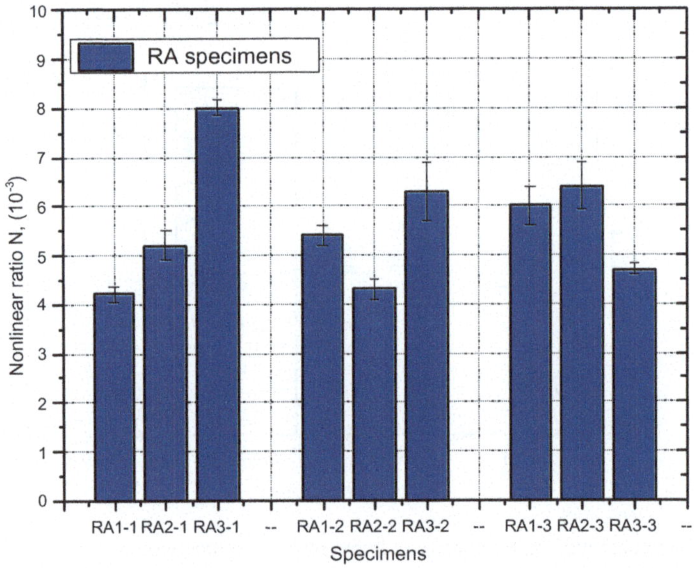

Fig. 4.24 Nonlinearity of each of the 3 specimens (−1, −2, −3) as a function of RA level of contamination

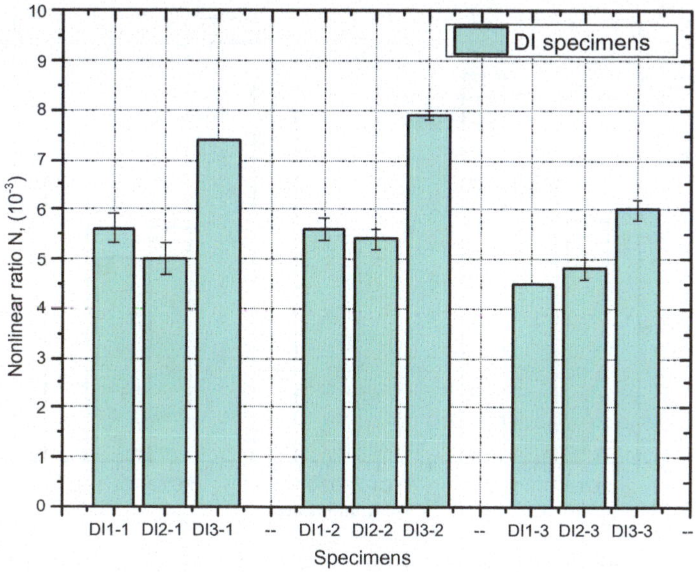

Fig. 4.25 Nonlinearity of each of the 3 specimens (−1, −2, −3) as a function of DI level of contamination

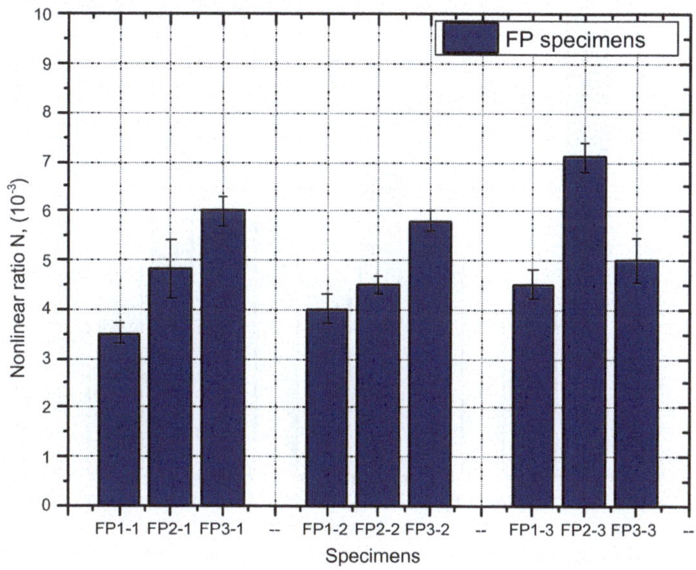

Fig. 4.26 Nonlinearity of each of the 3 specimens (−1, −2, −3) as a function of FP level of contamination

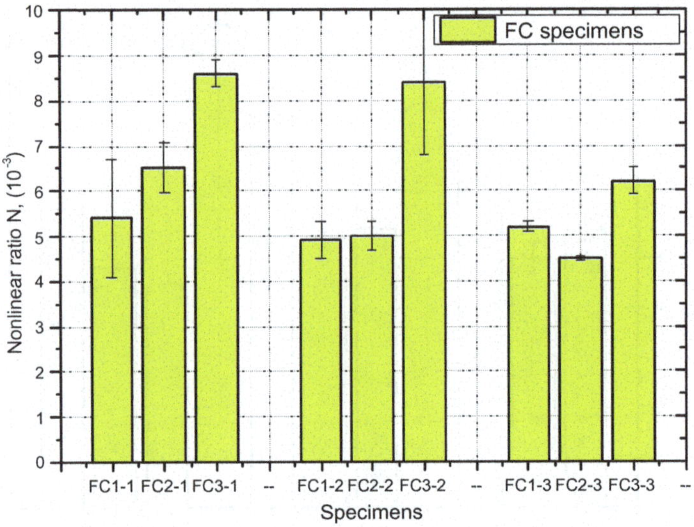

Fig. 4.27 Nonlinearity of each of the specimens (−1, −2, −3) as a function of FC level of contamination

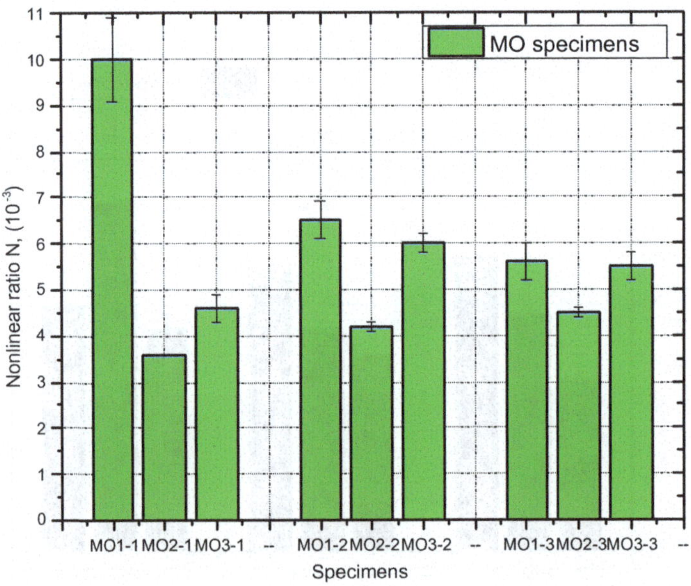

Fig. 4.28 Nonlinearity of the contaminated specimens (−1, −2, −3) as a function of MO level of contamination

MO case (Fig. 4.28) is not so convincing: with exception of the abrupt kick for specimen MO1-1, the nonlinearity does not change noticeably so that average N is ≈ 5 for all levels of contamination.

4.5.4 Evaluation of Bonding Quality in Realistic Aviation Components

The experiments of the previous section are concerned with a nonlinear material response for locally excited vibrations. An alternative approach based on the excitation of the standing wave in the specimen enables to probe the vibration nonlinear response over a total area of the specimen (except nodal lines). This methodology is applicable to mapping and imaging of the contaminated adhesion areas in large-size composite parts.

The large specimens tested are realistic airplane components shown in Fig. 4.29 [53]. Component N1 belongs to the production scenario with a contaminated part of the lower stringer (di-icing liquid + a few fingerprints, DI + FP). The CFRP patch in component N2 illustrates a repair scenario when a part of the patch is contaminated in a similar way. The exact location of the contaminated areas was not known before the experimental testing ("blind testing").

A piezo-stack transducer and high-power supply (Branson Ultrasonics) provided µm-range displacements at the fundamental frequency of 20 kHz in the source area (nonlinear regime). According to the above, the higher harmonics are supposed to be generated efficiently in the contaminated areas. (the unknown parts of the stringer in sample N1 and the patch in component N2). To visualize the nonlinear areas the

Fig. 4.29 The specimens tested: 80 × 80 cm^2 CFRP component (N1) with 2 stringers (production scenario, left); 100 × 80 cm^2 CFRP component (N2) with repair patch (20 cm diameter, right)

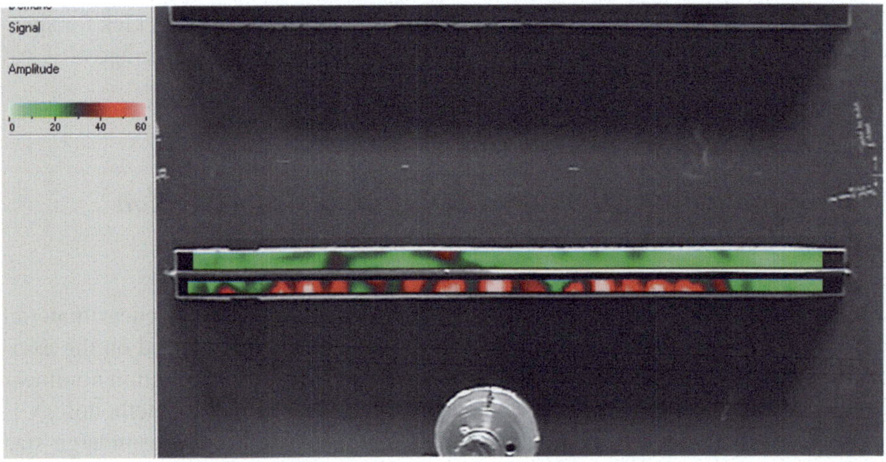

Fig. 4.30 Distribution of the fundamental frequency (20 kHz) ultrasonic amplitude along the stringer length in specimen N1

spectrum of the local vibrations over the specimen surface was probed with a laser vibrometer and the color-coded distributions of the higher harmonics were mapped.

In the experiment with specimen N1, the excitation transducer was positioned symmetrically in regard of the stringer to be tested (Fig. 4.30). Such a position provided symmetrical insonation 20 kHz field of the stringer in question as shown in Fig. 4.30. The second higher harmonics measured in specimen N1 demonstrates quite different distributions shown in Fig. 4.31: the higher efficiency of the local nonlinear generation in the left hand-side of the stringer is clearly seen and identifies a contaminated area with lower bonding quality.

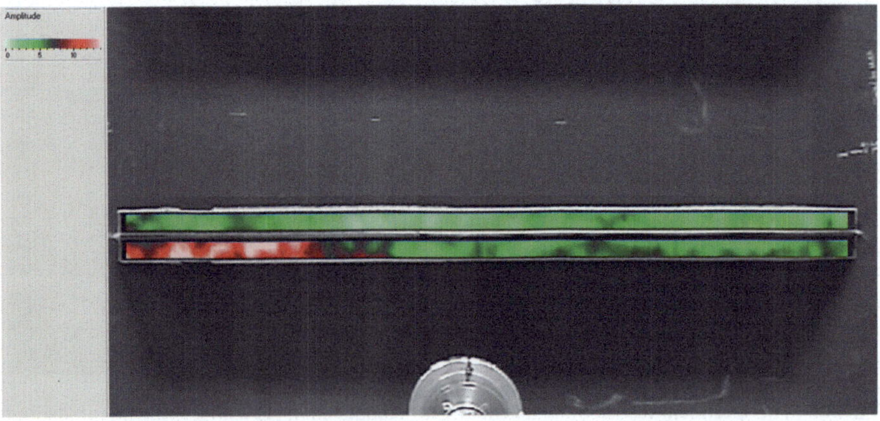

Fig. 4.31 Distribution of the second higher harmonic (40 kHz) along the stringer in component N1

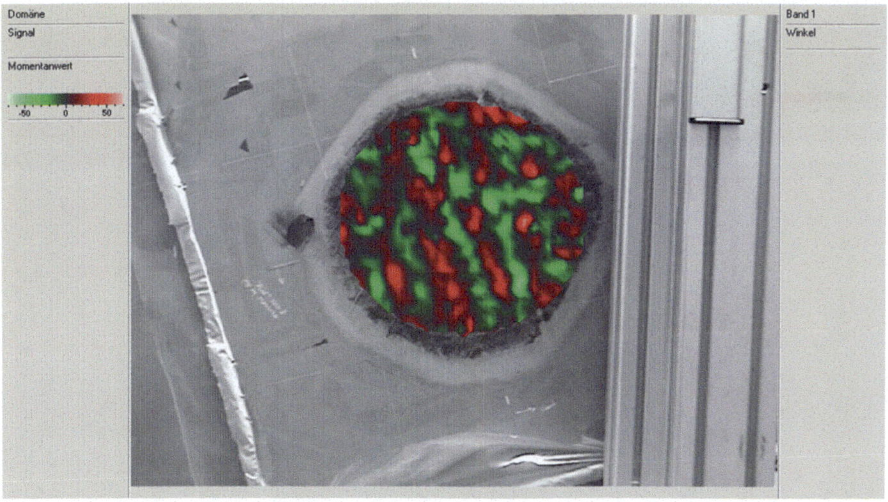

Fig. 4.32 Fundamental frequency field excited in the repair patch of specimen N2

In component 2, the fundamental frequency field visualized in Fig. 4.32 shows a rather conventional standing wave pattern over the total area of the repair patch. On the opposite, the higher harmonic fields measured in the patch (Fig. 4.33) demonstrate quite different nonlinear field distributions. They reveal a strong enhancement of the nonlinearity and thus lower bonding quality in the right hand-side and lower parts of the patch for the 2d and 3d higher harmonics (Fig. 4.33).

4.6 Local Defect Resonance (LDR) and Localized Nonlinearity

An evident way to enhance a local nonlinear response of a certain area is to increase its vibration amplitude. A novel methodology proposed recently for NDE and imaging of damage [10, 11] makes use of a mechanical resonance of the defect areas. It is based on the effect of LDR which develops when the defect is activated by sonic/ultrasonic excitation whose frequency matches its natural vibration frequency. Under the frequency match condition, the input energy is delivered and trapped selectively in the defect area that increases dramatically its vibration amplitude. Unlike conventional (classical) nonlinear inspection concerned with the nonlinear wave propagation the LDR-based methodology shifts to nonlinear vibration effects in particular isolated areas of a nonlinearly inhomogeneous material.

A direct way to experimentally reveal a local defect resonance is to measure an individual contribution of each point of the specimen in its overall vibrational response in a wide frequency range. In the experiment, the specimen vibrations are

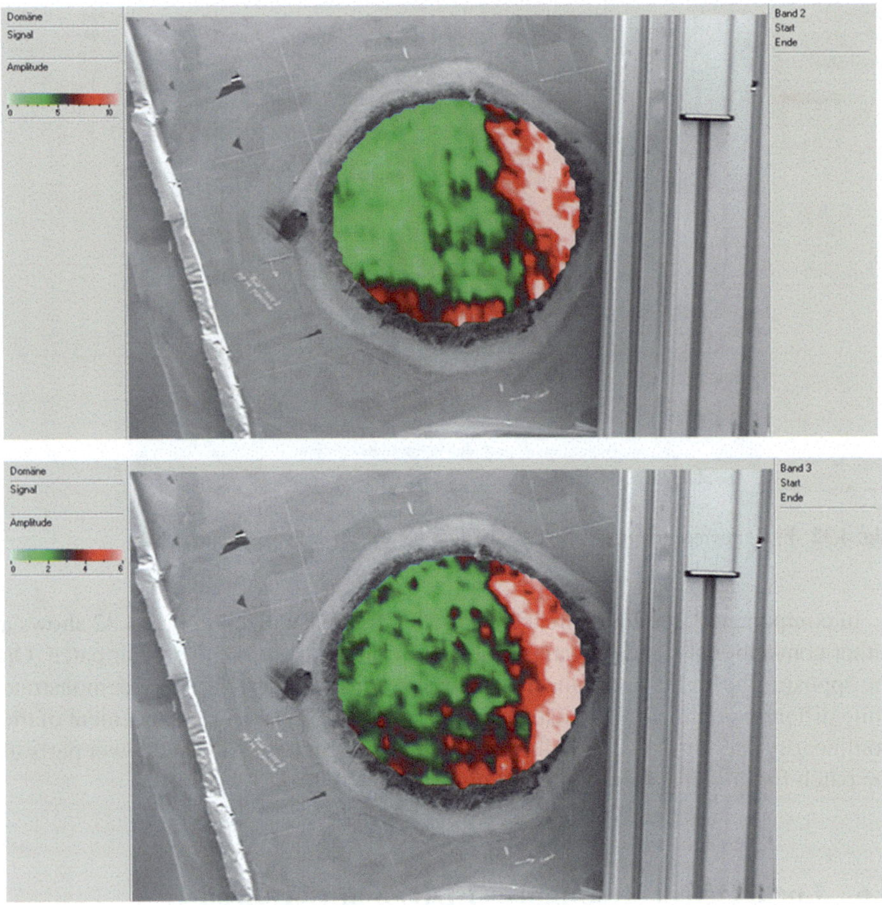

Fig. 4.33 The second harmonic (40 kHz, top) and the third harmonic (60 kHz, bottom) fields in the repair patch of specimen, N2

activated with a wide-band (400 Hz–100 kHz) piezoelectric transducer while a scanning laser vibrometer (Polytec 300, vibration velocity mode) is used for interrogation of the specimen frequency response. The scanning mode enables to probe and indicate all possible resonances in the vibration spectrum for every point of the specimen. The origin of each maximum is then verified by imaging the vibration pattern at the corresponding frequency. An example of application of such an approach is shown in Fig. 4.35 for a circular flat-bottomed hole (FBH, a circular disk of thickness 1 mm and radius 1 cm) in a PMMA plate (thickness 4.3 mm) (Fig. 4.34).

The spectrum obtained indicates the following resonance frequencies of interest: the fundamental LDR at $\omega = 7900$ Hz (in-phase vibration of the defect) accompanied by the second-order normal mode (one nodal diameter) at 16,250 Hz and the third-order resonance (two nodal diameters) at 26,400 Hz. The other maxima indicate

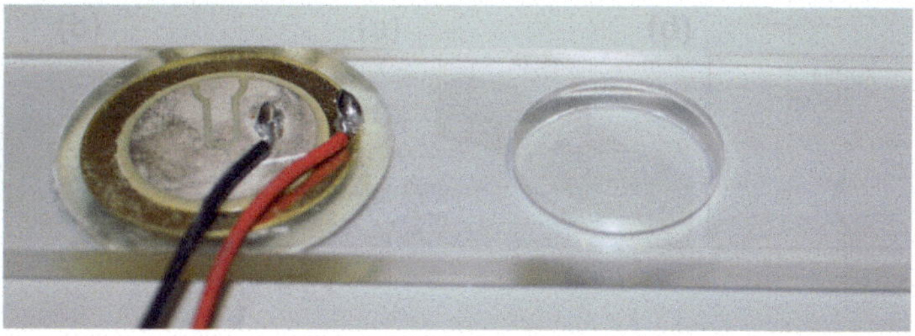

Fig. 4.34 The specimen used in the experiment: piezo-transducer (left) and circular FBH (right)

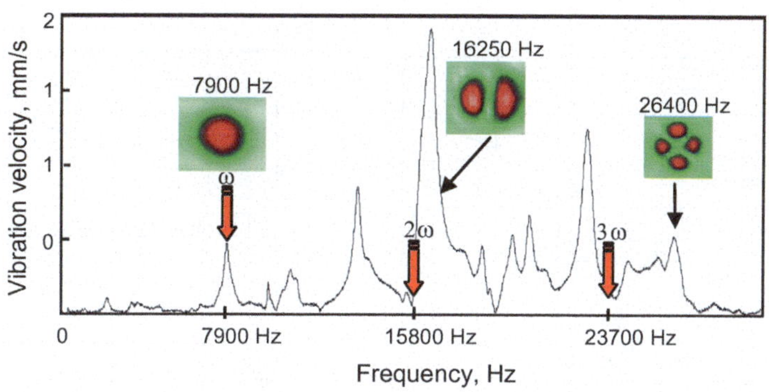

Fig. 4.35 Vibration spectrum and vibration fields of the fundamental and some of the higher-order LDR in FBH. Arrows indicate the higher harmonic positions for excitation at the fundamental LDR 7900 Hz

the resonances of the specimen; the amplitudes of all peaks are modulated by the frequency response of the excitation transducer (maximises in the range 10–24 kHz).

The 'amplification' of local defect vibrations depends on the LDR quality factor Q. A pair of LDR frequency responses measured for different defects in CFRP specimens (Fig. 4.36) demonstrates classical resonance frequency behaviour and shows that LDR can manifest a high Q resonance. The values of Q factors measured for other types of realistic defects in various materials ranged from ~10 to 100 so that the LDR induced amplification of local vibrations as high as ~20–40 dB was observed.

The local increase of the damage vibration on the background of a low-amplitude tremble of the rest of the specimen is readily detected with a scanning laser vibrometer (as well as via thermosonics and/or shearosonics) and brings a clear image of the damage (Fig. 4.36). As expected, it also enhances its nonlinear response: the higher harmonics increase as the defect is driven at the LDR frequency. However, due to dispersion the higher harmonics generated in a planar nonlinear resonator excited at a fundamental mode are, however, not matched to the higher-order resonances (see, Fig. 4.35) that prevents the maximal higher harmonic response of the resonator. To

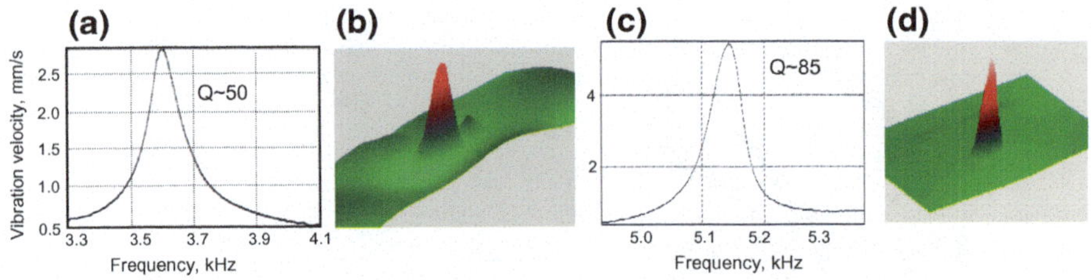

Fig. 4.36 LDR frequency responses and vibration patterns for impact-induced fiber loss (**a**, **b**) and impact damage (**c**, **d**) in 280 × 40 × 1 mm^3 CFRP plate

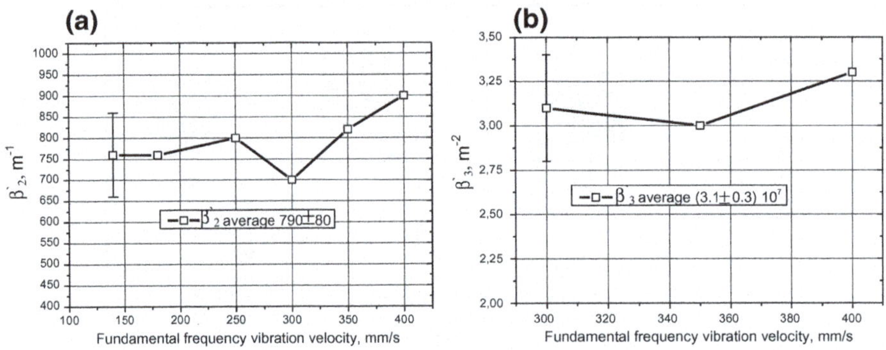

Fig. 4.37 Second (**a**) and third (**b**) harmonic nonlinearity parameters for fundamental LDR excitation

optimize the LDR higher harmonic response the excitation frequencies have to be changed in order the higher harmonic frequencies match the higher-order resonances.

The LDR efficiency of the harmonic generation can be quantified by the ratios $\beta_2' = u_{2\omega}/u_\omega^2$ and $\beta_3' = u_{3\omega}/u_\omega^3$ (u are the displacement amplitudes). These ratios characterize local nonlinearity and are, of course, different from the classical nonlinearity parameters β_n associated with the accumulation of the higher harmonic with distance and considered in Sect. 4.1. The nonlinearity parameters β_n' are measured for the FBH resonator shown in Fig. 4.34 in various excitation modes. The data for fundamental LDR frequency (7900 Hz) excitation (non-matched mode) are given in Fig. 4.37. The excitation frequencies are then changed for 8125 and 8800 Hz for the harmonics to match, correspondingly, the second- and third-order resonances in Fig. 4.35. The values of β_2', β_3' obtained (Fig. 4.38) indicate a dramatic rise in the higher harmonic generation efficiency: the second harmonic nonlinearity increases by more than an order of magnitude, while matching increases the third harmonic efficiency by almost three orders of magnitude.

With less information known about the higher-order LDR frequencies, the fundamental LDR mode can also be used for matching the higher harmonics when the fundamental LDR frequency is selected as an integer multiple of the input frequency. In this case, the HH are matched to the fundamental normal mode and use the benefit

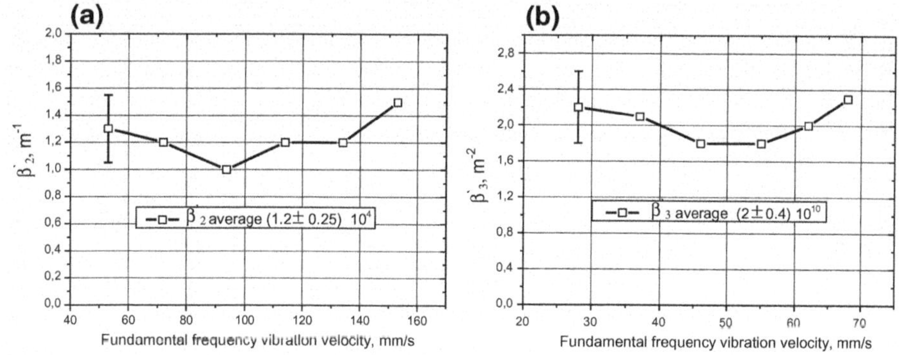

Fig. 4.38 Second (**a**) and third-order (**b**) nonlinearity parameters for matched HH generation

of the resonant gain provided by the so-called super-harmonic resonance (SHR). In the experiment, the SHR was applied to the second harmonic case so that the input frequency was placed around the fundamental LDR subharmonic at 3950 Hz. The results shown in Fig. 4.39 indicate further increase in β_2' whereas the vibration velocity of the second harmonic grows to more than 10% of the fundamental that makes the defect vibration noticeably nonlinear (Fig. 4.40).

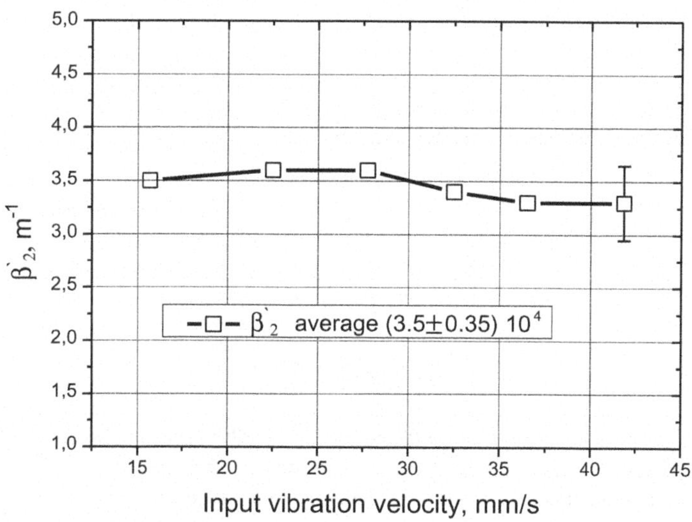

Fig. 4.39 Nonlinearity parameter for the second harmonic matched to SHR

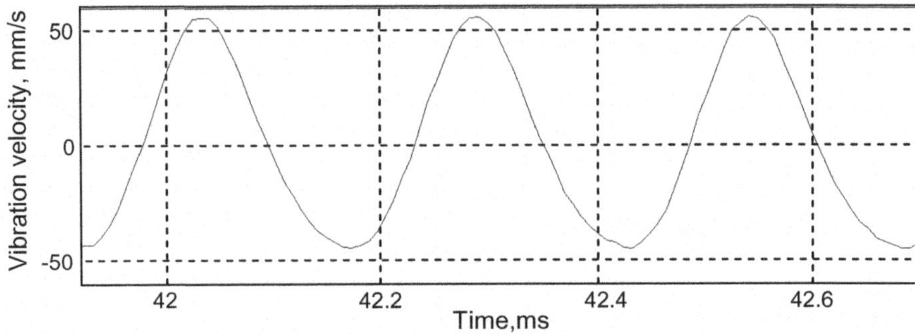

Fig. 4.40 FBH vibration patterns for the second harmonic matched to SHR

4.7 Enhancement of Nonlinearity in Various Types of Nonlinear Resonances

A low efficiency of conversion from fundamental frequency to nonlinear frequency components is a long-term bottleneck problem of nonlinear acoustic applications in NDE. The results presented above indicate an obvious progress obtained by using the localized resonant nonlinearity. Further experimental data given below for various types of nonlinear resonances validate the value of this approach for enhancement the efficiency of nonlinear effects.

4.7.1 Super-Harmonic Resonance

In the frame of phenomenological description, the effect of nonlinearity on resonant vibrations can be illustrated by using the model of a lumped nonlinear oscillator (natural frequency ω_0) with up to the third-order nonlinearity in the equation of its forced vibrations [54]:

$$\ddot{x} + 2\lambda\dot{x} + \omega_0^2 x + \alpha x^2 + \beta x^3 = F_0 \cos \nu t, \tag{4.19}$$

where λ is the dissipation factor, α and β are the corresponding factors in the series expansion of the stress-strain relation (parameters of nonlinearity).

For the super-harmonic resonance [20], the input frequency is taken as $\approx \omega_0/n$ and converted into ω_0 via the nth-order nonlinearity of the oscillator. The examples below correspond to $n = 2, 3$ and use the perturbation approach for finding solutions to Eq. (4.19).

For the driving frequency $\nu = (\omega_0/2) + \varepsilon$ and small λ, the solution of the first approximation is a non-resonant linear driven vibration:

$$x_1 = \frac{4F_0}{3\omega_0^2}\cos[(\omega_0/2) + \varepsilon)t] \tag{4.20}$$

In the second approximation, the resonance driving force will be developed directly via quadratic nonlinearity so that one readily obtains:

$$x_2^{2\omega} = -\frac{4\alpha F_0^2}{9\omega_0^5\sqrt{4\varepsilon^2 + \lambda^2}}\cos[(\omega_0 + 2\varepsilon)t] \tag{4.21}$$

A similar procedure for the input $\nu = (\omega_0/3) + \varepsilon$ leads to the solution for the third HH:

$$x_2^{3\omega} = -\frac{\beta(x_1^0)^3}{8\omega_0\sqrt{9\varepsilon^2 + \lambda^2}}\cos[(\omega_0 + 3\varepsilon)t], \tag{4.22}$$

where $x_1^0 = 9F_0/8\omega_0^2$.

Both resonant solutions (4.21) and (4.22) increase at $(\varepsilon, \lambda) \rightarrow 0$ that demonstrates the benefit of the resonance for the generation of the higher harmonics.

An example of the super-harmonic resonance in defects is demonstrated for the third-order resonance in impact damaged CFRP specimen with LDR around 5140 Hz in Figs. 4.41 and 4.42. One-third of the LDR frequency (1714 Hz) is selected for the excitation while the input voltage was increased up to 80 V. The spectrum (Fig. 4.41) measured in the defect area illustrate the dominance of the third harmonic vibration

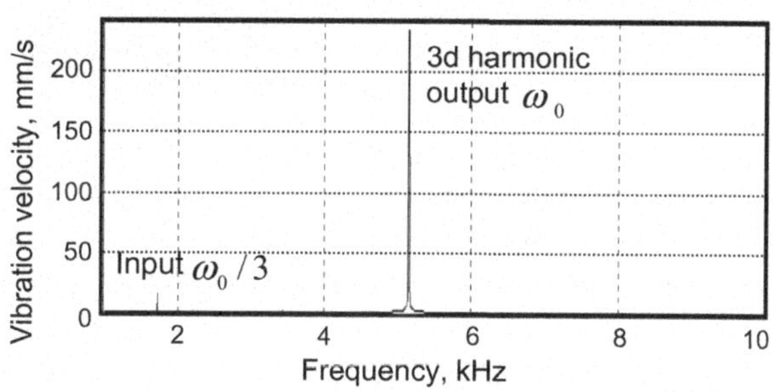

Fig. 4.41 Spectrum of the third-order superharmonic LDR in impact damaged CFRP plate

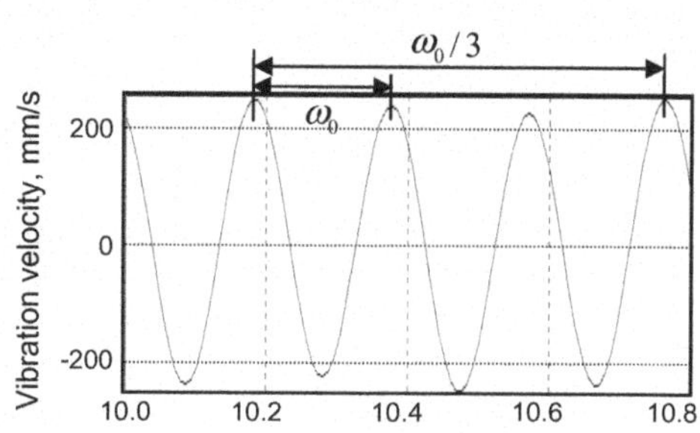

Fig. 4.42 Vibration pattern of the third-order superharmonic LDR in impact damaged CFRP plate

(25 dB higher than the fundamental, Fig. 4.41) also proved by quite "clean" third harmonic (5142 Hz) vibration pattern (period ~0.19 ms in Fig. 4.42).

4.7.2 Combination Frequency Resonance

A high quality factor of LDR can also be used as a resonant "amplifier" in the frequency mixing mode nonlinear NDE. The method is based on the nonlinear interaction of ultrasonic waves of different frequencies (ν_1, ν_2) that results in a combination frequency output: $\nu_\pm = \nu_1 \pm \nu_2$. For a bi-frequency excitation of a nonlinear defect, the equation of motion is:

$$\ddot{x} + 2\lambda\dot{x} + \omega_0^2 x + \alpha x^2 + \beta x^3 = F_1 \cos \nu_1 t + F_2 \cos \nu_2 t, \tag{4.23}$$

with the first-order (linear) solution:

$$x_1 = x_1^{\nu_1} \cos \nu_1 t + x_1^{\nu_2} \cos \nu_2 t, \tag{4.24}$$

where $x_{1,2}^{\nu_{1,2}} = \frac{F_{1,2}}{\nu_{1,2}^2[(\omega_0^2/\nu_{1,2}^2)-1)}$.

The second-order term in (4.23) yields the driving force of the combination frequencies that for the condition of frequency match $\nu_1 \pm \nu_2 = \omega_0 + \varepsilon$ brings the resonant solution:

$$x_2^\pm = -\frac{\alpha x_1^{\nu_1} x_1^{\nu_2}}{2\omega_0\sqrt{\varepsilon^2 + \lambda^2}} \cos[(\omega_0 + \varepsilon)t]. \tag{4.25}$$

An application of LDR as a resonant "frequency mixing amplifier" for realistic defects is illustrated then in Figs. 4.43 and 4.44 for an impact-induced damage (area ~ 5 × 5 mm^2) in a CFRP plate (280 × 40 × 1 mm^3). A linear LDR frequency response of the impact demonstrates a well-defined double-maxima peak around

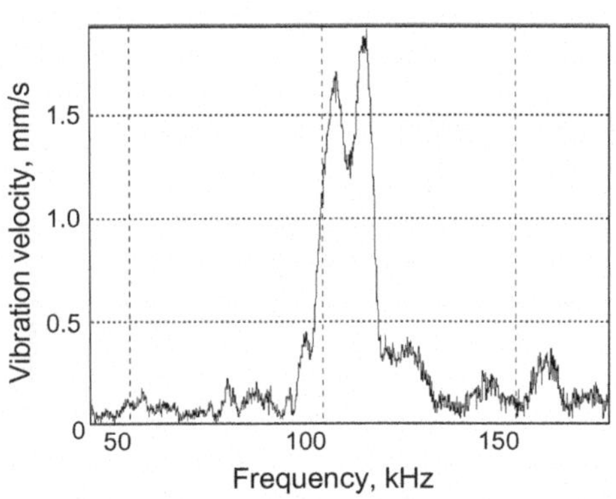

Fig. 4.43 LDR frequency response for an impact induced damage in a CFRP plate

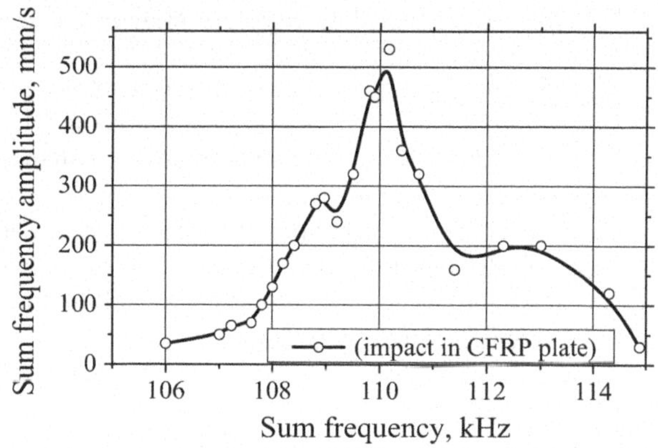

Fig. 4.44 LDR induced amplification at the sum frequency vibration for impact damage in a CFRP plate

110 kHz (Fig. 4.43). The frequency of one of the interacting flexural waves was fixed at $f_1 = 77.5$ kHz while the other was swept from $f_2 = 28.5$–37.5 kHz to provide the sum frequency variation around the LDR frequency.

The amplitude frequency response measured with laser vibrometer at the sum frequencies shown in Fig. 4.44 demonstrates unambiguously the effect of the resonance: more than 20 dB increase in the output is observed when the sum frequency matches the frequency of LDR.

4.7.3 *Subharmonic and Parametric Resonances*

Manifestation of parametric effects is due to the amplitude-dependent shift (modulation) of LDR frequency induced by the driving signal. The frequency shift induced by the driving signal causes a periodic modulation of the oscillator natural frequency that activates manifestation of the parametric resonance. The parametric approach enables further to reveal other features of the nonlinear resonance phenomena.

The nonlinear source of parametric resonance effects can be disclosed by applying the perturbation method to Eq. (4.19). For the driving frequency close to the resonance $\nu = \omega_0 + \varepsilon$, the solution of the first approximation is a linear vibration:

$$x_1 = \frac{F_0}{2\omega_0\sqrt{\varepsilon^2 + \lambda^2}} \cos \nu t. \tag{4.26}$$

In the next approximation, the quadratic nonlinearity in (4.19) apparently produces the third-order interaction term ~$2\alpha x_1 x_2$. By using (4.26) and taking into account this term only, one obtains:

$$\ddot{x}_2 + 2\lambda\dot{x}_2 + \omega_0^2[1 + \frac{\alpha F_0}{\omega_0^3\sqrt{\varepsilon^2 + \lambda^2}} \cos \nu t]x_2 = 0 \tag{4.27}$$

Equation (4.27) is a Mathieu's-type equation whose general form is [56]:

$$\ddot{x} + \omega_0^2(1 + h\cos\gamma t)x = 0 \tag{4.28}$$

and which is known to reveal parametric resonance and instability phenomena [55, 56].

The form (4.27) corresponds to the second-order parametric resonance at $\nu \approx \omega_0$. In this case, the frequencies of the solutions to (4.28) include the HH of the input signal: $\omega = n\nu (n = 1, 2, 3, \ldots)$ [57]. Beyond the input threshold, their amplitudes grow in time exponentially (instability).

In a similar way, for $\nu \approx 2\omega_0$ drive, the second approximation in (4.7) yields [54]:

$$\ddot{x}_2 + 2\lambda\dot{x}_2 + \omega_0^2[1 - \frac{2\alpha F_0}{3\omega_0^4}\cos[(2\omega_0 + \varepsilon)t]x_2 = 0, \tag{4.29}$$

which is a condition for a subharmonic (fundamental parametric) resonance. In this case, the solutions to (4.29) comprise unstable ultra-subharmonic outputs: $\omega = m\nu/2 (m = 1, 2, 3, \ldots)$.

The experimental proof for unstable parametric dynamics of the HH (the second-order parametric resonance) is given in Figs. 4.46 and 4.47 for a realistic defect of heat damage in CFRP plate ($300 \times 300 \times 4$ mm^3, Fig. 4.45). A local heating causes near-surface damage (burnout of epoxy and fibers) and induces delaminations between a few upper plies of CFRP.

LDR frequency for ~(8×12 mm^2) heat-induced delamination was found to be 10,500 Hz (Fig. 4.46, top) and used for excitation of plate waves in the specimen.

The instability is manifested in the vibration spectra (Fig. 4.46): a dramatic increase of the HH content in the range between 20 and 40 V (instability threshold) reveals development of the parametric resonance as a result of the combined action of the defect nonlinearity and resonance.

The frequency response (FR) of the defect measured for various input voltages of the wideband (chirp signal of a few kHz bandwidth around LDR frequency) input is shown in Fig. 4.47 and demonstrates a minor frequency shift but a strong LDR

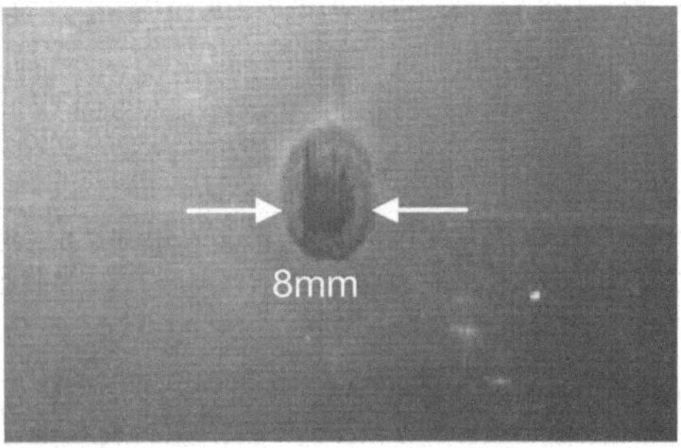

Fig. 4.45 Heat-induced damage in CFRP plate

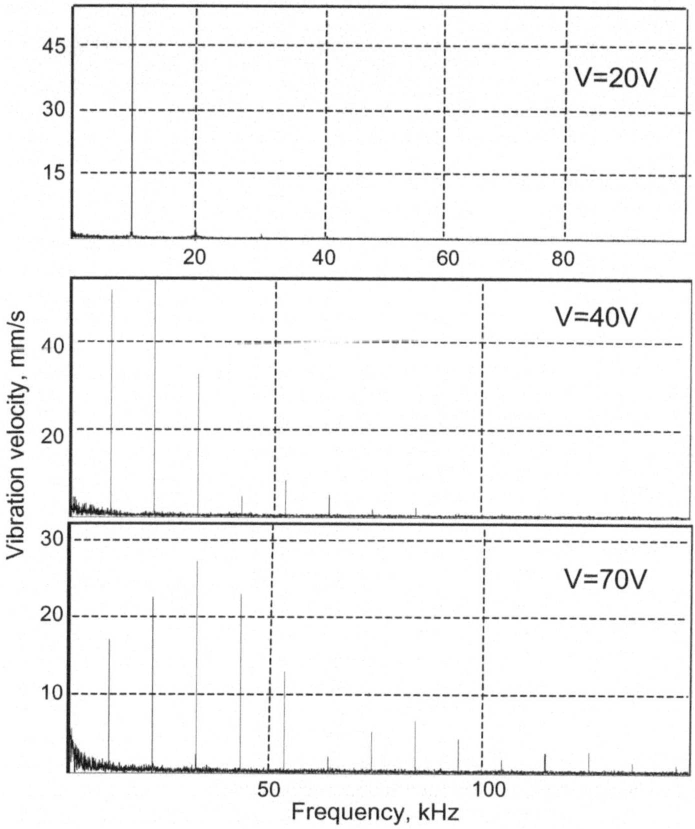

Fig. 4.46 LDR spectra of heat-induced damage in CFRP at different input voltages

amplitude decay. The decay sets in abruptly between 20 and 40 V inputs and leads to virtually total collapse of LDR at 80 V.

This spectral conversion throughout the LDR decay range of the inputs is quantified in Fig. 4.48 by plotting the ratio of the vibration energies at the fundamental frequency and the higher harmonics (nonlinear HH ratio): $E/E_{HH} = v_0^2/\sum_n v_n^2$, where v is the vibration velocity. The sharp decline in the fundamental frequency content (by about 4 orders of magnitude) in Fig. 4.48 due to nonlinear energy transfer to the higher harmonics is apparently the major reason for LDR decay and collapse. Total vibration energy is, of course, conserved so that the LDR decay does not have negative implications for its use in either nonlinear or/and thermosonic resonant methodologies of nondestructive evaluation.

A fundamental (subharmonic) parametric resonance is demonstrated in Fig. 4.50 for an impact damage in CFRP sample (Fig. 4.49). The excitation frequency is chosen at the second harmonic (10,280 Hz) of the fundamental LDR (5140 Hz, Fig. 4.50b). The threshold for the resonance was found to be ≈45 V. Beyond the

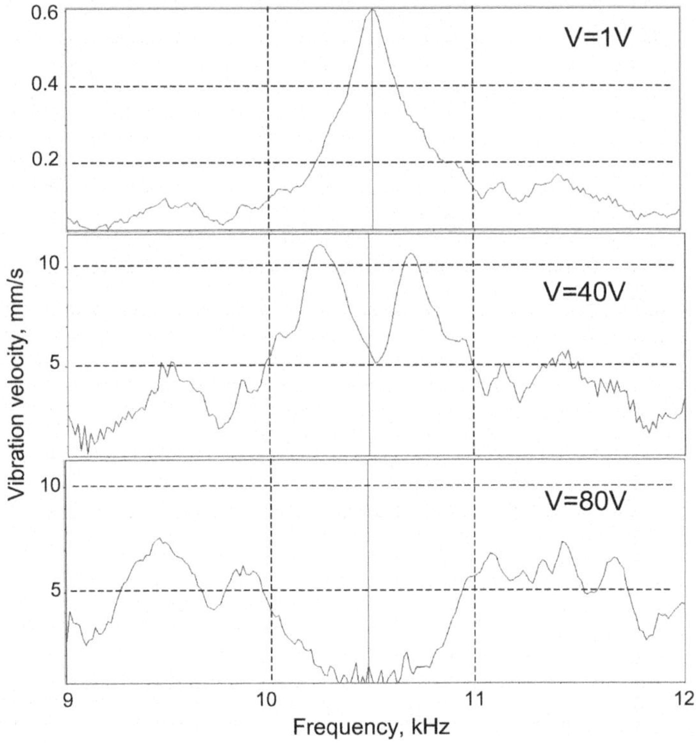

Fig. 4.47 FR of heat damage in CFRP at different input voltages

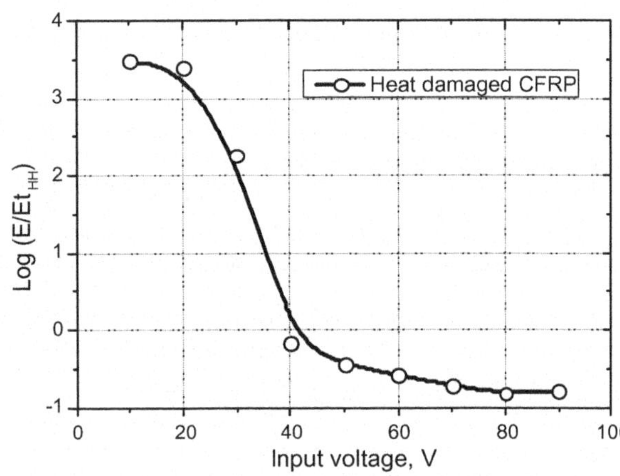

Fig. 4.48 Nonlinear HH ratio as a function of input voltage

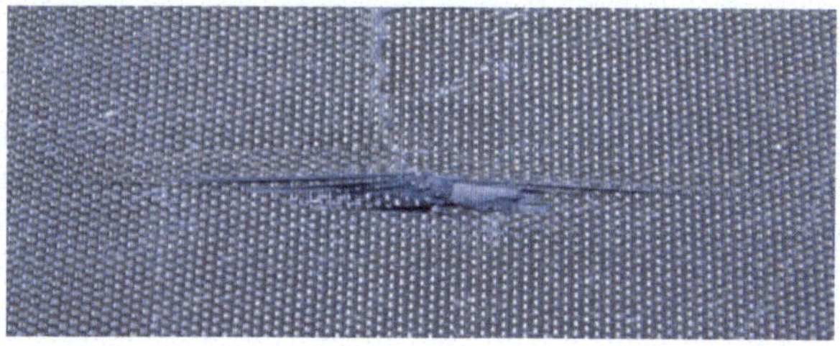

Fig. 4.49 Impact-induced damage in CFRP plate

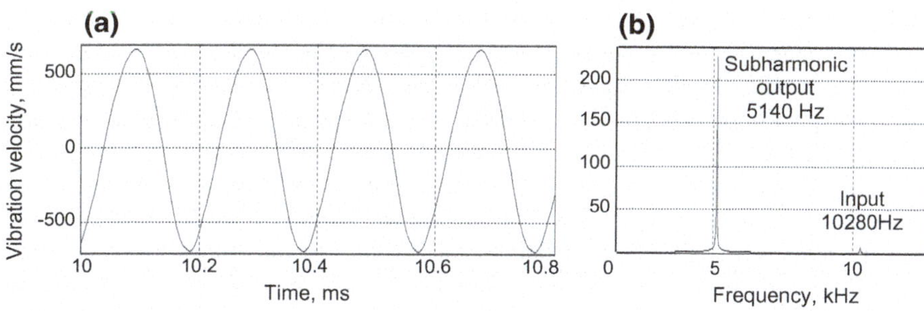

Fig. 4.50 Subharmonic LDR for impact damage in CFRP: vibration patterns (**a**) and spectrum (**b**) beyond threshold input voltage 45 V

threshold, the subharmonic component increases dramatically and prevails in the vibration (velocity) spectrum: $V_{\omega/2}/V_{\omega} \approx 30$ dB at 10,280 Hz input (Fig. 4.50b)). This complies with pure sinusoidal subharmonic vibration pattern in the impact area beyond the threshold (Fig. 4.50a)).

4.8 Linear and Nonlinear LDR for Non-contact Diagnostic Imaging of Defects

A selective acoustic activation of defects based on the concept of local defect resonance enables to enhance considerably the intensity of defect vibrations and makes it possible to reduce the input acoustic powers to the levels permissible for noncontact nondestructive inspection. Since for cm-size defects in composite materials, the LDR frequencies lie in the low kHz-range the resonant noncontact activation shifts to an audible frequency range and can be provided by conventional sonic equipment [58].

4.8.1 Experimental Methodology

Two sets of piezoelectric loudspeakers (CTS Corporation, model 232 (N1) and P5123 by Kemo Electronic GmbH (N2)) were combined in $\approx 20 \times 20\,\text{cm}^2$ arrays (Fig. 4.51) to maintain a wide insonation area of the specimens. To probe the sound field produced by the arrays the ACU laser vibrometry technique [59] was applied. Scanning laser vibrometer Polytec PSV 300 was also used for imaging of damage activated remotely with airborne sound waves.

The overall frequency response (FR) where the airborne radiation was measurable with the laser vibrometer covered the range from 2 to about 45 kHz with two distinct maxima around 4–5 kHz and 8–10 kHz for the arrays N1 and N2, correspondingly. In this frequency range, the speakers were driven either in single frequency or wide-band chirp modes (input voltage 20–40 V) generated by the HP 33120A arbitrary waveform generator combined with HVA 3/450 amplifier. The acoustic pressure produced by the arrays was measured with a ½ inch condenser microphone (B&K 4130, sensitivity 6 mV/Pa) positioned in the specimen location area and combined with 40 dB preamplifier type B&K 2642 and power supply B&K 2810. Most of the measurements were carried out at distances 1–1.5 m. The maximum sound pressure level (SPL) at these distances was measured to be within 10–20 Pa ($\leq$ 120 dB) in the range of the FR maxima indicated above and fell gradually by 30–40 dB to the edges of the overall FR.

Fig. 4.51 Loudspeaker arrays used in the experiments: 232 CTS (left) and P5123 Kemo (right)

4.8.2 Noncontact LDR Imaging Results

Figure 4.52a) shows a fundamental LDR (at 3750 Hz) image of a circular artificial delamination in CFRP plate for a single-sided access airborne activation. The delamination is a circular interlayer (self-healing PV5414 ionomer films (5 cm diameter, 500 μm thickness)) pressed between two plates (10 × 10 cm^2) of satin woven carbon fabrics (courtesy of TU Delft). A similar image (Fig. 4.52b) was obtained in airborne transmission mode in a close proximity to the specimen. The noncontact airborne imaging was vital to monitor the ionomer self-healing process by tracing LDR amplitude and frequency variations as a function of exposure time at elevated temperature [60].

A similar fundamental LDR case is shown in Fig. 4.53 for heat-induced delamination in CFRP specimen (300 × 300 × 4 mm^3). The fundamental LDR (9500 Hz) remote image (Fig. 4.53b) reveals a homogeneous vibration of a subsurface delamination of much larger size than that seen on the specimen surface (Fig. 4.53a).

A single frequency excitation at LDR frequency activates a fundamental vibration pattern which usually visualizes only some basic features of a defect. A wideband mode is required to reproduce all the details of the defect shape and its real size. Figure 4.54, a) shows a fundamental frequency (8.5 kHz) image of a simulated delamination (square 20 × 20 mm^2 Teflon insert) in CFRP plate (5 × 290 × 330 mm^3) which reveals a circular vibration pattern. As the input frequency increases, the higher-order LDR with additional nodal lines and near border-line of the defect localization are also seen (Fig. 4.54b, c). A superposition of various order vibration patterns in a wideband excitation mode (5–30 kHz) reproduces an entire square shape of the defect (Fig. 4.54d).

Another illustration of the full-scale remote defect selective imaging in the wideband mode is given in Fig. 4.55 for a more complex shape defect (subsurface Teflon

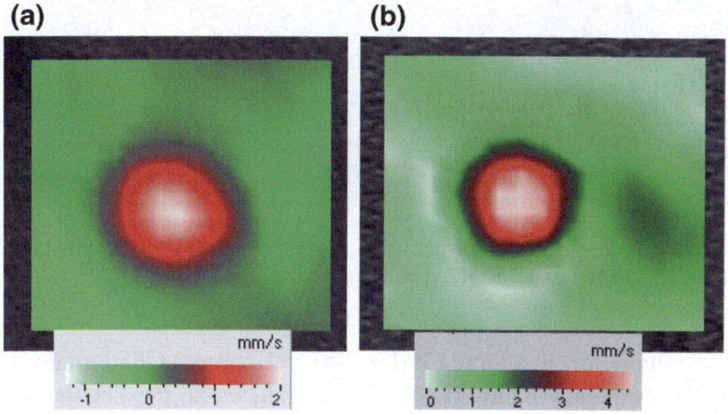

Fig. 4.52 **a**, **b** Remote imaging of self-healing ionomer interlayer in CFRP plate in: single-sided (**a**) and transmission mode (**b**)

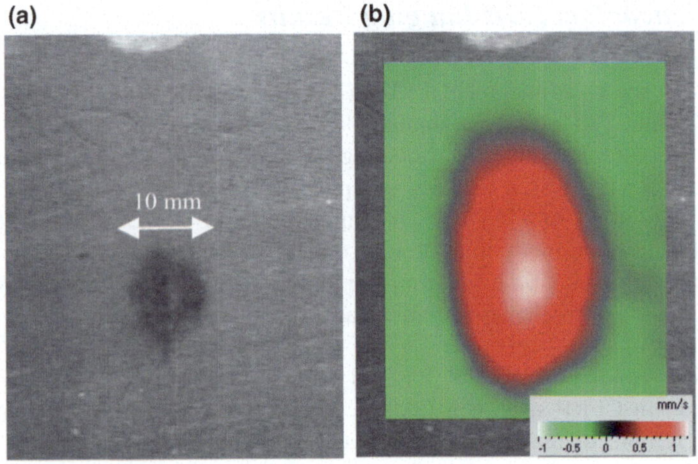

Fig. 4.53 Heat-induced damage in CFRP plate (**a**) and its remote LDR image at 9500 Hz (**b**)

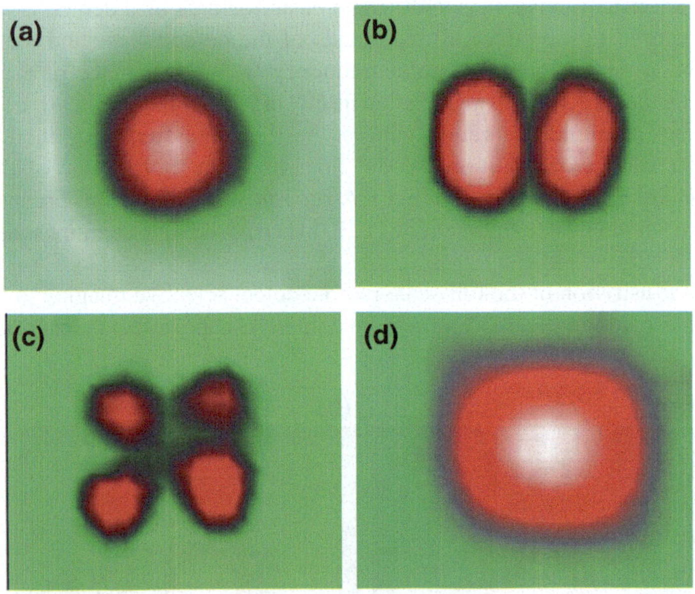

Fig. 4.54 Airborne images of various order LDR for square insert (20 × 20 mm^2) in CFRP plate at: **a** 8500 Hz, **b** 12,380 Hz, **c** 15,250 Hz, **d** wideband excitation

ring of 55 mm diameter) in a large (390 × 390 × 2.2 mm^3) CFRP plate. The image obtained in the frequency band 10–35 kHz reproduces all the details of the defect shape.

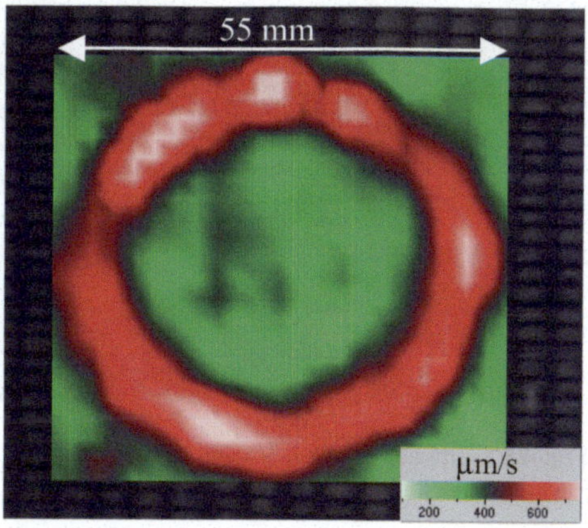

Fig. 4.55 Wideband (10–35 kHz) remote image of subsurface ring insert in CFRP plate

The strain developed in the tested composite specimen by remote airborne excitation (SPL 100–120 dB) subject to LDR is in the range typical for conventional contact ultrasonic NDE (10^{-7} to 10^{-6}). A few case studies given below show that it makes feasible to activate and visualize remotely the tight defects, like barely visible (BVID) and even virtually invisible (VIND) damage.

The BVID shown in Fig. 4.56 is the result of 20 J impact in the non-crimp fabric CFRP plate ($150 \times 100 \times 3$ mm^3). The damage is completely internal and on the reverse side (Fig. 4.56a) becomes apparent only due to barely noticeable bleaching of the stitching yarn (inside the dotted area in Fig. 4.56a). The remote image (Fig. 4.56b) reveals inhomogeneous damage which displays fractional LDR at 32,200 Hz. The maximum vibration area in the image is aligned along the upper roving layer and, thus, indicates near-surface damage of the fabric.

Figure 4.57 demonstrates an example of the lower energy impact-induced VIND: the damage is virtually invisible on the reverse side of the CFRP plate (170 × 50

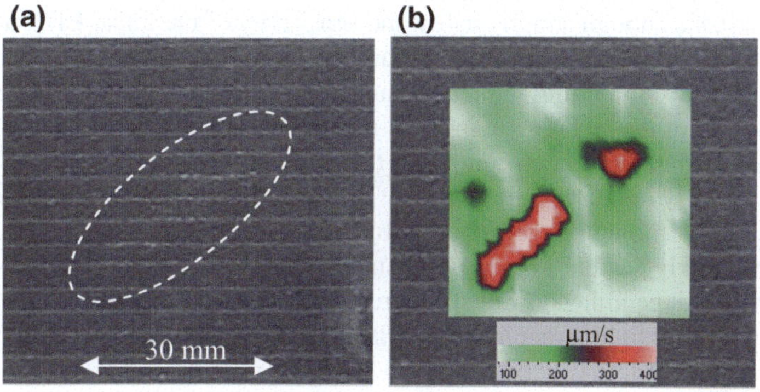

Fig. 4.56 20 J impact-induced BVD in non-crimp fabric CFRP plate (**a**) and its remote image at fractional LDR frequency 32,200 Hz (**b**)

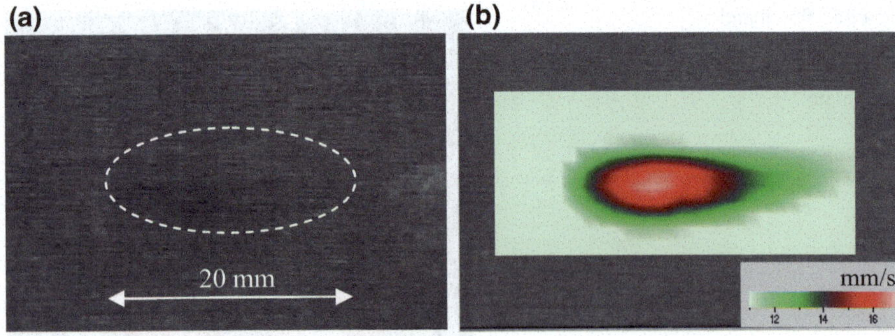

Fig. 4.57 Impact-induced VIND in CFRP plate (**a**) and its remote LDR image at 4500 Hz (**b**)

× 2 mm^3) (dotted area Fig. 4.57a). However, remote insonation reveals a rather extensive damaged area stretched along the 0° orientation of the surface layer of the fibres (Fig. 4.57b). The LDR image appears at quite low frequency of 4500 Hz and exhibits unexpectedly high vibration amplitude (see scaling in Fig. 4.57b) that might be an indication that it is facilitated by the resonance of the whole specimen.

4.8.3 Nonlinear LDR Diagnostic Imaging

Nonlinear acoustic NDE is a promising technology for diagnostics of damage based on the frequency conversion (e.g. generation of higher harmonics, mixing frequencies, etc.). The use of LDR, which "amplifies" local defect vibrations, enables to enhance the input-output frequency conversion and realize efficient contact and even noncontact versions of nonlinear NDE and defect imaging. The contact and noncontact (in transmission) imaging results were reported before [58]. Some new results on remote nonlinear imaging of defects in single-sided airborne version will be presented here [61]. Though a high quality and sensitivity of the linear LDR imaging is a "serious rival" to its nonlinear counterpart, the latter is recognized and valued as a sensitive indicator of a faint/incipient damage.

A feasibility of the HH remote mode is illustrated in Fig. 4.58 for an oval delamination in glass fibre-reinforced composite (GFRP). A linear LDR image (20,100 Hz, Fig. 4.58a) clearly indicated a loose part of the delamination which is also confirmed by the second harmonic image (Fig. 4.58b).

LDR in clearly visible (strong) realistic damages is found to manifest in two different ways depending on the structure of the damage induced. If the damage results mainly in a tight structure of its fragments or forms, on the contrary, a fully loose pattern, the LDR is activated as an in-phase vibration over a total damaged area at a single frequency (similar to the defect considered above). If the damage combines multiple tight and loose elements (with different stiffness) their weakly

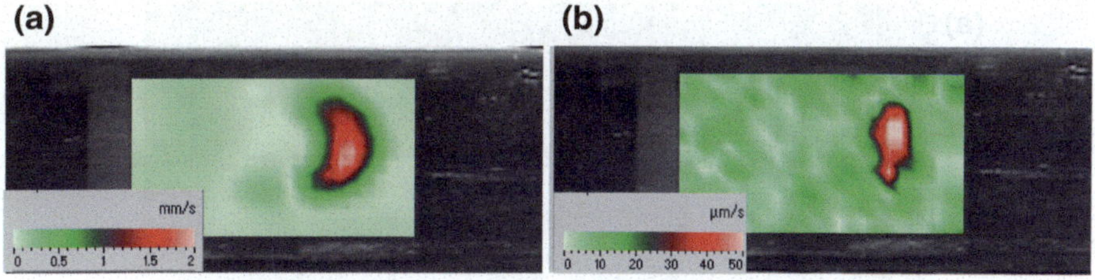

Fig. 4.58 Imaging of delamination in GFRP via linear (**a**) and second harmonic (**b**) LDR

coupled resonant vibrations will occur at different frequencies and, thus, LDR will be a combination of multiple *fractional* resonances.

The occurrence of the fractional LDR is demonstrated in Fig. 4.59 for an impact-induced loose crack in a thin CFRP plate (280 × 45 × 1.1 mm^3). The crack includes the fiber fracture (marked by the dotted line in Fig. 4.59a) which divides it in two different parts. The remote images of the damage (Fig. 4.59b, c) reveal two separate LDR frequencies (4150 and 5300 Hz) for the two parts of the defect. The presence of elastic coupling between them also provides LDR for the whole damaged area at lower frequency of 3500 Hz (Fig. 4.59d).

A remote excitation of the defect fractional resonance (5300 Hz, Fig. 4.59c) yields a nonlinear spectrum with multiple higher harmonics noticeably modulated by the *sinc*-function (Fig. 4.60a). The second harmonic images (Fig. 4.60b, c) validate the positions of the defect parts responsible for the corresponding fractured LDR.

The combination frequency resonance mode is applied for nonlinear imaging of an impact-induced damage in non-crimp fabric CFRP plate (250 × 180 × 1.1 mm^3, Fig. 4.61a) that manifests a single frequency LDR at 21,600 Hz (Fig. 4.61b). For dual-frequency airborne excitation (11,600 and 10,000 Hz), the spectrum of the damage vibration manifests a strong nonlinear interaction with multiple frequency mixing components (Fig. 4.62). Each of the side-lobe frequencies in Fig. 4.62 is an indicator of the defect presence and its position as seen from Fig. 4.61 where a couple of examples of the mixing frequency images are given. It is instructive noting that according to the images in Fig. 4.61c, d, the nonlinear part of the defect may indicate somewhat different fragments of a total damage area.

Remote excitation proves to be energetic enough to activate fractional LDR in tight defects. The presence of multiple defect resonances is clearly seen in the frequency range between 18 and 36 kHz in the vibration spectrum (Fig. 4.63a) of the compact impact-induced damage in CFRP specimen (170 × 50 × 2 mm^3) shown in Fig. 4.63b. A pair of the remote images obtained at the extremes of this frequency range (18,200 and 33,400 Hz, Fig. 4.63c, d) indicates that various resonances are attached to and visualize different fractions of the damaged area.

To test the feasibility of the super-harmonic imaging the excitation frequency is chosen at 16,700 Hz that is a subharmonic of the fractional LDR at 33,400 Hz (Fig. 4.64, a). The nonlinear spectrum measured indicates the highly-efficient frequency conversion in the super-harmonic resonance vibrations of the defect. This is

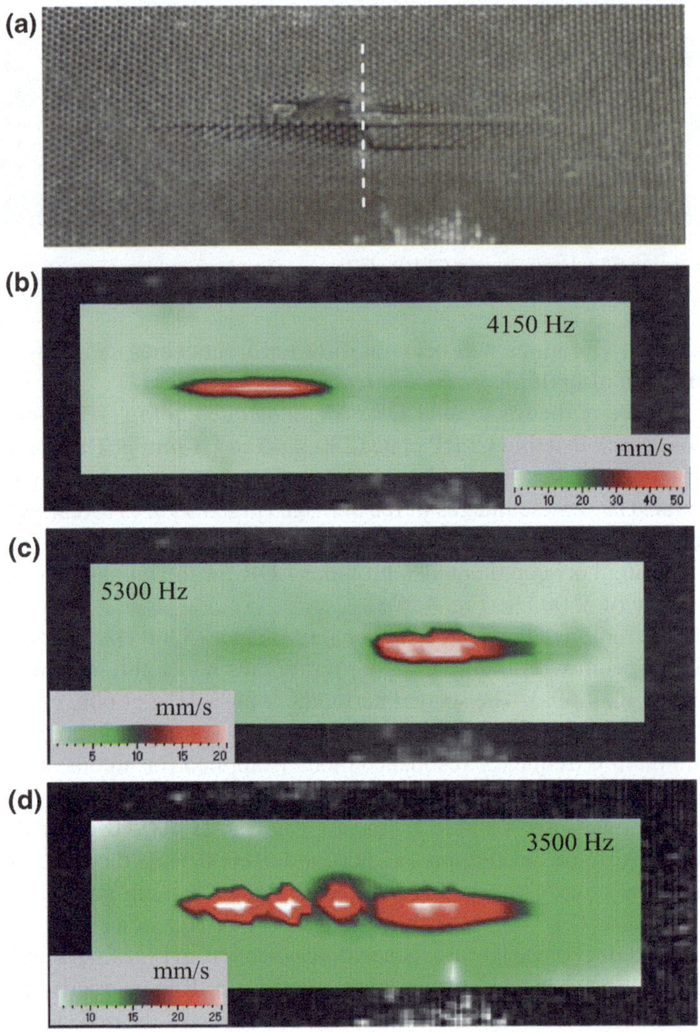

Fig. 4.59 Remote imaging of two-part crack in CFRP (**a**): fractional LDR for the left (**b**) and right (**c**) parts. Total LDR of the damaged area (**d**)

confirmed by the images in Fig. 4.64b, c: the signal of the nonlinear image (c) is of the higher amplitude (see, the colour scale) while the linear image (b) is fully out of resonance and completely unrelated to the defect image.

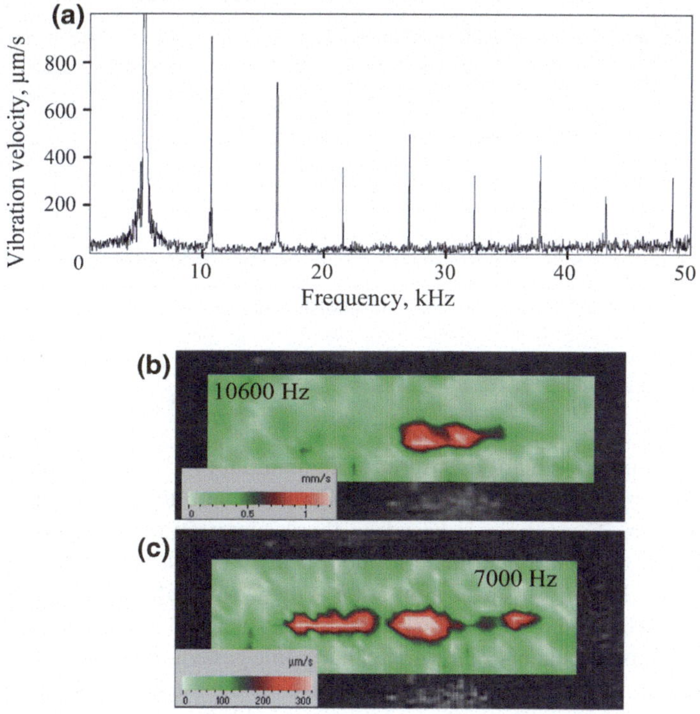

Fig. 4.60 Higher harmonic spectrum (**a**) and second harmonic image (**b**) for fractional LDR at 5300 Hz. Second harmonic image of total damage LDR at 3500 Hz (**c**)

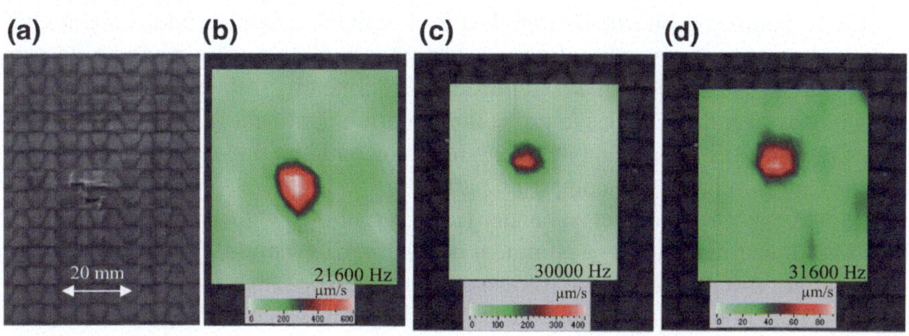

Fig. 4.61 Impact-induced damage in CFRP plate (**a**), its remote fundamental LDR image (**b**), and combination frequency images (**c, d**)

4.9 Conclusions

For over than 50 years, nonlinear acoustics of solids conceived by its masterminds to be a delicate methodology of solid state physics has changed to a new applied

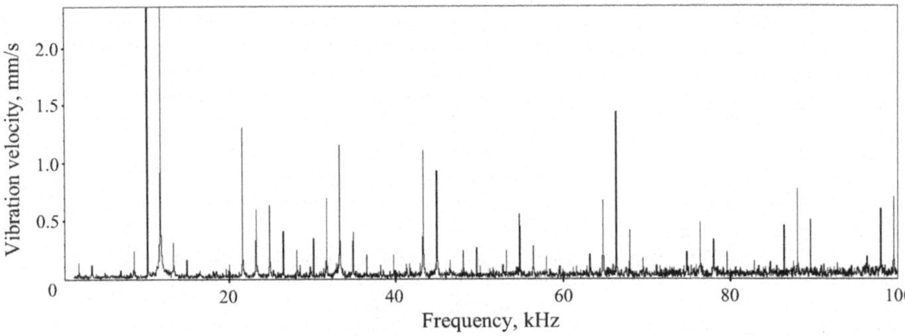

Fig. 4.62 Vibration spectrum of impact-induced damage in Fig. 4.61a, for dual frequency (11,600 and 10,000 Hz) excitation

technology for monitoring of deterioration in material properties and diagnostics of damage. The application field now includes both the nonlinear wave and nonlinear vibration modes. The former is based on the assumption and is applicable to the case studies of the distributed material nonlinearity (variation due to fatigue, heat treatment, etc.). It profits from accumulation of the wave nonlinear response along the propagation distance and in practical terms relies solely on the second and the third harmonic signals. The theory of the higher harmonic generation is well developed and documented for different types of acoustic waves while the nonlinear measurements require some experimental precaution and specific high-power instrumentation.

A strong nonlinear response of non-bonded interfaces in planar defects, like cracks, delaminations, debondings, etc. introduces the nonlinearity localized in the defect area where the vibration nonlinearity steps up. Activated by an incident acoustic wave it manifests abnormally high levels of both the quadratic and cubic nonlinearity and provides a severe nonlinear distortion in both transmission and reflection modes which can be used to read out the defect nonlinearity (related to its gravity). A standing wave version enables to probe the vibration nonlinear response over a total area of the specimen and is applicable to mapping and imaging of the contaminated adhesion areas in large-size composite parts.

The concept of local defect resonance (LDR) combined with its nonlinearity identifies a nonlinear inclusion as a nonlinear oscillator. Conventional nonlinear practice of its excitation at a fundamental frequency appears to be rather inefficient for planar defects. Their nonlinear response raises considerably when the higher harmonics match the oscillator higher-order normal modes. The LDR also brings about quantitatively and qualitatively different dynamic and frequency scenarios (nonlinear and parametric resonances) in vibration nonlinear phenomena. In particular, the parametric resonance provides dramatic efficiency of the frequency conversion into either the subharmonic or the higher harmonics. In the latter case, the LDR collapses due to nonlinear depletion caused by the energy pumping into the higher harmonics (the effect similar to nonlinear wave case).

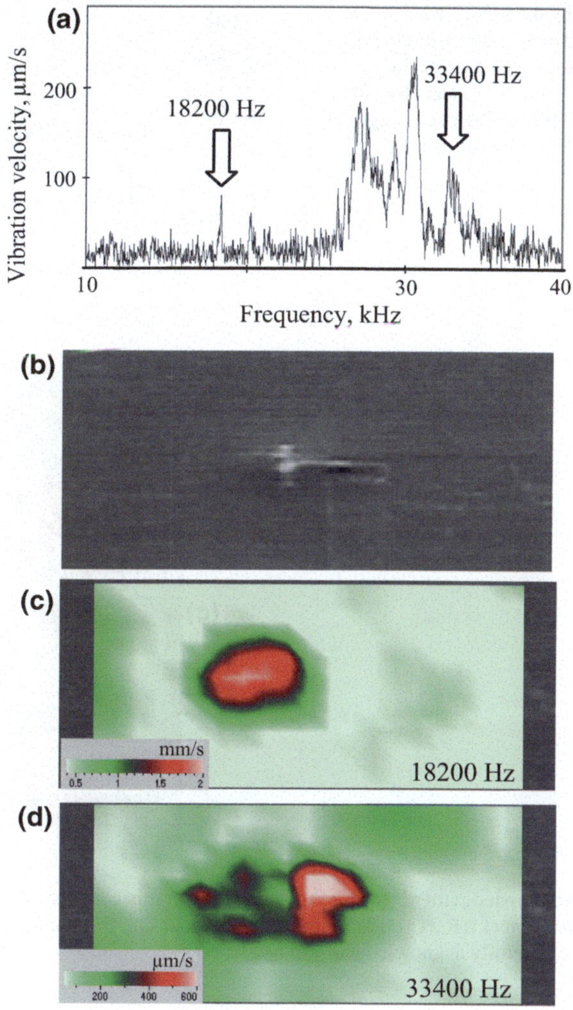

Fig. 4.63 Fractional LDR spectrum (**a**) and fractional LDR images (**c, d**) for a solid impact-induced damage in CFRP plate (**b**)

The LDR-induced trapping of the nonlinearity in the defect area has its pros and contras. A direct acoustic read-out of the defect nonlinearity by the driving wave is barely possible and one has to use e.g. a laser vibrometer or a thermal camera instead. On the other hand, it generates a defect-selective nonlinearity and conditions for diagnostic imaging of damage. The LDR-induced boost in efficiency provides feasibility for noncontact nonlinear imaging by using airborne activation of damage. It employs commercial instrumentation and applicable to remote sonic inspection of different materials and various scale components.

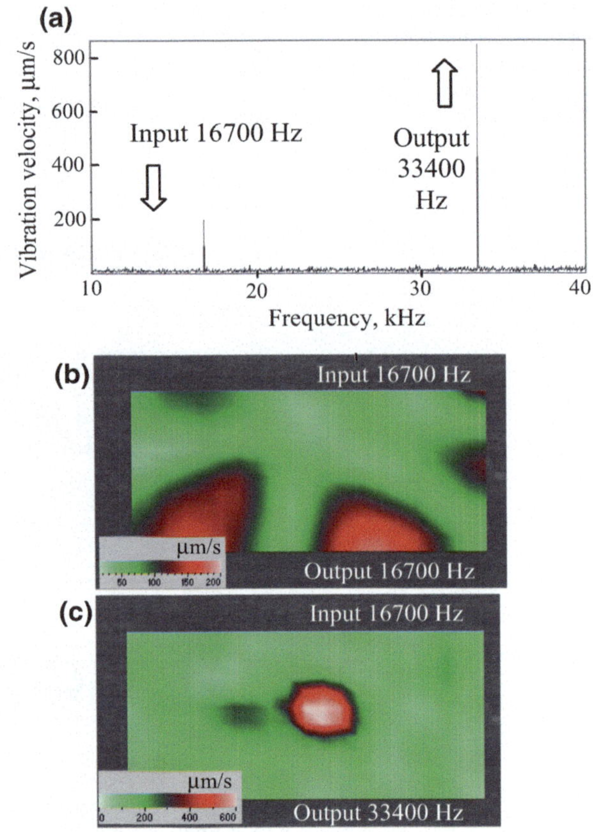

Fig. 4.64 Super-harmonic resonance spectrum (**a**) of the impact damage in Fig. 4.63b. Linear (**b**) and nonlinear SHR (**c**) images at 16,700 Hz excitation

Acknowledgements A part of the results in this Chapter has been obtained in the framework of the project KR 2131/12-1 funded by the Deutsche Forschungsgemeinschaft (DFG) whose support is gratefully acknowledged.

References

1. L.K. Zarembo, V.A. Krasilnikov, *Vvedenie v nelineinuyu akustiku (Introduction to Nonlinear Acoustics)* (Nauka, Moskva, 1966)
2. M.A. Breazeale, J. Ford, J. Appl. Phys. **36**, 3488 (1965)
3. A.A. Gedroitz, V.A. Krasilnikov, Sov. Phys. JETP **16**, 1122 (1963)
4. W.T. Yost, J.H. Cantrell, Rev. Progr. Quant. Nondestr. Eval. **9**, 1669 (1990)
5. J.H. Cantrell, W.T. Yost, W.T.J., Appl. Phys. **81**, 2957 (1997)
6. K.S. Len, F.M. Severin, IYu. Solodov, Sov. Phys. Acoust. **37**, 610 (1991)
7. I. Solodov, N. Krohn, G. Busse, Ultrasonics **40**, 621 (2002)
8. I.Yu. Solodov, Ultrasonics **36**, 383 (1998)
9. I. Solodov, J. Bai, S. Bekgulyan, G. Busse, Appl. Phys. Lett. **99**, 211911 (2011)
10. I. Solodov, J. Bai, G. Busse, J. Appl. Phys. **113**, 223512 (2013)

11. I. Solodov, G. Busse, Appl. Phys. Lett. **102**, 061905 (2013)
12. J. Hettler, M. Tabatabaeipour, S. Delrue, K.V.D. Abeele, J. Nondestr. Eval. **36**, 2 (2017)
13. M. Rahammer, M. Kreutzbruck, NDT E Internat. **86**, 83 (2017)
14. G.P.M. Fierro, D. Ginzburg, F. Ciampa, M. Meo, J. Nondestr. Eval. **36**, 4 (2017)
15. L. Pieczonka, L. Zietek, A. Klepka, W.J. Staszewski, F. Aymerich, T. Uhl, Damage imaging in composites using nonlinear vibro-acoustic wave modulations. Struct. Contr. Health Monit. **25**, 2e2063 (2018)
16. J. Segers, M. Kersemans, S. Hedayatrasa, J.A.C. Tellez, W. Van Paepegem, NDT E Internat. **98**, 130 (2018)
17. B. Roy, T. Bose, K. Debnath, J. Sound Vibr. **443**, 703 (2019)
18. Ch. Andreades, P. Mahmoodi, F. Ciampa, Characterisation of smart CFRP composites with embedded PZT transducers for nonlinear ultrasonic applications. Comp. Struct. (2018). https://doi.org/10.1016/j.compstruct.2018.08.083
19. F. Ciampa, G. Scarselli, M. Meo, J. Acoust. Soc. Am. **141**, 2364 (2017)
20. I. Solodov, J. Nondestr. Eval. **33**, 252 (2014)
21. R.N. Thurston, in *Physical acoustics*, vol. 1, chap. 1. (Academic, New York, 1964)
22. J. Na, M. Breazeale, J. Acoust. Soc. Am. **95**, 3213 (1994)
23. K. Van Den Abeele, M. Breazeale, J. Acoust. Soc. Am. **99**, 1430 (1996)
24. Ch. Kube, A. Arquelles, J. Acoust. Soc. Am. **142**, EL224 (2017)
25. D. Gerlich, M. Breazeale, J. Appl. Phys. **67**, 3278 (1990)
26. K. Naugolnykh, L. Ostrovskii, *Nonlinear Wave Processes in Acoustics*. (Cambridge University Press, 1998)
27. R. Beyer, J. Acoust. Soc. Am. **32**, 719 (1960)
28. E.M. Ballad, B.A. Korshak, V.G. Mozhaev, I. Solodov, Moscow Univ. Bull. Phys. **3**, 44 (2001)
29. M. Breazeale, D. Thompson, Appl. Phys. Lett. **3**, 77 (1963)
30. K. Matlack, J.-Y. Kim, L. Jacobs, J. Qu, J. Nondestr. Eval. **34**, 273 (2015)
31. V.E. Nazarov, Akust. Zh. **37**, 150 (1991)
32. P. Hess, A. Lomonosov, A.P. Mayer, Ultrasonics **54**, 39 (2014)
33. V.A. Krasilnikov, V.E. Ljamov, I. Solodov, Izv. AN SSSR Phys. **35**, 944 (1971)
34. Y. Shui, I. Solodov, J. Appl. Phys. **64**, 6155 (1988)
35. Y. Zheng, R. Maev, I. Solodov, Can. J. Phys. **77**, 927 (1999)
36. A.V. Porubov, *Lokalizacija nelinejnyh voln deformacii (Localisation of Nonlinear Deformation Waves)* (Fizmatgiz, Moscow, 2009)
37. B.A. Konyukhov, G.M. Shalashov, J. Appl. Math. Theor. Phys. **4**, 125 (1974)
38. I. Solodov, D. Sci. Dissertation, Moscow State University, 1987
39. I.B. Yakovkin, D.V. Petrov, *Difrakcija sveta na poverkhnostnykh akusticheskikh volnakh (Light Diffraction on Surface Acoustic Waves)* (Nauka, Novosibirsk, 1979)
40. Y. Shui, I. Yu. Solodov, in *Proceedings II WESTPA*. (Polytech. Institute Hong Kong, 1985), pp 188
41. A.P. Brysev, V.A. Krasilnikov, A.A. Podgornov, IYu. Solodov, Fiz. Tverd. Tela **26**, 2204 (1984)
42. I.Yu. Solodov, C. Wu, Acoust. Phys. **39**, 476 (1993)
43. K. Sel Len, F.M. Severin, I.Yu. Solodov, Sov. Phys. Acoust. **37**, 610 (1991)
44. R. Maev, I. Solodov, Rev. Progr. Quant. Nondestr. Eval. **19**, 1409 (2000)
45. R. Maev, I. Solodov, IEEE Ultrason. Symp. Proc. 707 (1998)
46. I. Yu. Solodov, A.F. Asainov, K. Sel Len, Ultrasonics **31**, 91 (1993)
47. I. Solodov, B.A. Korshak, Phys. Rev. Lett. **88**, 014303 (2002)
48. J.D. Achenbach, O.K. Parikh, Rev. Progr. Quant. Nondestr. Eval. **10B**, 1837 (1991)
49. C. Bermes, J.Y. Kim, J. Qu, L.J. Jacobs, Appl. Phys. Lett. **90**, 021901 (2007)
50. T.-H. Lee, I.-M. Choi, K.-Y Jhang, Mod. Phys. Lett., B. **22**(11), 1135 (2018)
51. J. Zhao, V.K. Chillara, B. Ren, H. Cho, J. Qiu, C.J. Lissenden, J. Appl. Phys. **119**, 064902 (2016)
52. I. Solodov, D. Segur, M. Kreutzbruck, Res. Nondestr. Eval. **30**, 1 (2019)
53. I. Solodov, D. Segur, M. Kreutzbruck, in Proceedings of 12th ECNDT, Gothenburg, (2018)
54. L.D. Landau, E.M. Lifshitz, *Mechanics* (Pergamon Press Ltd., Oxford-London-Paris, 1960)

55. N.W. McLachlan, *Theory and Applications of Mathieu Functions* (University Press, Oxford, 1951)
56. N. Minorsky, *Nonlinear Oscillations* (D. Van Nostrand Co., Inc., Princeton, 1962)
57. F.K. Kneubuehl, *Oscillations and Waves* (Springer, Berlin, 1997)
58. I. Solodov, M. Rahammer, N. Gulnizkij, M. Kreutzbruck, J. Nondestr. Eval. **35**, 47 (2016)
59. I. Solodov, D. Döring, G. Busse, Appl. Optics **48**, C33 (2009)
60. W. Post, M. Kersemans, I. Solodov, K. Van Den Abeele, S.J. Garsia, S. van der Zwaag, Compos. Part A: Appl. Sci. Manuf. **101**, 243 (2017)
61. I. Solodov, M. Kreutzbruck (2019). https://doi.org/10.1016/j.ndteint.2019.102146

Chapter 5
Nonlinear Ultrasonic Phased Array for Measurement of Closed-Crack Depth

Yoshikazu Ohara, Tsuyoshi Mihara and Kazushi Yamanaka

Abstract This chapter summarizes the comprehensive review of the nonlinear ultrasonic phased array (PA) methods for the measurement of closed-crack depth. Various nonlinear ultrasonic PA methods are categorized into four groups: (I) subharmonics, (II) parallel and sequential transmission, (III) all-elements, odd-elements, and even-elements transmission, and (IV) utilization of thermal stress. Each method is described in the order of principles, experimental conditions and imaging results, key features.

5.1 Introduction

To ensure the safety and reliability of aging structures, mechanical components, and manufactured materials, the reliable detection of cracks and their accurate measurement by nondestructive testing (NDT) are important. Specifically, crack depth is one of the critical factors determining material strength, where crack depth is defined as the size of surface-breaking cracks in the thickness direction. Among the various NDT methods, ultrasonic testing (UT) has been widely used because of its high sensitivity to cracks and its ease of use [1–3]. When cracks are irradiated by ultrasound, the ultrasound is strongly scattered at a crack tip with an air gap, as illustrated in Fig. 5.1a. Such a crack can be referred to as an open crack. The depth of open cracks can be nondestructively measured by UT. This has been carried out by waveform measurement using monolithic transducers, where crack depths are estimated from the time of flight of crack tip echo, the wave speed, and the geometric relationship [1–3].

Electronic supplementary material The online version of this chapter (https://doi.org/10.1007/978-981-15-1461-6_5) contains supplementary material, which is available to authorized users.

Y. Ohara (✉) · T. Mihara · K. Yamanaka
Department of Materials Processing, Tohoku University, Sendai, Japan
e-mail: ohara@material.tohoku.ac.jp

K.-Y. Jhang et al. (eds.), *Measurement of Nonlinear Ultrasonic Characteristics*, Springer Series in Measurement Science and Technology,
https://doi.org/10.1007/978-981-15-1461-6_5

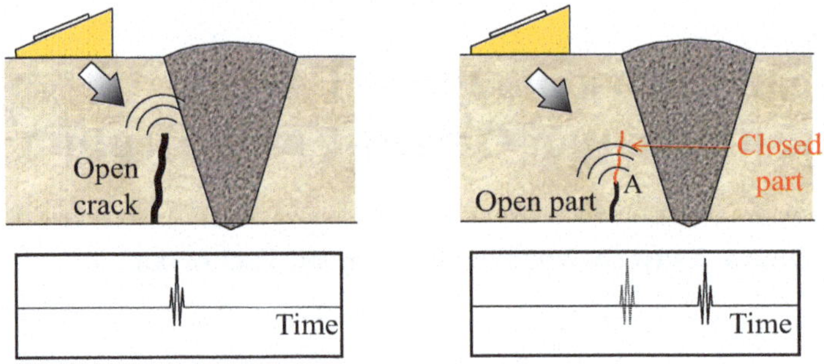

(a) Crack with an open tip (Open crack)

(b) Crack with a closed tip (Closed crack)

Fig. 5.1 Ultrasonic testing (UT) of cracks with open and closed tips

On the other hand, as a real-time ultrasonic imaging method, ultrasonic phased array (PA) has been commonly used in NDT [4, 5]. PA was originally proposed to diagnose the human body [6]. PA usually uses a linear (one-dimensional (1D)) array transducer. A linear array transducer is typically composed of multiple (16–128) rectangular individual elements. With an appropriately designed array transducer [7, 8], PA can produce a B-scan image, which is a cross-sectional image of a sample, in real time by electronic scanning without the mechanical scanning of the array transducer [9]. Furthermore, a data acquisition method called full matrix capture (FMC), which acquires a data set comprising a matrix of signals from every transmit–receive element pair, and a post-processing algorithm, such as the total focusing method (TFM), for the data set acquired by FMC are also commercially available [10]. PA has a strong merit that it can provide ultrasonic images, from which NDT engineers can intuitively recognize defects. Furthermore, such images have been more valuable as evidence than the waveforms (A-scans) obtained by conventional UT, since it is relatively easy for NDT engineers to explain such results to non-experts. Thus, PA has been widely used for NDT in the engineering field, although the hardware cost of PA is still higher than that of conventional UT hardware for waveform measurement.

The aforementioned UT methods including PA have been widely used to measure the depth of open cracks. On the other hand, cracks are often closed because of compressive residual stress due to various mechanisms [11] and/or the generation of oxide debris between the crack faces [11, 12]. When cracks are closed, ultrasound penetrates through them [13–15]. This results in serious problems such as the underestimation or nondetection of cracks. This may result in catastrophic accidents.

It should be noted here that the definition of closed cracks in UT may be different from that in fracture mechanics, in which the phenomenon of crack closure was first found by Elber [16] in 1970. Subsequently, various studies have been carried out because crack closure stress is one of the factors affecting the crack growth rate. Crack closure has mainly been explained by three mechanisms, namely, plasticity-, roughness-, and oxide-induced crack closure (PICC, RICC, and OICC, respectively)

[11]. PICC, OICC, and RICC are caused by residual plastic strains in the wake of propagating fatigue cracks [16], the generation of oxide debris between crack faces [17, 18], and the contact of rough (inconsistent) crack faces [19], respectively. The reason why crack closure can reduce the crack growth rate is that a tensile stress that is repeatedly applied to cracks can be mitigated by a closure stress [15]. Thus, in fracture mechanics, cracks are regarded as being closed when a compressive stress acts on crack faces.

We next look at crack closure in a different way. From the microscopic viewpoint, crack faces are not perfectly in contact with each other even when the crack is closed. This means that a compressive stress acts on the crack faces through multiple contact points between them. That is, the closed crack defined in fracture mechanics should have numerous microscopic air gaps between the crack faces. Note that ultrasound can be scattered at such air gaps. Hence, it is possible that UT can be used to measure the closed cracks defined in fracture mechanics, which are not a problem in UT. On the other hand, the closed crack defined in UT means that the crack cannot be measured by UT since the ultrasound penetrates through the crack. In other words, the closed crack should have crack faces that are mostly in contact with each other and thereby do not generate observable scattered waves. The closed crack definition in UT may partially overlap with that in fracture mechanics. To avoid any confusion on the definition, the term "closed cracks" is used as defined in UT hereafter in this chapter.

Given the above definition, closed cracks cannot be measured by UT. How can we determine whether a crack is closed? A simple way is to apply a tensile stress to a closed crack while measuring the crack by UT [13–15]. An example of this is shown in Fig. 5.2. Here, two transducers were used for transmission and reception in a time-of-flight diffraction (TOFD) configuration. A fatigue crack was introduced in

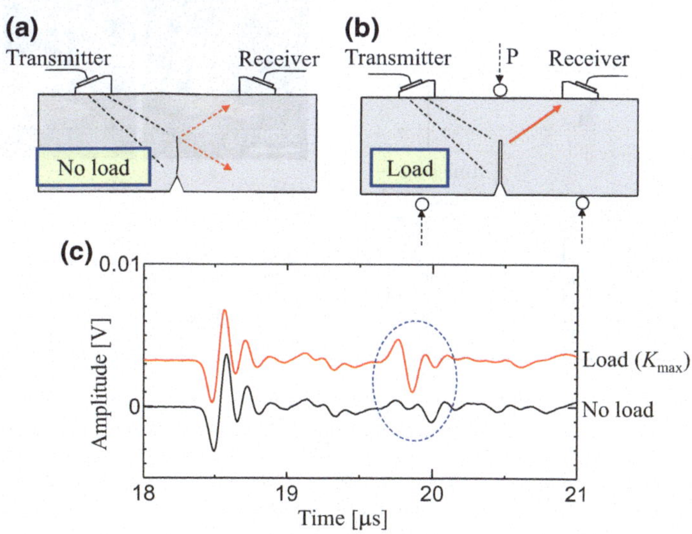

Fig. 5.2 Crack tip echoes with and without tensile loading

a specimen made of A7075 aluminum alloy with fatigue conditions of a maximum stress intensity factor of $K_{max} = 4.3$ MPa $m^{1/2}$ and a minimum stress intensity factor of $K_{min} = 0.6$ MPa $m^{1/2}$. A crack tip echo was examined when a load was applied to impose a tensile bending stress on the crack. When no load was applied (Fig. 5.2a), the crack tip echo was invisible in the lower waveform in Fig. 5.2c, where the temporal region surrounded by the blue dashed elliptical circle corresponds to the arrival time of the crack tip echo. On the other hand, when the tensile load was applied (Fig. 5.2b), the crack tip echo appeared in the upper waveform of Fig. 5.2c. This is because the crack was opened by the application of the tensile load. That is, the crack was originally closed.

This can also be confirmed by PA. Figure 5.3 shows the experimental configuration and the imaging results obtained by PA [15]. As shown in Fig. 5.3a, a 32-element array transducer (5 MHz, 0.5 mm pitch) manufactured by Imasonic (Voray-sur-l'Ognon, France) was positioned on the top surface of the specimen, and was operated by commercial PA hardware (PAL) produced by KJTD (Tokyo, Japan). A fatigue crack was introduced in a compact tension (CT) specimen made of A7075 aluminum alloy with fatigue conditions of $K_{max} = 9.0$ MPa $m^{1/2}$ and $K_{min} = 0.6$ MPa $m^{1/2}$. The region around the crack was imaged in real time under the loading conditions of $K = 0$ and $K = 7.0$ MPa $m^{1/2}$, as respectively shown in Fig. 5.3b, c. At $K = 0$ (Fig. 5.3b), the crack was not visualized while the starting notch was visualized. In contrast, at $K = 7$ MPa $m^{1/2}$ (Fig. 5.3c), the crack tip was clearly visualized. This is because the crack was opened by the application of the tensile load, as illustrated in Fig. 5.3d, e. That is, the crack was originally closed.

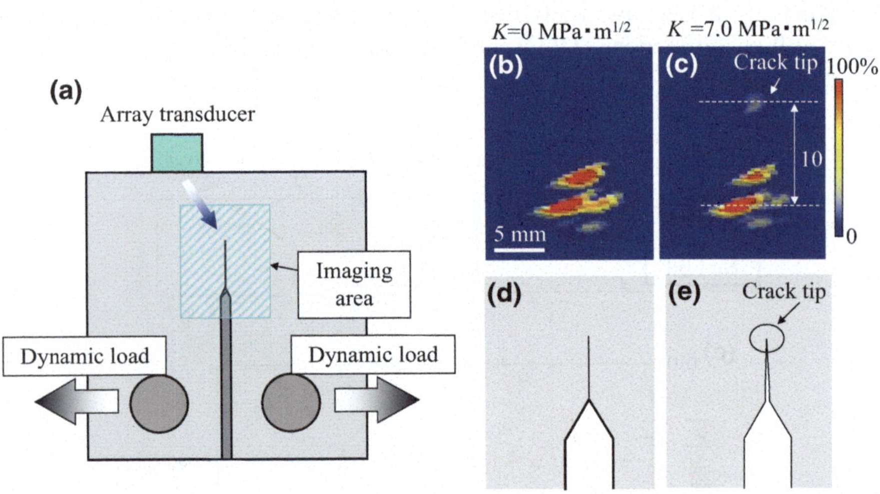

Fig. 5.3 Ultrasonic phased array images of crack tips without and with an applied tensile stress. A 32-element array transducer (5 MHz) was used. **a** Schematic illustration of experimental configuration and PA images at **b** $K = 0$ and **c** $K = 7$ MPa $m^{1/2}$. Schematic illustrations of the crack state at **d** $K = 0$ and **e** $K = 7$ MPa $m^{1/2}$ to interpret the images in (**b**) and (**c**), respectively. Modified from [15], with permission from Elsevier

Furthermore, in the same experimental configuration, the dynamic change in the closed crack was monitored by PA in real time under sinusoidal loading with $K = 0$ to 7 MPa $m^{1/2}$ at a frequency of 0.1 Hz. In the lower graph in Fig. 5.4, the intensity of the response at the crack tip is plotted as a function of time at 0.25 s intervals, while K, which is proportional to the tensile stress applied to the crack, is also similarly plotted in the upper graph in Fig. 5.4. The intensity did not change at $K \leq 0.5$ MPa $m^{1/2}$ (region I). Between $K = 0.5$ and 2.4 MPa $m^{1/2}$ (region II), the intensity increased linearly with K. Above $K = 2.4$ MPa $m^{1/2}$ (region III), the intensity saturated. This shows that the crack was closed in region I, gradually opened (was partially closed) in region II, and was completely open in region III. Accordingly, the crack in region III can be measured by conventional UT, whereas the crack in regions I and II may be underestimated or overlooked in UT.

Thus, a closed crack can be confirmed by UT by applying an external tensile load to the crack with, for example, a servohydraulic testing machine. Such a testing machine, however, can only be used in lab experiments and is not suitable for practical application to actual structures and materials. Hence, a UT method without the use of such a testing machine is required to solve the problem of closed cracks.

To overcome this difficulty, nonlinear ultrasonics has been the most promising approach and has therefore been widely studied, as described in review articles [20–22] and a book [23]. Let us start with a brief introduction to the study of nonlinear ultrasonics for solids. Classically, nonlinear ultrasonics was first studied with the aim of measuring uniform elastic nonlinearity such as a third-order elastic constant [24],

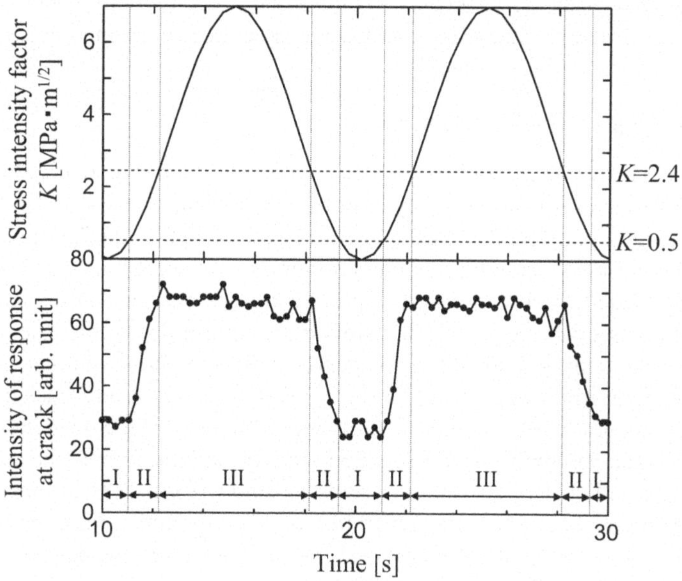

Fig. 5.4 Dynamic load measured by a load cell attached to a servohydraulic fatigue testing machine and the intensity of the response at the crack tip. Modified from [15], with permission from Elsevier

which is related to the thermal expansion coefficient arising from the anharmonicity of atomic interactions. The nonlinearity has mainly been investigated by measuring the second harmonic wave with a frequency of $2f$ generated during the propagation of an incident wave with a frequency of f in a solid, since the second harmonic amplitude is proportional to the nonlinearity, which has been recently referred to as "classical" nonlinearity [23].

On the other hand, an epoch-making discovery was made by Buck et al. [25] in the late 1970s. They found that a contact interface, such as one composed of crack faces, can behave like "two hands clapping" because of ultrasonic irradiation and thereby can generate second harmonic waves. This means that the crack can exhibit a local elastic nonlinearity through the nonlinear interaction with ultrasound. Since this discovery, nonlinear ultrasonics has been extensively studied for NDT. Thus far, various nonlinear phenomena including ones that are completely different from classical nonlinearity have been reported [20–23]. Hence, the nonlinearity arising from the contact vibrations of an interface has recently been referred to as "nonclassical" nonlinearity [26–30] or contact acoustic nonlinearity (CAN) [26, 27].

Research on nonclassical nonlinearity can be categorized into two methods, A and B, in terms of how the contact vibration of the interface is induced. Method A involves the use of a large-amplitude probe wave in a MHz-order frequency range. That is, the ultrasonic wave is used to induce the contact vibration and also to measure the effect. Method B involves the use of a large-displacement pump wave (or vibration) in a kHz-order (or lower) frequency while the effect of the contact vibration is measured by a small-amplitude probe wave. That is, a low-frequency pump wave (or vibration) and a high-frequency probe wave are used to induce the contact vibration and to measure the effect, respectively. Methods A and B are separately explained below in more detail.

Method A: Large-amplitude probe wave

As mentioned above, in the late 1970s, it was found by Buck et al. that a closed crack can exhibit strong elastic nonlinearity, resulting in the generation of second harmonics ($2f$) for a large-amplitude incidence with a frequency of f [25]. The generation of nonlinear components including second harmonics can be explained as follows. By irradiating a closed crack with large-amplitude ultrasound, one can induce the contact vibrations of crack faces, which results in waveform distortion. The waveform distortion in the temporal domain corresponds to the generation of higher harmonic components with frequencies (e.g., $2f$, $3f$, ...) that are integral multiples of the fundamental frequency, corresponding to the center frequency of the incident wave [25, 31–33]. The contact vibration has been theoretically explained under the assumption of phenomena called clapping, rubbing, popping, and so forth. [23]. Interestingly, the contact vibration can result in the generation of not only higher harmonics but also other nonlinear components. For instance, by increasing the incident wave amplitude, subharmonics with a frequency of $f/2$ can be generated in addition to higher harmonics. Higher harmonics can be generated by both classical and nonclassical nonlinearities, whereas subharmonics is generated only by nonclassical nonlinearity.

Note that subharmonics has an advantage of higher selectivity than higher harmonics because of their different generation mechanisms [34–38].

Method B: Small-amplitude probe wave with large-displacement pump wave (or vibration)

Method B is based on the combination of a probe wave and a pump wave (or vibration) [39–42]. This method takes advantage of both a high frequency of a probe wave and a low frequency of pump wave, namely, a high spatial resolution and a large displacement, respectively. Low-frequency pump excitation is used to induce the contact vibration of crack faces, which can result in changes in the transmission and reflection coefficients of a probe wave. Consequently, amplitude modulation occurs in the response of the probe wave. Here, the amplitude modulation in the time domain corresponds to the generation of side bands (i.e., sum and difference frequencies) in the frequency domain.

The aforementioned studies on nonlinear ultrasonics were based on waveform observation and frequency analysis. Given the practical applicability of nonlinear ultrasonics and the progress in imaging techniques (e.g., PA), their combination is very promising. In this context, a novel closed-crack imaging technique combining nonlinear ultrasonics with PA was proposed by Ohara et al. [43], and three more techniques have emerged recently [44–46]. These techniques can be referred to as nonlinear ultrasonic PA. The efficacy of each technique for closed-crack imaging has been demonstrated. On the other hand, each technique has its own merits and demerits. Although it is important to understand the fundamental principles and characteristics of these techniques for their use in research and practical applications, it may not be easy to disentangle all of them in terms of similarities, differences, advantages, drawbacks, ease of use, implementability, and so forth, because of their complexity.

In this chapter, we comprehensively review the nonlinear ultrasonic PA methods that have been reported so far. To this end, we first categorize them into four methods:

(I) Subharmonics (Sect. 5.2)
(II) Parallel and sequential transmission (Sect. 5.3)
(III) All-elements, odd-elements and even-elements transmission (Sect. 5.4)
(IV) Utilization of thermal stress (Sect. 5.5)

(I) is based on the measurement of a specific nonlinear component, i.e., subharmonics, generated using intense ultrasonic incidence [43]. Both (II) and (III) are based on the measurement of fundamental components to indirectly measure all nonlinear components, whereas they utilize different transmission modes and post-processing algorithms [44, 45]. (IV) is based on the combination of a pump wave (i.e., thermal stress induced by cooling sprays) and a probe wave (i.e., PA) [46]. In summary, (I)–(III) and (IV) belong to methods A and B, respectively.

This chapter starts with a brief summary of nonlinear ultrasonics, PA, the definition of closed cracks, and nonlinear ultrasonic PA (Sect. 5.1). In Sect. 5.2, nonlinear ultrasonic PA (I) will be described in the order of principles, experimental conditions, and imaging results. Sections 5.3 and 5.4 focus on nonlinear ultrasonic PAs (II) and (III), respectively, which utilize fundamental components based on different transmission modes and various post-processing algorithms. This is followed by the method of utilizing thermal stress (IV) in Sect. 5.5.

5.2 Subharmonics

Nonlinear ultrasonics generally measures a specific one of the nonlinear components generated by the contact vibration of crack faces induced by large-amplitude ultrasonic incidence. Among various nonlinear components, higher harmonics ($2f$, $3f$, …) for a large-amplitude incident wave with a frequency of f have been measured in many studies [20–23]. Higher harmonics, however, can be generated not only at a closed crack but also in liquid couplant, a transducer, electric circuits, and so forth. This will make it difficult to identify the generation source of higher harmonics and to realize the high-selectivity imaging of closed cracks. On the other hand, it has been reported that the subharmonics ($2f$, $3f$, …) for a large-amplitude incident wave (f) can be generated at a closed crack [34, 37, 38]. Note that the mechanism of subharmonic generation seems to be explainable by both contact vibration and resonance phenomena, while the mechanism of higher harmonic generation can only be explained by the contact vibration. Because of this difference, subharmonics cannot be generated at liquid couplant, transducers, or electric circuits; hence, in principle, the method of using subharmonics can achieve higher-selectivity imaging of closed cracks than that of using higher harmonics. On the basis of this knowledge, the first nonlinear ultrasonic PA was proposed in a method of using subharmonic waves. This method is called subharmonic phased array for crack evaluation (SPACE) [12, 43, 47–61]. SPACE is based on a large-amplitude short burst input for subharmonic generation and a PA algorithm with frequency filtering.

For the implementation of SPACE, the following points should be considered. First, it is pivotal in SPACE to establish a means of generating a large-amplitude short burst wave, because the incident wave amplitude required for subharmonic generation is generally higher than that for higher harmonic generation [62]. It is also important to ensure that the incident wave does not contain any subharmonic components (as expected in principle). Moreover, the length of a short burst should be appropriate, because a bandwidth of the short burst must be carefully treated in frequency filtering to extract fundamental (linear) and subharmonic (nonlinear) components, separately. In this section, we will start with the fundamental principle of SPACE and its imaging algorithm. Subsequently, experimental conditions will be described along with a way of examining the suitability of incident wave for SPACE. Last, a couple of imaging results obtained in closed-crack specimens will be exemplified.

5.2.1 Principles

A schematic illustration of SPACE is shown in Fig. 5.5. Figure 5.5a shows the first implementation of SPACE that uses a $LiNbO_3$ single-crystal (LN) transmitter and an array transducer as a receiver. To irradiate cracks with the intense ultrasound required to generate subharmonics at closed cracks, a 36° Y-cut LN transmitter with a wedge made of polyimide was fabricated. LN was selected as the piezoelectric material for the transmitter because it has a high dielectric breakdown voltage. Hence, the LN transmitter can withstand the high excitation voltage required to achieve a large-amplitude incidence. A sinusoidal wave with a frequency of f produced by a function generator is amplified using a gated amplifier. Subsequently, it is applied to the LN transmitter, and thereby, a large-amplitude short burst is irradiated on a crack. When the crack has open parts and a closed tip, linear and nonlinear scattering at fundamental (f) and subharmonic ($f/2$) frequencies will occur at the open parts and closed tip, respectively. The linearly and nonlinearly scattered waves are received by a piezoelectric array transducer, which should be selected to have a bandwidth that covers both the fundamental and subharmonic frequencies. Note here that the LN transmitter and array transducer should be synchronized when they are operated by different apparatus. Before performing delay-and-sum (DAS) processing [4–6, 9], which is a kind of PA imaging algorithms, the received waves are treated separately after dividing them into the fundamental and subharmonic components using band-pass filters (BPFs). Here, the bandwidth of the scattered waves is inversely proportional

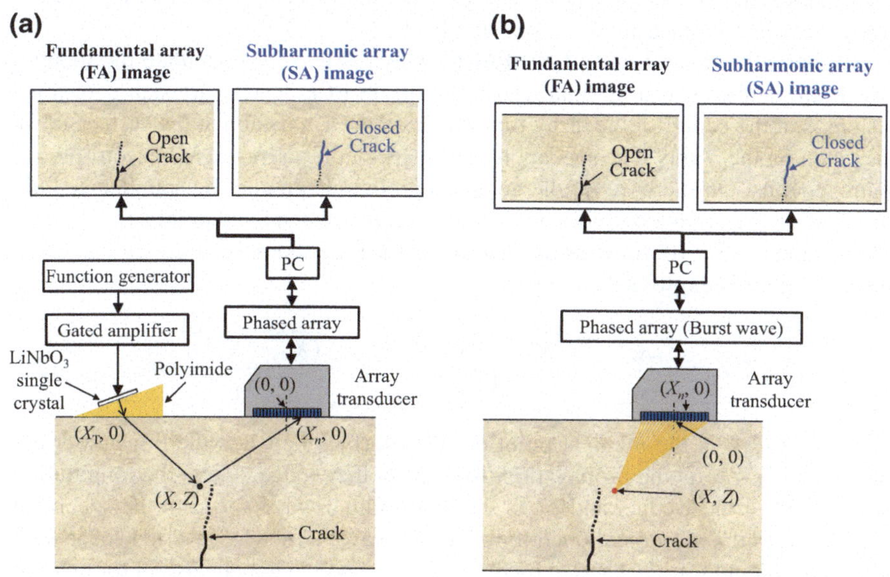

Fig. 5.5 Schematic illustration of SPACE. **a** SPACE with a monolithic transmitter and an array transducer (LN SPACE) and **b** SPACE with a single-array transducer (single-array SPACE)

to the number of cycles of the incident burst wave. Hence, the bandwidths of both the BPFs and the incident burst wave should be optimized. Regarding the selection of the BPFs, although analog filters have classically been used, digital filters are currently adopted in many cases because of their flexibility [9]. To implement the filters in PA hardware e.g. via a field programmable gate array (FPGA) for fast real-time imaging, a finite impulse response (FIR) or infinite impulse response (IIR) filter may be useful, since the BPFs can be implemented on an FPGA immediately after an analog-to-digital converter (ADC). On the other hand, a digital filter for post-processing after data transfer to a personal computer (PC) can also be used thanks to recent progress in the speed of data transfer between PA hardware and a host PC. Many types of filters, such as a fast Fourier transform (FFT) filter, a wavelet filter, an FIR filter, and an IIR filter, are available. Subsequently, DAS processing for a focal point on reception is applied to the received waves filtered around the fundamental and subharmonic frequencies, respectively. Details of the DAS processing will be described later with the formulation. By repeating the post-processing over an imaging area (i.e., the region of interest (ROI)), the fundamental array (FA) and subharmonic array (SA) images are produced. These images show the open and closed parts of cracks, respectively. This is the fundamental principle of SPACE (Fig. 5.5a) using a transmitter and an array receiver, which can be called LN SPACE [43, 47–51, 59–61]. As another option, SPACE (Fig. 5.5b) using a single array transducer for transmission and reception was proposed, which is called single-array SPACE [12, 52–55, 57, 58, 60]. Single-array SPACE is similar to LN SPACE but with the large-amplitude incidence achieved by an ultrasonic focusing effect on transmission under high-voltage burst excitation. The following are the imaging algorithms of both methods, which partly include nonlinear mode conversion [61].

A schematic illustration of LN SPACE, which is used to formulate the imaging algorithm, is shown in Fig. 5.5a. Here, the origin of the x–z Cartesian coordinate system is at the center of the array transducer on the x-axis and at the surface of the specimen on the z-axis. The specimen to be inspected is assumed to be isotropic and homogeneous. On the basis of the geometric relationship and the sound speed, the propagation time of a longitudinal (LL) wave scattered at a focal point $F(X, Z)$ to the nth element $E_n(X_n, 0)$ of the array transducer for longitudinal wave incidence from the transmitter is expressed as

$$t_{\mathrm{L}}(X, Z, n) = \frac{L_{\mathrm{W}}}{V_{\mathrm{W}}} + \frac{\sqrt{(X_{\mathrm{T}} - X)^2 + Z^2}}{V_{\mathrm{L}}} + \frac{\sqrt{(X_n - X)^2 + Z^2}}{V_{\mathrm{L}}}, \tag{5.1}$$

where $T(X_{\mathrm{T}}, 0)$ is the incident point on the interface between the transmitter and the specimen, L_{W} is the propagation distance in the wedge, V_{W} is the longitudinal wave speed in the wedge, and V_{L} is the longitudinal wave speed in the specimen. In addition to the aforementioned longitudinal wave, a mode-converted shear wave will be simultaneously generated by the scattering, which may be useful for imaging [60, 61]. Hence, the imaging algorithm for the mode-converted shear wave is also formulated. The propagation time of a mode-converted shear (LS) wave scattered at

$F(X, Z)$ to $E_n(X_n, 0)$ for longitudinal wave incidence from the transmitter is expressed as

$$t_{\mathrm{S}}(X, Z, n) = \frac{L_{\mathrm{W}}}{V_{\mathrm{W}}} + \frac{\sqrt{(X_{\mathrm{T}} - X)^2 + Z^2}}{V_{\mathrm{L}}} + \frac{\sqrt{(X_n - X)^2 + Z^2}}{V_{\mathrm{S}}}, \tag{5.2}$$

where V_{S} is the shear wave speed in the specimen. The propagation time differences $\Delta t_{\mathrm{L}}(X, Z, n)$ and $\Delta t_{\mathrm{S}}(X, Z, n)$ between the first and nth elements for LL and LS scattered waves are given by

$$\Delta t_{\mathrm{L}}(X, Z, n) = t_{\mathrm{L}}(X, Z, n) - t_{\mathrm{L}}(X, Z, 1), \tag{5.3}$$

$$\Delta t_{\mathrm{S}}(X, Z, n) = t_{\mathrm{S}}(X, Z, n) - t_{\mathrm{S}}(X, Z, 1), \tag{5.4}$$

which correspond to delay laws. Assuming that the signals received at the nth element and then filtered around the fundamental and subharmonic frequencies are $u_{n,f}(t)$ and $u_{n,f/2}(t)$, respectively, the fundamental and subharmonic signals obtained by DAS processing using Eqs. (5.1) and (5.3) for the LL mode can be expressed by

$$U_{f,\mathrm{L}}(X, Z, t) = \frac{1}{N} \sum_{n=1}^{N} u_{n,f}(t - \Delta t_{\mathrm{L}}(X, Z, n)), \tag{5.5}$$

$$U_{f/2,\mathrm{L}}(X, Z, t) = \frac{1}{N} \sum_{n=1}^{N} u_{n,f/2}(t - \Delta t_{\mathrm{L}}(X, Z, n)), \tag{5.6}$$

where N is the total number of elements of the array transducer. The fundamental and subharmonic scattering intensities can be calculated as the root-mean-square (RMS) values of Eqs. (5.5) and (5.6), respectively, as

$$I_{f,\mathrm{L}}(X, Z) = \left(\frac{1}{\Delta\tau} \int_{t_{\mathrm{L}}(X,Z,1)}^{t_{\mathrm{L}}(X,Z,1)+\Delta\tau} U_{f,\mathrm{L}}^2(X, Z, t)dt \right)^{1/2}, \tag{5.7}$$

$$I_{f/2,\mathrm{L}}(X, Z) = \left(\frac{1}{\Delta\tau} \int_{t_{\mathrm{L}}(X,Z,1)}^{t_{\mathrm{L}}(X,Z,1)+\Delta\tau} U_{f/2,\mathrm{L}}^2(X, Z, t)dt \right)^{1/2}, \tag{5.8}$$

where $\Delta\tau$ is the time window for the RMS calculation.

Likewise, the fundamental and subharmonic signals obtained by DAS processing using Eqs. (5.2) and (5.4) for the LS mode can be expressed by

$$U_{f,\mathrm{S}}(X, Z, t) = \frac{1}{N}\sum_{n=1}^{N} u_{n,f}(t - \Delta t_{\mathrm{S}}(X, Z, n)), \tag{5.9}$$

$$U_{f/2,\mathrm{S}}(X, Z, t) = \frac{1}{N}\sum_{n=1}^{N} u_{n,f/2}(t - \Delta t_{\mathrm{S}}(X, Z, n)). \tag{5.10}$$

The fundamental and subharmonic scattering intensities are calculated as the RMS values of Eqs. (5.9) and (5.10), respectively, as

$$I_{f,\mathrm{S}}(X, Z) = \left(\frac{1}{\Delta\tau}\int_{t_{\mathrm{S}}(X,Z,1)}^{t_{\mathrm{S}}(X,Z,1)+\Delta\tau} U_{f,\mathrm{S}}^2(X, Z, t)dt\right)^{1/2}, \tag{5.11}$$

$$I_{f/2,\mathrm{S}}(X, Z) = \left(\frac{1}{\Delta\tau}\int_{t_{\mathrm{S}}(X,Z,1)}^{t_{\mathrm{S}}(X,Z,1)+\Delta\tau} U_{f/2,\mathrm{S}}^2(X, Z, t)dt\right)^{1/2}. \tag{5.12}$$

By plotting $I_{f,\mathrm{L}}(X, Z)$, $I_{f/2,\mathrm{L}}(X, Z)$, $I_{f,\mathrm{S}}(X, Z)$, and $I_{f/2,\mathrm{S}}(X, Z)$ at each pixel over an imaging area (i.e., ROI), the FA and SA images are produced on the basis of LL and LS scattered waves in LN SPACE.

In almost the same manner, the imaging algorithm of single-array SPACE (Fig. 5.5b) is formulated as follows. The propagation time from $E_n(X_n, 0)$ to $F(X, Z)$ is given by

$$t_{\mathrm{T}}(X, Z, n) = \frac{\sqrt{(X - X_n)^2 + Z^2}}{V_{\mathrm{L}}}. \tag{5.13}$$

The propagation time difference $\Delta t_{\mathrm{T}}(X, Z, n)$ between the first and nth elements is expressed as

$$\Delta t_{\mathrm{T}}(X, Z, n) = t_{\mathrm{T}}(X, Z, n) - t_{\mathrm{T}}(X, Z, 1), \tag{5.14}$$

which corresponds to the delay laws for focusing on transmission. This is used with a high-voltage burst excitation to focus ultrasound at $F(X, Z)$ to achieve a large-amplitude incidence.

The propagation time of the LL scattered wave from the center of the array transducer (i.e., (0, 0)) through $F(X, Z)$ to $E_n(X_n, 0)$ is given by

$$t_{\mathrm{R}}(X, Z, n) = \frac{\sqrt{X^2 + Z^2} + \sqrt{(X - X_n)^2 + Z^2}}{V_{\mathrm{L}}}. \tag{5.15}$$

Hence, the propagation time difference between the first and nth elements is expressed as

$$\Delta t_{\mathrm{R}}(X, Z, n) = t_{\mathrm{R}}(X, Z, n) - t_{\mathrm{R}}(X, Z, 1), \tag{5.16}$$

which corresponds to the delay laws for focusing on reception. The fundamental and subharmonic signals obtained by DAS processing using Eqs. (5.15) and (5.16) for the LL mode can be expressed by

$$U_{f,\mathrm{L}}(X, Z, t) = \frac{1}{N}\sum_{n=1}^{N} u_{n,f}(t - \Delta t_{\mathrm{R}}(X, Z, n)), \tag{5.17}$$

$$U_{f/2,\mathrm{L}}(X, Z, t) = \frac{1}{N}\sum_{n=1}^{N} u_{n,f/2}(t - \Delta t_{\mathrm{R}}(X, Z, n)). \tag{5.18}$$

The fundamental and subharmonic scattering intensities are calculated as the RMS values of Eqs. (5.17) and (5.18), respectively, as

$$I_{f,\mathrm{L}}(X, Z) = \left(\frac{1}{\Delta\tau}\int_{t_{\mathrm{R}}(X,Z,1)}^{t_{\mathrm{R}}(X,Z,1)+\Delta\tau} U_f^2(X, Z, t)dt\right)^{1/2}, \tag{5.19}$$

$$I_{f/2,\mathrm{L}}(X, Z) = \left(\frac{1}{\Delta\tau}\int_{t_{\mathrm{R}}(X,Z,1)}^{t_{\mathrm{R}}(X,Z,1)+\Delta\tau} U_{f/2}^2(X, Z, t)dt\right)^{1/2}. \tag{5.20}$$

By plotting $I_f(X, Z)$ and $I_{f/2}(X, Z)$ at each pixel over an imaging area, the FA and SA images are produced in single-array SPACE.

5.2.2 Experimental Conditions

In this subsection, four experimental conditions are summarized to exemplify imaging results obtained by LN SPACE and single-array SPACE to be shown in Sect. 5.2.3. First, to evaluate the fundamental performance of SPACE, two well-defined fatigue crack specimens made of A7075 aluminum alloy ($V_{\mathrm{L}} = 6230$ m/s) were formed by performing a three-point bending fatigue test (Fig. 5.6). To introduce open and closed fatigue cracks, the fatigue conditions were selected to be $K_{\max} = 5.3$ MPa m$^{1/2}$ and $K_{\min} = 0.6$ MPa m$^{1/2}$, and $K_{\max} = 4.3$ MPa m$^{1/2}$ and $K_{\min} = 0.6$ MPa m$^{1/2}$, respectively. Each crack was extended from the top of a starting notch to a depth of approximately 20 mm. An LN transmitter with a wedge made of polyimide was fabricated, where the center frequency of the LN was 7 MHz. The excitation voltage used was a three-cycle tone burst with a frequency of 7 MHz, which was amplified by a gated amplifier (GA5000) produced by Ritec (Warwick, USA). As the receiver, a 32-element array transducer (5 MHz, 0.5 mm pitch) was used to receive both the

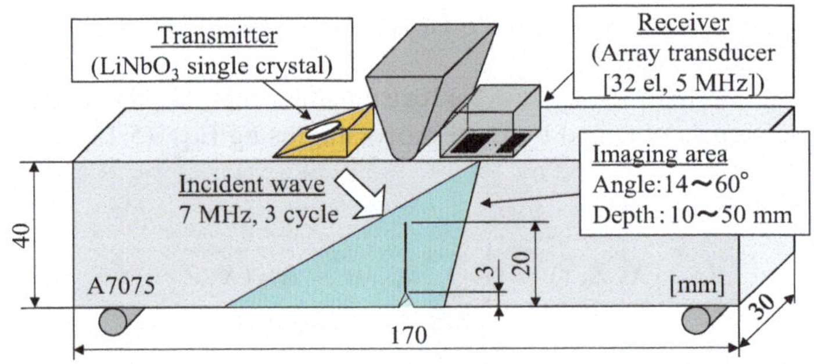

Fig. 5.6 Experimental configuration of LN SPACE for imaging open and closed fatigue-crack specimens (A7075)

fundamental (7 MHz) and subharmonic (3.5 MHz) components, and was operated by PA hardware (PAL) manufactured by KJTD (Tokyo, Japan). The sampling rate was 50 MS/s. The DAS processing was carried out with 1 mm steps in depth and 1° steps in angle. IIR BPFs with bandwidths of 6–8 and 2.5–4.5 MHz were implemented on an FPGA to filter the received signals around the fundamental and subharmonic frequencies, respectively.

For this experiment, a means of examining the suitability of the incident wave for SPACE is described here as an example. It is critically important for SPACE to use an appropriate incident wave that has a large amplitude, an appropriate bandwidth (i.e., burst length), and single-frequency excitation (without any nonlinear components). To confirm whether an incident wave meets the requirements, the incident wave should be examined in the specimen after passing through the liquid couplant. Specifically, it is best to examine the incident wave just before it arrives at the crack. It is, however, impossible to measure the incident wave at the crack position in the fatigue-crack specimens. To realize such measurement, a specimen made of the same material (A7075) was machined to give it an oblique surface, as shown in Fig. 5.7a, so that the longitudinal incident wave could be measured at the crack position using a laser Doppler vibrometer, which can measure an out-of-plane displacement on a free surface with a very broad and flat bandwidth. The measured displacement waveform is shown in Fig. 5.7b. The peak-to-peak displacement amplitude was achieved to be more than 20 nm. The wavelet analysis of the displacement waveform clearly shows that the incident wave does not contain any nonlinear components including subharmonics as theoretically expected. The bandwidth can be seen to be sufficiently narrow to separate the received waves into the fundamental (7 MHz) and subharmonic (3.5 MHz) components in Fig. 5.7c while the short burst length required to obtain a high spatial resolution in the imaging is also retained in Fig. 5.7b. This also suggests an optimum bandwidth for frequency filtering. In this case, IIR BPFs with bandwidths of 6–8 and 2.5–4.5 MHz were chosen for the fundamental and subharmonic components, respectively, on the basis of the above preliminary experimental results (Fig. 5.7). Thus, it can be confirmed whether an incident wave is appropriate

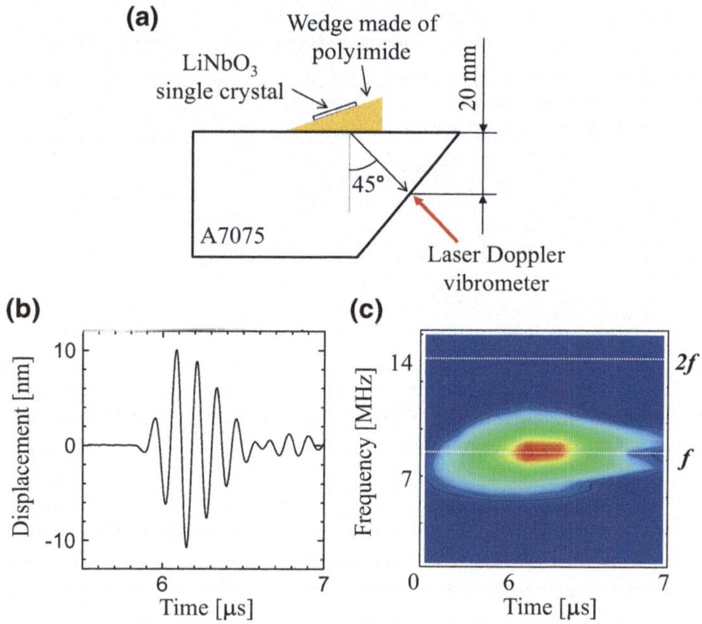

Fig. 5.7 Measurement of an incident wave at the position of the crack in Fig. 5.6 using a laser Doppler vibrometer. **a** Experimental configuration, **b** displacement waveform, and **c** wavelet analysis of **b**

for SPACE. In order to obtain meaningful results, such confirmation should be carried out not only for SPACE but also for all experiments using nonlinear ultrasonics.

After the fundamental experiments in the open and closed fatigue cracks, the crack-closure dependence of SPACE images was investigated in a fatigue-crack specimen made of SUS316L austenitic stainless steel ($V_L = 5940$ m/s) with a thickness of 40 mm, which is employed for the recirculation pipes of nuclear power plants. Under the fatigue conditions of $K_{max} = 18.6$ MPa m$^{1/2}$ and $K_{min} = 0.6$ MPa m$^{1/2}$, a crack was extended from a starting notch to a depth of approximately 8 mm on the side surface. To image closed cracks with different closure stresses in a single specimen, the measurement was carried out while applying a static load, which induces a tensile bending stress that acts on the crack to relieve an original closure stress, as illustrated in Fig. 5.8. Here, the nominal bending stress was estimated from the applied static load, elastic properties, density, and geometry under the condition of a crack-free beam. The nominal bending stress was adopted as a measure of the stress relieved from the closure stress, because the rigorous calculation of the closure stress is difficult because of the complexity of the stress field around the crack. The measurement was conducted at nominal bending stresses of 19, 84, and 112 MPa, and the other experimental conditions, such as the transmitter, array transducer, excitation voltage, and imaging area, selected for this experiment were the same as those for the previous experiment (Fig. 5.6).

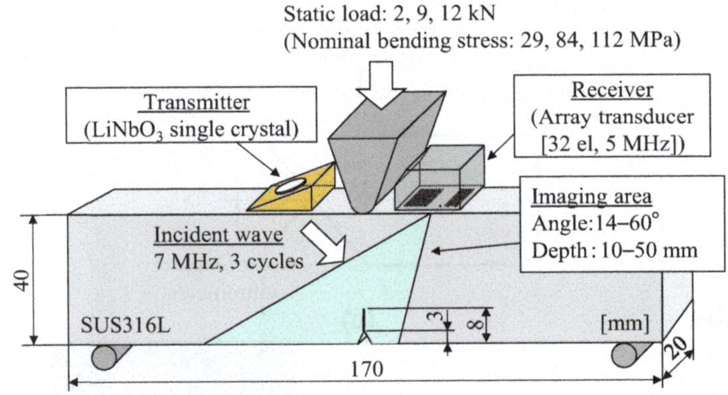

Fig. 5.8 Experimental configuration of LN SPACE for imaging SUS316L fatigue-crack specimen while simulating various crack closure stresses

The next experiment was carried out on a more realistic specimen to simulate the inspection in actual nuclear reactors. Here, single-array SPACE was applied to a specimen with a stress corrosion cracking (SCC). The SCC was introduced in the heat affected zone (HAZ) of a sensitized SUS304 stainless steel ($V_L = 5900$ m/s) pipe under a residual stress in high-temperature pressurized water at 280 °C in an autoclave. After the SCC was formed over more than a year, the specimen was cut and machined to a rectangular parallelepiped for UT. Note that this specimen has not only welds but also a coarse-grained base metal, which will lead to strong linear scattering at the coarse grain boundaries. In other words, this is a very challenging specimen in terms of crack measurement. The experimental configuration is shown in Fig. 5.9. A 32-element array transducer (5 MHz, 0.5 mm pitch) was employed for transmission and reception and was operated by PA hardware (MultiX LF) manufactured by M2M (Les Ulis, France). The excitation voltage was a three-cycle tone burst with a frequency of 7 MHz. The excitation frequency was selected so that both

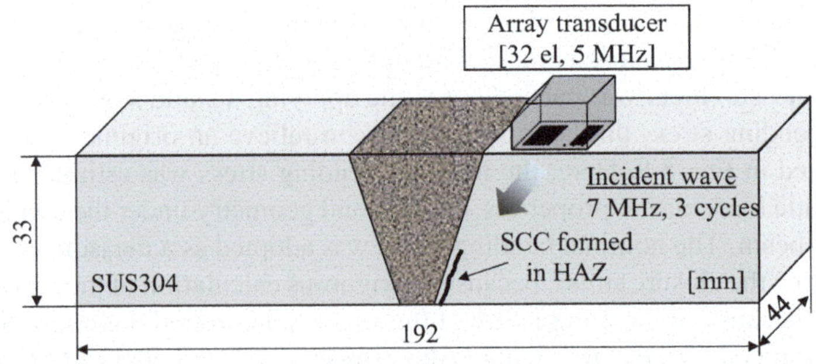

Fig. 5.9 Experimental configuration of single-array SPACE for imaging stress corrosion cracking (SCC) formed in the heat affected zone (HAZ) of SUS304 austenitic stainless steel specimen in high-temperature pressurized water at 280 °C

fundamental (7 MHz) and subharmonic (3.5 MHz) components could be received by the array transducer. FFT BPFs with bandwidths of 6–8 and 2.5–4.5 MHz were used to separate the received signals into the fundamental and subharmonic components, respectively. An appropriate position for the array transducer was searched for by manual mechanical scanning.

Finally, LN SPACE was used again to explore the possibility of multi-mode SPACE imaging. Here, a complexly branched SCC was prepared by an acceleration formation method [46, 47, 56]. It is not easy to form complexly branched SCCs with a depth of more than 10 nm in the laboratory owing to their low growth rates. To solve this problem, a fatigue precrack was first extended by performing a three-point bending fatigue test from a starting notch to a depth of approximately 10 mm in SUS304 austenitic stainless steel ($V_L = 5800$ m/s), which had been sensitized at 600 °C for 4 h. Here, the fatigue conditions were $K_{max} = 28$ MPa m$^{1/2}$ and $K_{min} = 0.6$ MPa m$^{1/2}$. Subsequently, the specimen was immersed in a cell with a corrosive environment, and then, an SCC was introduced from the tip of the fatigue precrack under a static bending stress. The corrosive environment was a solution of 30 wt% $MgCl_2$ at 90 °C, and a nominal bending stress of 124 MPa was applied to the crack for 1300 h (i.e., 54 days) to form a very complexly branched SCC. The experimental configuration is shown in Fig. 5.10. The LN transmitter was excited by a three-cycle burst with a frequency of 7 MHz at 220 Vp-p. As the receiver, a 64-element array transducer (5 MHz, 0.5 mm pitch) was used to receive both forward-

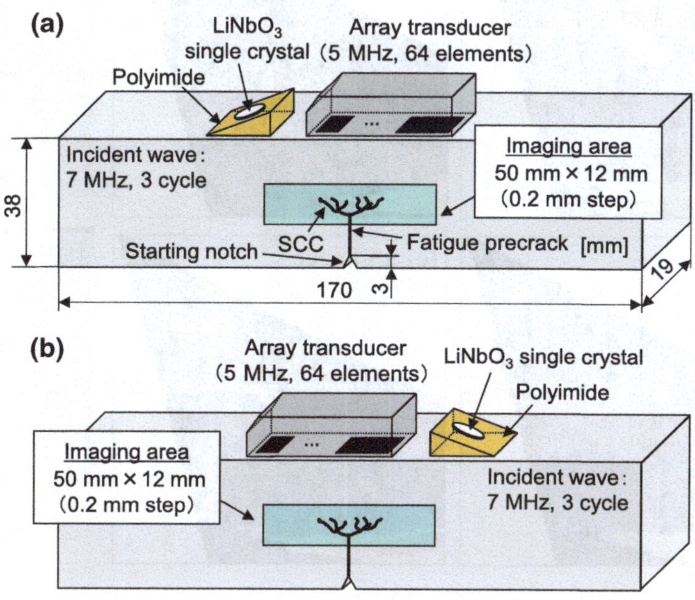

Fig. 5.10 Two experimental configurations of multi-mode SPACE used to image a branched SCC. **a** Longitudinal wave incidence from the left of the SCC. **b** Longitudinal wave incidence from the right of the SCC. Modified from [61] (Copyright (2019). The Japan Society of Applied Physics)

and backscattered waves from the SCC and was operated by PA hardware (MultiX LF) manufactured by M2M (Les Ulis, France). The imaging area around the SCC was 50 mm × 12 mm (0.2 mm steps). FFT BPF with bandwidth of 2.5–4.5 MHz was used to extract the subharmonic component (3.5 MHz) from the received waves. Furthermore, the experiment was performed in two configurations to simulate actual inspections that use the ultrasonic incidence from the left and right of the crack, as shown in Fig. 5.10a, b, respectively.

5.2.3 Imaging Results

To evaluate its fundamental performance, LN SPACE was applied to two types of A7075 fatigue-crack specimens: the open and closed fatigue cracks. The experiments were carried out in the experimental configuration shown in Fig. 5.6 using the incident wave whose suitability was verified in Fig. 5.7. The imaging results are shown in Fig. 5.11. In the open-crack specimen, the crack tip was clearly visible in the FA image

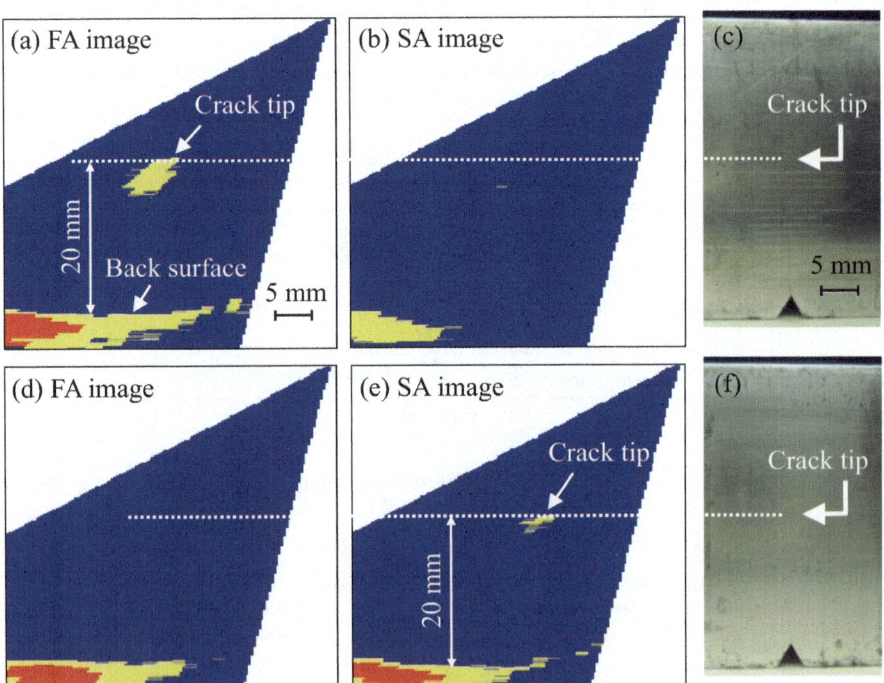

Fig. 5.11 Comparison of crack tip images obtained by SPACE for A7075 open- and closed-crack specimens. **a** FA image, **b** SA image, and **c** photograph of open-crack specimen. **d** FA image, **e** SA image, and **f** photograph of closed-crack specimen. Modified from [43] (Copyright © 2007 American Institute of Physics)

(Fig. 5.11a), whereas the SA image did not show any crack responses (Fig. 5.11b). It was also found that the location of the crack tip in the obtained image agreed with that of the actual crack tip on the side surface shown in the photograph (Fig. 5.11c). This shows that an open crack can be visualized in an FA image, corresponding to a linear PA image. In the closed-crack specimen, the crack tip was invisible in the FA image in Fig. 5.11d, in contrast to that in Fig. 5.11a. On the other hand, the crack tip was clearly observed in the SA image (Fig. 5.11e). The depth measured in the SA image was in good agreement with the actual crack depth (Fig. 5.11f). Thus, it was demonstrated that SPACE is useful for visualizing closed cracks that cannot be visualized by conventional (i.e., linear) UT.

After evaluating the fundamental performance of SPACE for open and closed fatigue cracks, the crack-closure dependence of SPACE images was investigated for the fatigue-crack specimen made of SUS316L austenitic stainless steel, which is employed in nuclear reactors. In the experimental configuration (Fig. 5.8), the crack closure stress was controlled by applying different nominal bending stresses. FA and SA images of the fatigue crack obtained while the nominal bending stresses of 19, 84, and 112 MPa were applied are shown in Fig. 5.12.

Under the bending stress of 19 MPa, the FA image (Fig. 5.12a) showed only the notch part C (Fig. 5.12c). This suggests that the boundary C between the notch and the closed crack had a distinct discontinuity and thus C became a linear scattering source. On the other hand, the SA image (Fig. 5.12b) showed not only C but also the crack tip A and the middle part B. This shows that the SPACE implemented here can image various parts of a crack, including a closed crack tip.

When the bending stress was increased to 84 MPa, the FA image (Fig. 5.12d) showed the middle part B. This shows that B was a distinct boundary between the open and closed regions and thus became a linear scattering source. Interestingly, the SA image (Fig. 5.12e) showed both A and B. The intensity at B was lower than that in Fig. 5.12b. This shows that the SPACE used here can image a crack tip, irrespective of the closure stress.

Upon increasing the bending stress further to 112 MPa, the crack became invisible in the FA image (Fig. 5.12g). This was surprising, because the indication of B observed in Fig. 5.12d was lost, even though the crack was more open. The following assumption may explain this phenomenon: the boundary between the open and closed regions at B became ambiguous with increasing bending stress; and therefore, the linear scattering source diminished. In contrast, the SA image (Fig. 5.12h) showed A. This reveals that A was still closed, which was favorable for subharmonic generation.

The most striking one among the aforementioned findings is that the crack had a very high closure stress that was only partly relieved by applying the nominal maximum bending stress of more than 100 MPa. Nevertheless, the change in the crack state was successfully visualized at various closure stresses. Moreover, the SA images always gave an accurate crack depth, in contrast to the FA images, from which the crack depth was underestimated. Thus, it was demonstrated that the LN SPACE implemented in this experiment is highly promising as an inspection technique to

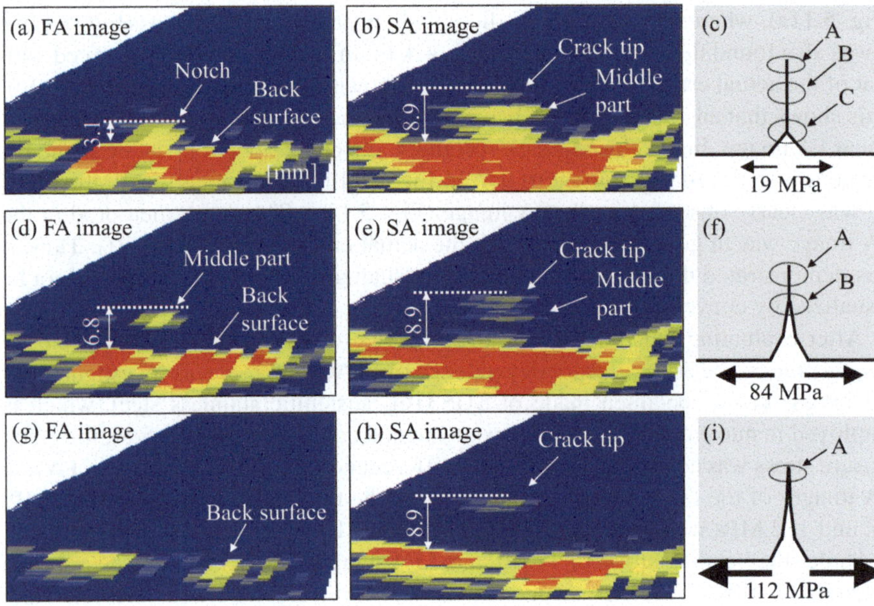

Fig. 5.12 Dependence of crack visualization on static bending stress applied to SUS316L fatigue-crack specimen. **a** FA and **b** SA images for nominal bending stress of 19 MPa. **c** Schematic illustration used for interpretation of (**a**) and (**b**). **d** FA and **e** SA images for nominal bending stress of 84 MPa. **f** Schematic illustration used for interpretation of (**d**) and (**e**). **g** FA and **h** SA images for nominal bending stress of 112 MPa. (**i**) Schematic illustration used for interpretation of (**g**) and (**h**). In **c**, **f**, and **i**, **A**–**C** denote the crack tip, middle part, and notch part, respectively, which are fundamental and/or subharmonic scattering sources. Modified from [43] (Copyright © 2007 American Institute of Physics)

satisfy the reliability requirements of materials and structural components in nuclear power plants.

The next experiment was carried out on a more realistic SCC specimen [52], since an SCC is one of the critical defects in power plants. In the experimental configurations (Figs. 5.9, 5.13a, and 5.13b), the SCC in the HAZ was visualized by single-array SPACE. In the FA image (Fig. 5.13c), the SCC was not visualized because of the numerous responses appearing over the imaging area. These responses are attributed to the strong linear scattering at the coarse grain boundaries in the base metal. On the other hand, in the SA image (Fig. 5.13d), the SCC was selectively visualized as two bright points, which are surrounded by white dashed ellipses. This suggests that the SCC was closed, presumably because of the generation of oxide films between the crack faces through the corrosion induced by the high-temperature pressurized water [12].

In this experiment, further frequency analyses were performed to demonstrate the subharmonic generation at the SCC in another way; the DAS waveforms at points A and B, which are positions with strong responses in the FA and SA images,

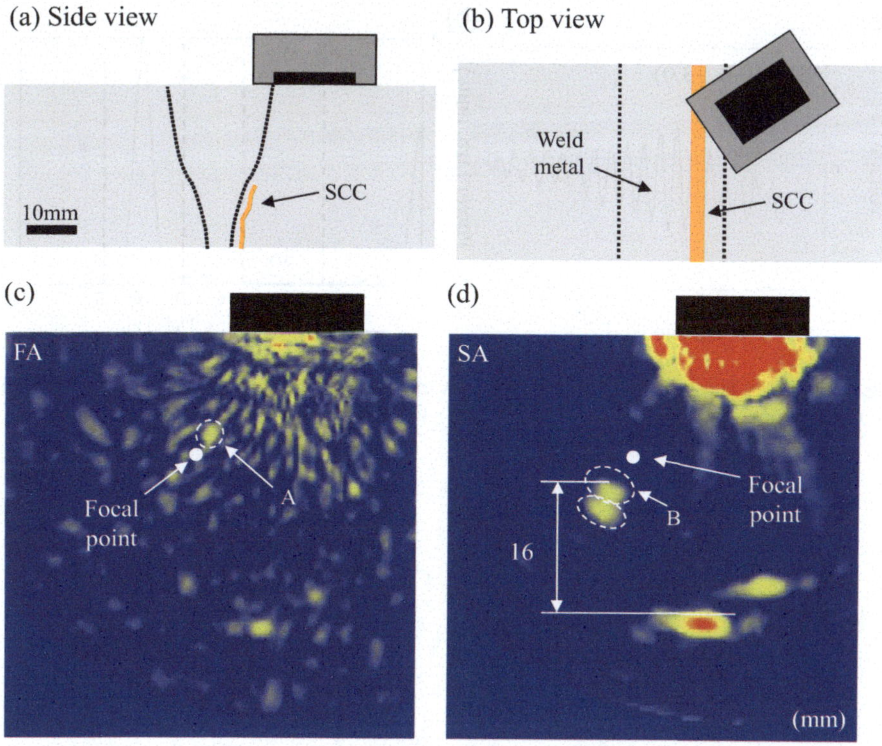

Fig. 5.13 Schematic illustrations showing array positions on SCC specimen and FA and SA images. **a** Side and **b** top views showing array positions. Here, the array transducer was mechanically scanned to optimize its position. **c** FA and **d** SA images of SCC obtained by single-array SPACE in the array position shown in (**a**) and (**b**).

respectively, were analyzed by an FFT. The DAS waveform and its power spectrum at A, which is a position with strong linear scattering (Fig. 5.13c), are shown in Fig. 5.14. No waveform distortion was observed in the DAS waveform (Fig. 5.14a) in the time range of 5.5–6 μs, which was used for point A in the imaging. In the power spectrum (Fig. 5.14b) of the DAS waveform in the time range of 5.5–6 μs, a clear peak was only observed around the fundamental frequency (7 MHz). This is reasonable because the response at A was the linear response from a coarse grain boundary. This also implies that the incident wave used in this experiment did not contain any nonlinear components. Next, the DAS waveform and its power spectrum at B, which is a position with strong subharmonic scattering (Fig. 5.13d), are shown in Fig. 5.15. In contrast to Fig. 5.14a, distortion was clearly observed in the DAS waveform (Fig. 5.15a) in the time range of 8.5–9 μs, which was used for point B in the imaging. In the power spectrum (Fig. 5.15b) of the DAS waveform in the time range of 8.5–9 μs, a clear peak appeared around the subharmonic frequency (3.5 MHz) in addition to multiple peaks around the fundamental frequency (7 MHz), which can be understood by assuming that the scattered waves from multiple coarse grains

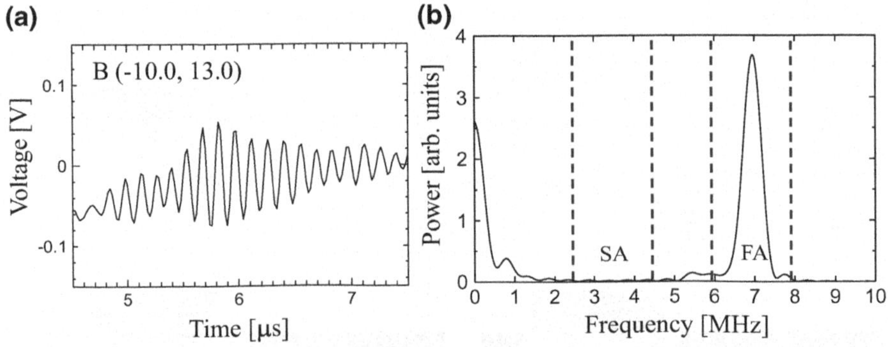

Fig. 5.14 Delay-and-sum (DAS) waveform in A of Fig. 5.13 and its power spectrum. Modified from [52] (Copyright (2011) The Japanese Society for Non-Destructive Inspection)

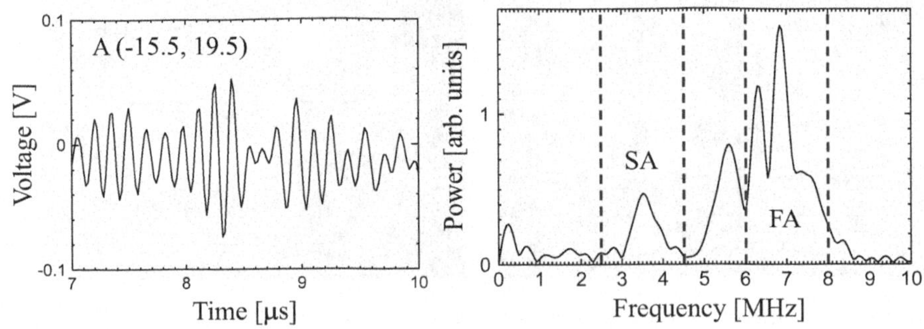

Fig. 5.15 Delay-and-sum (DAS) waveform at B in Fig. 5.13 and its power spectrum. Modified from [52] (Copyright (2011) The Japanese Society for Non-Destructive Inspection)

interfered with each other in a complex manner. Importantly, the linear scattering occurred throughout the imaging area, whereas the subharmonic generation was localized at the closed SCC. Thus, the above results show that single-array SPACE is useful for the high-selectivity imaging of closed cracks even in coarse-grained materials.

As the final example, LN SPACE was employed again to explore the efficacy of multi-mode SPACE [61] in the complexly branched SCC, which is commonly generated in aging power plants. Although the above experiments involved the measurement of longitudinal subharmonic waves, the nonlinear scattering is known to generate mode-converted shear subharmonic waves as well as longitudinal subharmonic waves. Multi-mode SPACE utilizes both longitudinal (LL) and mode-converted shear (LS) subharmonic waves for longitudinal wave incidence, providing SA images with the LL and LS modes. In the experimental configuration (Fig. 5.10a), SA images with the LL and LS modes were obtained as respectively shown in Fig. 5.16a, b. These images were produced using Eqs. (5.1)–(5.12). In the SA image with the LL mode (Fig. 5.16a), two responses were visualized. The appearance of multiple responses

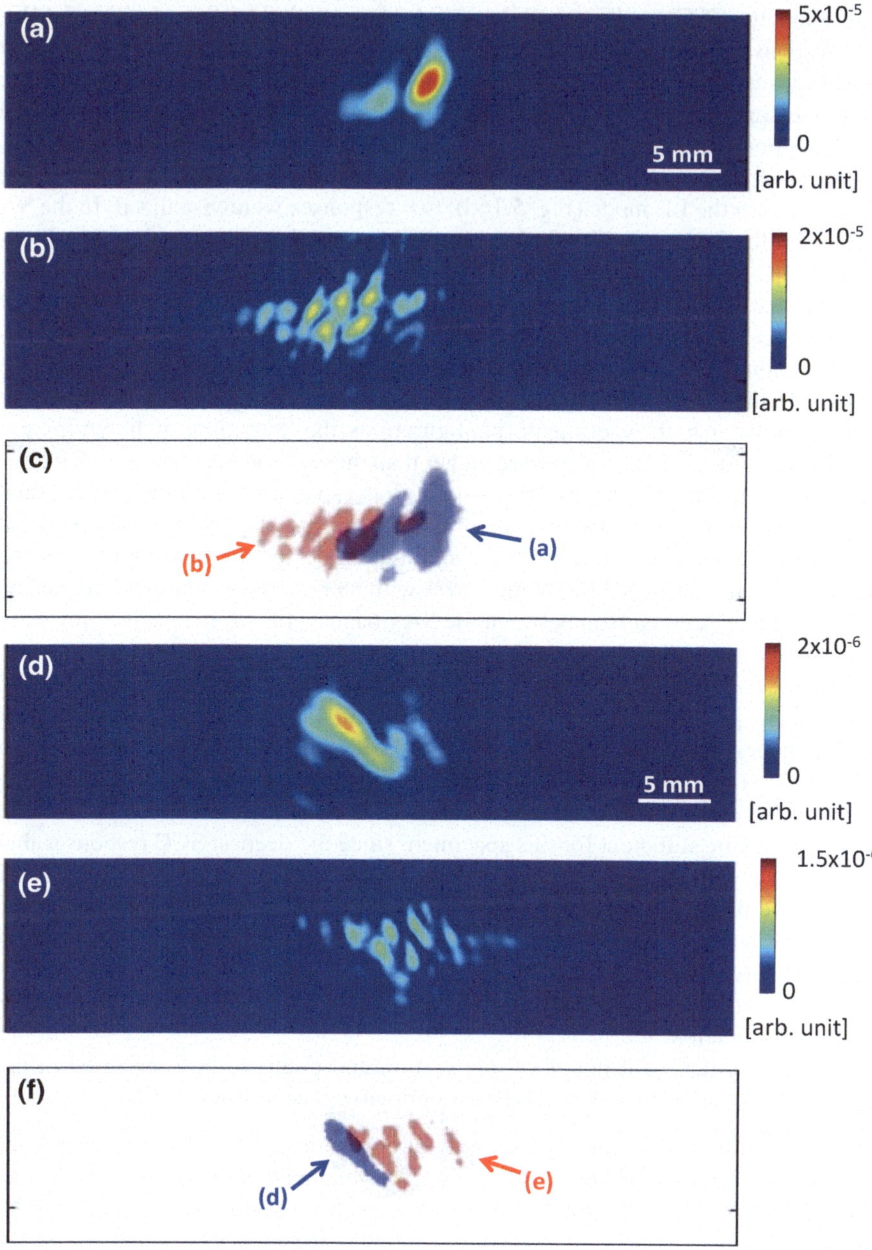

Fig. 5.16 Longitudinal (LL) and mode-converted shear (LS) SA images of complexly branched SCC. **a** LL and **b** LS SA images obtained in the experimental configuration of Fig. 5.10a. **c** Superimposition of (**a**) and (**b**) after their color bars were redefined to be blue and red at the maximum, respectively, and white at the minimum. **d** LL and **e** LS SA images obtained in the experimental configuration of Fig. 5.10b. **f** Superimposition of (**d**) and (**e**). The color bars of (**d**) and (**e**) were redefined to be blue and red at the maximum, respectively, and white at the minimum [61] (Copyright (2019) The Japan Society of Applied Physics)

implies that the SCC was branched. In the SA image with the LS mode (Fig. 5.16b), more responses than those in Fig. 5.16a were clearly visualized. This suggests that the SCC branched over a wider area than that indicated by the SA image with the LL mode. To examine the effect of the incident direction, the experiment was performed with the opposite arrangement of the LN transmitter and array transducer (Fig. 5.10b). Figure 5.16d, e show the SA images with the LL and LS modes, respectively. In the SA image with the LL mode (Fig. 5.16d), two responses were visualized. In the SA image with the LS mode (Fig. 5.16e), more responses than those in the LL subharmonic image were clearly visualized. These tendencies were similar, irrespective of the incidence direction, as shown in Fig. 5.16. On the other hand, the positions of the responses appearing in each image were almost opposite for the opposite incident directions. This suggests that the responses of a complexly branched SCC are sensitive to the incident angle.

Importantly, in both experimental configurations, the resolutions of the SA images with the LS mode (Fig. 5.16b, e) were higher than those of the SA images with the LL mode (Fig. 5.16a, d). This is reasonable because the spatial resolution is proportional to the wavelength, and the wavelength of a shear wave is half of that of a longitudinal wave with the same frequency. This can compensate for the deterioration in the spatial resolution due to subharmonic waves with half of the fundamental frequency.

For a detailed comparison between the SA images with the LL and LS modes in each configuration, the superimpositions of Fig. 5.16a, b, d, e are shown in Fig. 5.16c, f, where the color bars were converted to blue and red at the maximum for the LL and LS modes, respectively, and to white at the minimum. Figure 5.16c clearly shows that the SA images with the LL and LS modes indicated different parts of the branched SCC. Surprisingly, Fig. 5.16f shows the same feature, although the positions of the responses are symmetric. For the crack depth measurement, the SA image with the LL mode may be sufficient for this specimen, since the deepest SCC responses that appeared were at almost the same position. On the other hand, it is important that many of the branched parts that were not visualized in the SA image with the LL mode were successfully visualized in the SA image with the LS mode for both experimental configurations. This results suggest that both images can complement each other and that the use of both modes will make LN SPACE more robust for practical application.

Finally, the important points of this section and points to be considered in the application of the discussed methods are summarized as follows:

- Higher harmonics is generated not only at closed cracks but also in a transducer and liquid couplant, whereas subharmonics is generated only at closed cracks. By using subharmonics, the nonlinearity of a transducer and liquid couplant can be eliminated, in contrast to the case of using higher harmonics.
- Large-amplitude incidence is required to generate subharmonics at closed cracks through the contact vibration of crack faces, since the threshold for subharmonic generation is higher than that for higher harmonic generation.
- In the case of using a monolithic transmitter with a different array receiver, a transmitter with a high dielectric breakdown voltage and a high-power amplifier

may be required to realize the large-amplitude incidence above a threshold of subharmonic generation. The two devices for transmission and reception must be accurately synchronized.

- An array transducer with a broad reception bandwidth is required to receive both the fundamental and subharmonic components since they are used to produce both linear and nonlinear images.
- Comparison between FA (linear) and SA (nonlinear) images is very important to clarify the necessity of nonlinear ultrasonic PA and to gain physical insight.
- Single-array SPACE can be applied to a surface smaller than that for LN SPACE, whereas the maximum incident wave amplitude may be limited in single-array SPACE.
- An LN transmitter can be replaced with a higher-output transmitter [56, 63]. This could extend the applicable fields.
- An incident wave should be checked in advance to verify that it has a large amplitude, an appropriate bandwidth (i.e., burst length) for frequency filtering, and single-frequency excitation (without any nonlinear components before it arrives at cracks). To this end, a laser Doppler vibrometer that can measure a displacement with a very broad and flat bandwidth should be used. The results of this measurement should be taken into account when selecting an appropriate bandwidth for BPFs and/or an appropriate excitation voltage.

5.3 Parallel and Sequential Transmission

5.3.1 Principles

Most techniques in nonlinear ultrasonics are based on the measurement of a specific nonlinear component generated by the contact vibration of crack faces. The nonlinear component will be one of various components (e.g., higher harmonics, subharmonics, and so forth.). Such measurement may lose much information contained in the nonlinear components other than a specific component. If all the nonlinear components were measured, very high sensitivity to closed cracks would be achieved. It is, however, impossible for a piezoelectric array transducer to directly measure all the nonlinear components with diverse frequencies (e.g., $f/2$, $f/3$, $f/4$, $2f$, $3f$, $4f$, …), since an array transducer has a finite bandwidth around its center frequency. Furthermore, higher harmonics with a high frequency may diminish before arriving back at the array transducer because of high attenuation.

On the other hand, a method of indirectly measuring all the nonlinear components by utilizing the amplitude dependence of the response mainly in the fundamental wave was proposed. This concept was initially employed on the basis of the waveform measurement with monolithic transducers and is referred to as the scaling subtraction method (SSM) [64–67]. The primary reason that the response of a fundamental wave from contact interfaces (e.g., closed cracks) can nonlinearly depend on the incident wave amplitude is that part of the energy of the fundamental wave is consumed by the contact vibration of the crack faces to produce nonlinear components, which are nonlinearly dependent on the incident wave amplitude. In other words, the energies of all the nonlinear components generated are supplied from that of the fundamental wave. Thus far, this has mainly been verified in highly attenuative materials such as concrete, since it is almost impossible to directly measure higher harmonics in such materials because of very high attenuation.

Recently, a nonlinear ultrasonic PA using the incident-wave-amplitude dependence of the fundamental component was proposed to enhance the selectivity of closed cracks. This can be referred to as the fundamental wave amplitude difference (FAD) [68, 69], and it was proposed that the nonlinearity may also originate from the threshold behavior of the fundamental wave. In addition to the generation of nonlinear components, the loss of the fundamental component can be attributed to the energy dissipation due to heat generation arising from the contact vibration and/or friction of the crack faces [70–74]. Note that the responses at the fundamental frequency are obtained not only from closed cracks but also from other linear scatterers such as the bottom of the specimen, coarse grains, open cracks, and so forth. Importantly, these phenomena are nonlinearly dependent on the incident wave amplitude, whereas linear scattering phenomena are linearly dependent on the incident wave amplitude. Hence, the nonlinear phenomena resulting from the contact vibration of crack faces can be inferred by utilizing the incident-wave-amplitude dependence of fundamental responses, since the energy of the fundamental wave consumed at closed cracks nonlinearly changes with increasing the incident wave amplitude while the fundamental responses at linear scatterers linearly change with the incident wave amplitude. This concept is not based on the measurement of a nonlinear component and requires only the measurement of fundamental components. This allows us to use a bandwidth only around the center frequency of an array transducer for both transmission and reception, although the bandwidth of the array transducer must be carefully considered in the methods of measuring a specific nonlinear component, as described in Sect. 5.2.

To realize this concept in experiments, the easiest means of changing the incident wave amplitude is to vary the excitation voltage of a pulser. By selecting two (or more) different excitation voltages, the incident wave amplitude can be readily changed. This approach has been employed in the SSM [64–67] and FAD, which is called different-voltage FAD [68, 69]. However, the variation in excitation voltages affects the generation of higher harmonics due to the transducer and liquid couplant. This may hide the nonlinear response arising from cracks in metallic samples, although this may be negligible in highly nonlinear materials such as sandstone, rock, and concrete [64–67].

Is it possible to realize different incident wave amplitudes without varying excitation voltage? The answer is yes. Recently, a very sophisticated means was proposed in Ref. [44]. This is based on two focusing approaches via an array transducer with multiple individual elements [44, 75–78]. The first focusing approach is carried out using sequential transmission, where each element of an array transducer is sequentially excited to input ultrasound into a specimen, and all elements are employed for reception. This acquisition is called full matrix capture (FMC) [4, 10]. Subsequently, the data set comprising a matrix of signals from every transmit–receive element pair is post-processed on the basis of a delay law for transmission focusing, that is, physical focusing in a specimen is not carried out. This focusing approach is used as a small-amplitude incidence. The second focusing approach is carried out using parallel transmission, where all elements of an array transducer are excited following the delay law used for the post-processing of the sequential transmission, and all elements are employed for reception. Therefore, physical focusing is carried out, in contrast to the case of sequential focusing. Therefore, the actual incident wave amplitude for parallel transmission is larger at the focal point in a specimen than that for sequential transmission. Importantly, these two focusing approaches are linearly equivalent, whereas they are not nonlinearly equivalent because of the difference in the absolute incident wave amplitude at the focal point. In consequence, when a closed crack exists around the focal point, a larger energy is transferred from the fundamental component to other nonlinear components for parallel transmission than for sequential transmission. Note that the nonlinear effects of piezoelectric array elements and liquid couplant are identical for the two types of transmission since the excitation voltage is fixed. In this section, nonlinear ultrasonic PA using parallel and sequential transmission is described.

The first implementation of nonlinear ultrasonic PA using parallel and sequential transmission is based on the measurement of a diffuse field to infer the nonlinear phenomena at the focal point [44], where the relative energy around the fundamental frequency within the diffuse field is measured for parallel and sequential focusing. As illustrated in Fig. 5.17, the sound field in a specimen after the transmission for parallel and sequential transmission transits from a coherent field to a diffuse field over time. Although most PAs form an image on the basis of backscattered waves in a coherent field, they cannot utilize the other modes such as the transmitted and scattered waves that propagate away from the array transducer. In the proposed method, a diffuse field is utilized since the energy sampled at any point is proportional to the total energy in the system in a diffuse field. This means that this technique can indirectly infer all the nonlinear components in all the wave modes. Thus, very high sensitivity to closed cracks is expected to be achieved.

On the other hand, a diffuse field may not be observed in large structures and highly absorptive materials, because an observable diffuse field can only be formed in a specimen that can confine the acoustic energy. Hence, the second implementation was proposed to utilize a coherent field for parallel and sequential transmission [77]. This has the advantage of applicability to large structures in which a diffuse field cannot be formed, whereas the ADC of the PA hardware used in this implementation requires a sufficiently high dynamic range to receive a significantly different level of coherently scattered waves for parallel and sequential transmission.

The schematic illustration used to formulate the imaging algorithm when using diffuse fields for parallel and sequential transmission is shown in Fig. 5.17, where the origin of the x–z Cartesian coordinate system is at the center of the array transducer on the x-axis and at the surface of the specimen on the z-axis. The specimen to be inspected is assumed to be isotropic and homogeneous. On the basis of the geometric

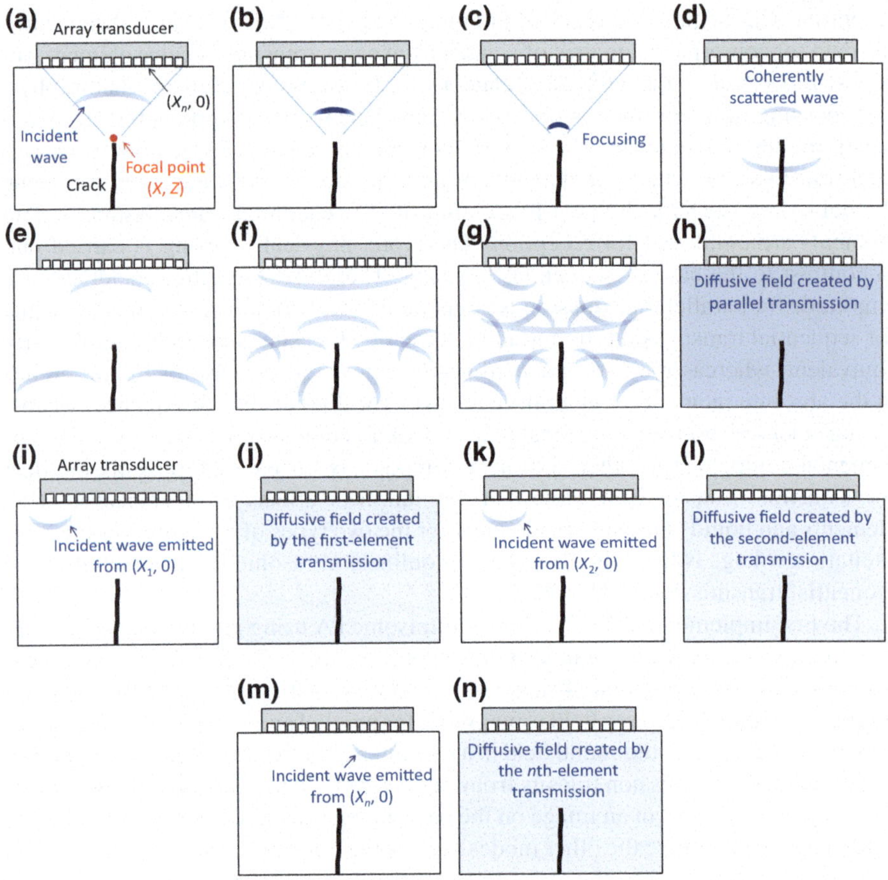

Fig. 5.17 Schematic illustration of nonlinear ultrasonic PA using parallel (**a**–**h**) and sequential (**i**–**n**) transmission. **a**–**h** Transition from coherent field to diffuse field for parallel transmission. **a** All elements are excited at a timing following the delay laws to focus the incident wave at a focal point (X, Z). **b** and **c** Incident wave approaching the focal point with increasing incident wave amplitude. **d** Coherently scattered wave and transmitted waves generated by the interaction between the focused wave and the crack. **e**–**g** Coherent field gradually approaching a diffuse field owing to multiple scattering and reflections. **h** Diffuse field with uniform energy at any position formed after a sufficient amount of time. **i**–**n** Transition from coherent field to diffuse field for sequential transmission. **i** and **j** Coherent and diffuse fields for the first-element transmission, respectively. **k**, **l** Coherent and diffuse fields for the second-element transmission, respectively. **m** and **n** Coherent and diffuse fields for the *n*th-element transmission, respectively

relationship and the sound speed, the propagation time from the *n*th element E_n $(X_n, 0)$ to the focal point $F(X, Z)$ is given by

$$t_T(X, Z, n) = \frac{\sqrt{(X_n - X)^2 + Z^2}}{V_L}, \tag{5.21}$$

where V_L is the longitudinal wave speed in the specimen and X_n is the coordinate of the nth element of the array transducer. The transmission delay law used to focus ultrasound at $F(X, Z)$ can be calculated by subtracting a reference propagation time (i.e., from the origin (0, 0) to F) from Eq. (5.21):

$$\Delta t_T(X, Z, n) = \frac{\sqrt{(X_n - X)^2 + Z^2} - \sqrt{X^2 + Z^2}}{V_L}, \tag{5.22}$$

which is used for the delayed excitation of each element in parallel transmission and for the post-processing in sequential transmission.

Assuming that the signals received at the kth element for the sequential transmission of element j and the parallel focusing at $F(X, Z)$ are $v_{j,k}(t)$ and $w_k(X, Z, t)$, respectively, the amplitude spectra of $v_{j,k}(t)$ and $w_k(X, Z, t)$ within the time window between t_d and $t_d + \Delta t_d$ are given by

$$V_{j,k}(\omega) = \int_{t_d}^{t_d+\Delta t_d} v_{j,k}(t)e^{-i\omega t}dt, \tag{5.23}$$

$$W_k(X, Z, \omega) = \int_{t_d}^{t_d+\Delta t_d} w_k(t)e^{-i\omega t}dt, \tag{5.24}$$

where ω is the angular frequency, and t_d and Δt_d are the reception delay and the time window for the measurement of the diffuse field, respectively. When an N-element array transducer is used, the diffuse acoustic kinetic energies in the parallel and sequential focusing are respectively calculated as

$$E_P(X, Z) = \sum_{k=1}^{N}\left(\int_{\omega_0-\frac{\Delta\omega}{2}}^{\omega_0+\frac{\Delta\omega}{2}} \omega^2|W_k(X, Z, \omega)|^2 d\omega\right), \tag{5.25}$$

$$E_S(X, Z) = \sum_{k=1}^{N}\left(\int_{\omega_0-\frac{\Delta\omega}{2}}^{\omega_0+\frac{\Delta\omega}{2}} \omega^2\left|\sum_{j=1}^{N} V_{j,k}(\omega)e^{i\omega\Delta t_T(X,Z,j)}\right|^2 d\omega\right), \tag{5.26}$$

where ω_0 is the center (i.e., fundamental) frequency and $\Delta\omega$ is the bandwidth of the BPF used to extract only the fundamental component. This bandwidth should be appropriately selected by considering the bandwidth of the fundamental wave. To infer the relative energy around the fundamental frequency within the diffuse field for parallel and sequential focusing, a nonlinear metric γ is defined as the normalized difference in the diffuse acoustic kinetic energies (i.e., Eqs. (5.25) and (5.26)) of the sequential and parallel transmission:

$$\gamma(X, Z) = \frac{E_S(X, Z) - E_P(X, Z)}{E_S(X, Z)}, \tag{5.27}$$

which represents the proportional energy loss from the fundamental frequency band owing to the nonlinear effects. By plotting $\gamma(X, Z)$ at each pixel over an imaging area, a diffuse nonlinear image are produced.

In a similar manner, the imaging algorithm when using coherent fields for parallel and sequential transmission is formulated below. A schematic illustration is shown in Fig. 5.17. The transmission modes and received waveforms for both types of transmission are identical to those for the diffuse field measurement. To focus the ultrasound at $F(X, Z)$, Eq. (5.21) can be used as the transmission delay. In addition, for coherent field measurement, the reception delay is required for image formation, which is expressed as

$$\Delta t_{\mathrm{R}}(X, Z, k) = \frac{\sqrt{(X_k - X)^2 + Z^2} - \sqrt{X^2 + Z^2}}{V_{\mathrm{L}}}. \tag{5.28}$$

The amplitude spectra of $v_{j,k}(t)$ and $w_k(X, Z, t)$ for coherent field measurement are given by

$$S_{j,k}(\omega) = \int_0^{t_r} v_{j,k}(t) e^{-i\omega t} dt, \tag{5.29}$$

$$P_k(X, Z, \omega) = \int_0^{t_r} w_k(X, Z, t) e^{-i\omega t} dt, \tag{5.30}$$

respectively, where t_r is the reception end time. Note that the only difference between the above equations for the diffuse- and coherent-field measurements is the integral range. For an N-element array transducer, the coherently scattered waves A_{S} and A_{P} for the sequential and parallel transmission are calculated as

$$A_S(t) = \int_{\omega_0 - \frac{\Delta\omega}{2}}^{\omega_0 + \frac{\Delta\omega}{2}} \left(\sum_{m=1}^{N} \left(\sum_{n=1}^{N} S_{j,k}(\omega) e^{i\omega \Delta t_{\mathrm{T}}(X,Z,n)} \right) e^{i\omega \Delta t_{\mathrm{R}}(X,Z,m)} \right) e^{-i\omega t} d\omega, \tag{5.31}$$

$$A_P(t) = \int_{\omega_0 - \frac{\Delta\omega}{2}}^{\omega_0 + \frac{\Delta\omega}{2}} \left(\sum_{m=1}^{N} P_k(X, Z, \omega) e^{i\omega \Delta t_{\mathrm{R}}(X,Z,m)} \right) e^{-i\omega t} d\omega. \tag{5.32}$$

The amplitude intensities for the sequential and parallel transmission are given by

$$I_S(X, Z) = A_S(t_f), \tag{5.33}$$

$$I_P(X, Z) = A_P(t_f), \tag{5.34}$$

where t_f is focal time for $F(X, Z)$ and I_S is equivalent to an image obtained by a conventional linear TFM. Finally, a nonlinear metric, $\zeta(X, Z)$, is calculated by subtracting the parallel and sequential amplitudes as follows:

$$\zeta(X, Z) = |I_S(X, Z) - I_P(X, Z)|. \tag{5.35}$$

By plotting $\zeta(X, Z)$ at each pixel over an imaging area, the coherent nonlinear image are produced.

5.3.2 *Experimental Conditions*

This subsection starts with a summary of the experimental conditions for diffuse-field measurement, followed by that for coherent-field measurement. Imaging experiments employing the method based on diffuse-field measurement for the parallel and sequential transmission were exemplified on two different fatigue-crack specimens. Following the ASTM (American Standard of Testing of Materials) standard E647, high-cycle fatigue cracks were grown in compact specimens (A2014 aluminum alloy ($V_L = 6270$ m/s)) under cyclic loading. The maximum load was 6 kN, which corresponds to K_{max} at the starting notch of a third of the fracture toughness K_{IC}. After 75,000 cycles, the fatigue cracks had reached a depth of approximately 15.5 mm. Subsequently, the specimens were machined to dimensions of 180 × 30 × 25 mm^3 with a flat top surface. One was used as a specimen with an isolated fatigue crack, as shown in Fig. 5.18a. The other one was used to simulate an early-stage fatigue crack grown from a geometric feature such as a fixing hole. To this end, a hole with a diameter of 5 mm was machined behind the crack tip of a similar fatigue specimen. As a reference, an identical hole with the same depth was machined at a distance of from the first hole in the specimen (Fig. 5.18b).

A 64-element array transducer (5 MHz, 0.63 mm pitch) manufactured by Imasonic (Voray-sur-l'Ognon, France) was operated with a Micropulse FMC model array controller produced by Peak NDT Ltd. (Derby, UK). Each element was excited by a pulse with 80 ns length and a voltage of 200 V. The array transducer was placed on the top surface and positioned centrally over the crack. To show the difference between coherent and diffuse acoustic fields, examples of acquired time traces are shown in Fig. 5.19, where the coherent and diffuse fields were acquired with reception delays of 0.01 and 1 ms, respectively. In this experiment, the received waves were acquired with a reception delay of $t_d = 1$ ms during a time window of $\Delta t_d = 120$ μs at a sampling rate of 25 MS/s. This choice was a compromise between maximizing the amplitude and the diffuse-field convergence. Here, the bandwidth for Eqs. (5.25) and (5.26) was set from $2\omega_0/3$ to $4\omega_0/3$.

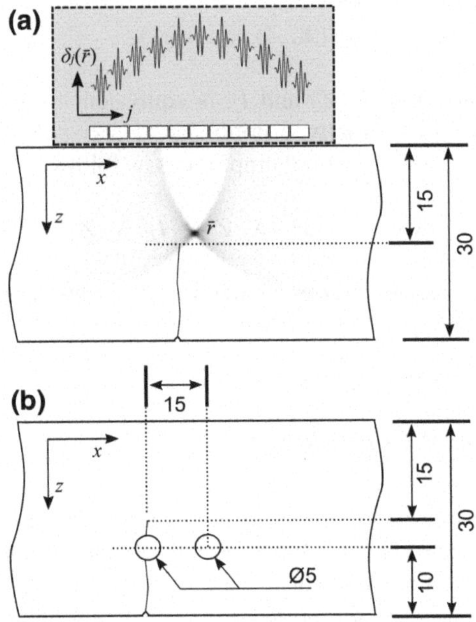

Fig. 5.18 Experimental configurations for imaging fatigue cracks by nonlinear ultrasonic PA with parallel and sequential transmission modes. **a** Isolated fatigue crack and **b** fatigue crack and linear scatterers. Dimensions are in mm [44] (Copyright (2014) American Physical Society)

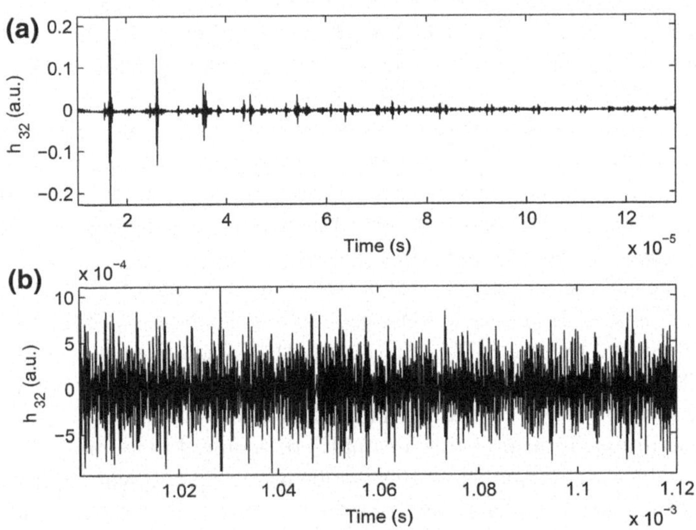

Fig. 5.19 Example of received waveforms for reception delays of **a** $t_d = 0.01$ ms and **b** $t_d = 1$ ms [44] (Copyright (2014) American Physical Society)

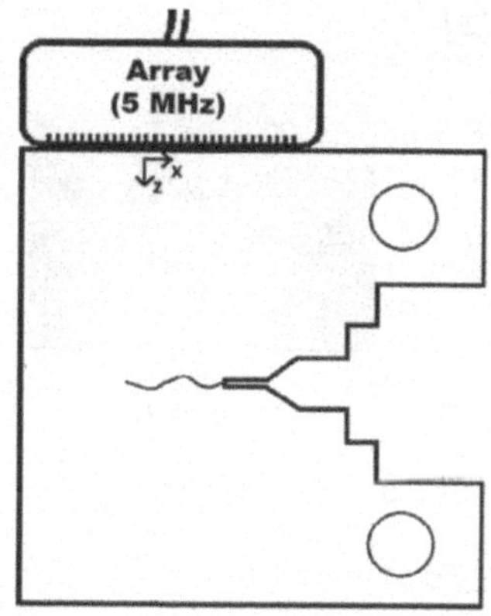

Fig. 5.20 Schematic diagram of pulse-echo configuration with array transducer placed on the side of the compact tension (CT) specimen [77] (Copyright (2018) IOP Publishing Ltd.)

For the sake of extending applicable objects, the method based on coherent-field measurement for the parallel and sequential transmission was verified through the monitoring of fatigue-crack growth. Following ASTM standard E647-05, a high-cycle fatigue crack was grown in a CT specimen (ASTM A36 mild steel, V_L = 5924 m/s) under a cyclic tensile load varying in the range of 2 and 15 kN [70, 71]. During the fatigue test, the crack was monitored by observing linear and nonlinear images. Furthermore, the crack length was periodically measured using an optical microscope during the fatigue test for comparison with the imaging results.

As shown in Fig. 5.20, a 64-element array transducer (5 MHz, 0.63 mm pitch) manufactured by Imasonic (Voray-sur-l'Ognon, France) was placed on the side surface parallel to the crack faces to obtain a large reflection from the crack, although the accessible surface on site will be restricted to that perpendicular to the crack faces as selected in the last experiment of Ref. [77]. The same PA controller as that in the previous experiments was used. Here, the bandwidth for Eqs. (5.31) and (5.32) was set from $5\omega_0/6$ to $7\omega_0/6$.

5.3.3 *Imaging Results*

The method of measuring the diffuse field for parallel and sequential transmission was applied to the isolated fatigue crack extending from the back wall, as shown in Fig. 5.18a. As a reference, a linear imaging method based on FMC and TFM [10] was employed for the specimen. As shown in Fig. 5.21d, the crack was visualized as a consequence of weak scattering at the middle part while the back wall was clearly visualized. The signal-to-noise ratio (SNR), however, was not high. Moreover, the crack tip was not visualized, resulting in underestimation of the crack depth.

Figure 5.21a, b show the diffuse acoustic kinetic energies obtained by parallel and sequential transmission modes on the basis of Eqs. (5.25) and (5.26), respectively. The images are seen to be identical. To produce a nonlinear image, the normalized

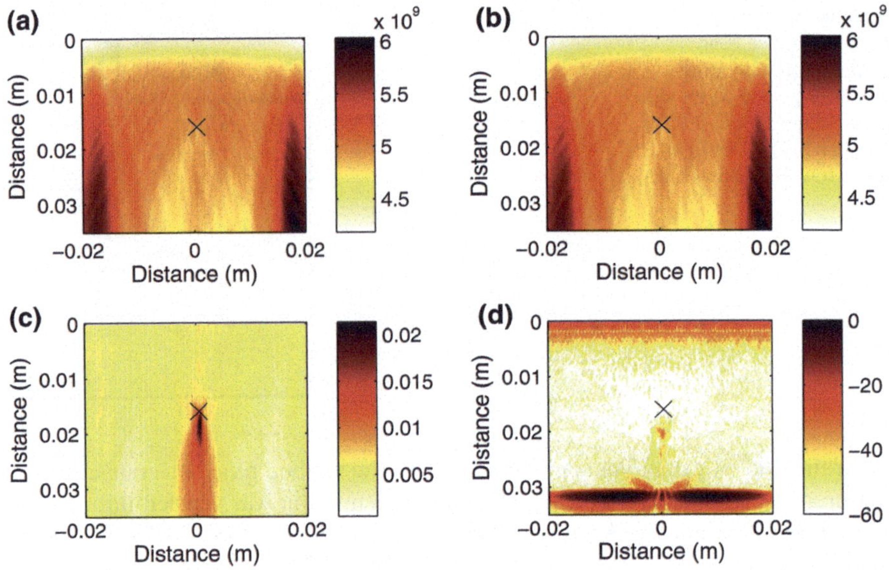

Fig. 5.21 Imaging results of isolated fatigue crack. **a** Sequential transmission energy (E_S), **b** parallel transmission energy (E_P), **c** diffuse nonlinear transmission energy (γ), and **d** linear TFM image (dB) [44] (Copyright (2014) American Physical Society)

difference between the absolute energy images for the parallel and sequential transmission modes was taken on the basis of Eq. (5.27). In the diffuse nonlinear image (Fig. 5.21c), the crack is clearly visible with an excellent SNR. The amplitude of the nonlinear metric increased locally at the crack. This signifies a much higher loss of the fundamental wave in the parallel transmission than in the sequential transmission. Note that the back wall was successfully canceled in the diffuse nonlinear image, although it was the most dominant feature in the linear image. That is, no artifacts due to the strong linear responses appeared in the nonlinear image. This shows the very high separation between linear and diffuse nonlinear images.

As a further demonstration, another experiment was carried out on the specimen with two holes, where the left-hand-side hole had the fatigue crack, which simulated an early-stage fatigue crack grown from, for example, a fixing hole. In the linear image (Fig. 5.22b), only the two holes were visualized in addition to the back wall. Note that the crack was not visualized in the linear image. This will lead to the overlooking of cracks. In the diffuse nonlinear image (Fig. 5.22a), the crack was visualized with an excellent SNR, in contrast to the linear image (Fig. 5.22b). Importantly, the very strong linear features such as the holes and the back wall were successfully canceled. This shows that the diffuse nonlinear image obtained with the parallel and sequential transmission is useful for visualizing early-stage cracks even in the vicinity of strong linear scatterers.

Thus, the efficacy of the method based on diffuse-field measurement for the parallel and sequential transmission was demonstrated using the two fatigue-crack

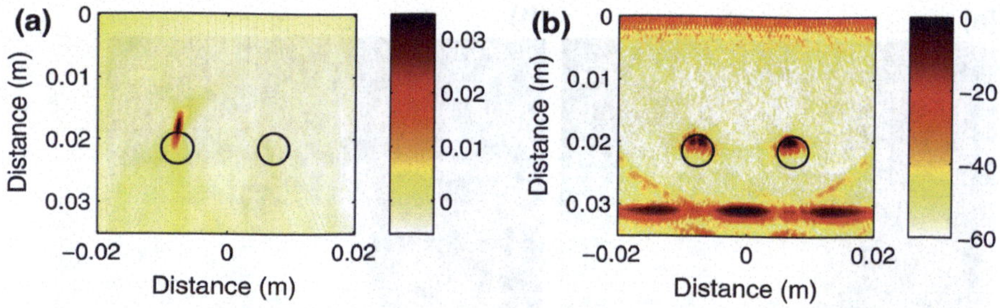

Fig. 5.22 Imaging results of fatigue crack and linear scatterers. **a** Diffuse nonlinear image (γ) and **b** linear TFM image (dB) [44] (Copyright (2014) American Physical Society)

specimens. On the other hand, this method cannot be used for large structures and absorptive materials because a diffuse field cannot be formed in such objects.

To overcome this difficulty, a method based on the coherent-field measurement for the parallel and sequential transmission was proposed and was applied to crack growth monitoring (Fig. 5.20). As an example, the imaging results obtained at 30,000 fatigue cycles are shown in Fig. 5.23. Figure 5.23a, b show the sequential and parallel images obtained using Eqs. (5.33) and (5.34), respectively. The images are seen to be almost identical. On the other hand, the coherent nonlinear image (Fig. 5.23c), which was obtained using Eq. (5.35), can be seen to suppress the linear geometric features. This reveals that the maximum nonlinear response appeared at a position around the crack tip.

On the other hand, residue can be seen around the notch, which is a linear scatterer that does not generate any nonlinear components. To understand this, multiple linear points selected from the back wall were examined [77]. It was found that the linear points exhibited the small differences in the phase and amplitude for sequential and parallel transmission. This can be attributed to the classical nonlinearity of the bulk material and, more significantly, the nonlinearity of the instrumentation. Presumably, the instruments do not produce perfectly identical waveforms in both the parallel and sequential transmission modes. To compensate for these effects on the basis of this finding, the relative amplitude and phase of the baseline measurements were first averaged for multiple points on the back wall. Subsequently, the baseline response was approximated through a linear fit in the frequency domain. Finally, a correction for the relative phase and amplitude arising from the instrumentation was applied to the sequential focused amplitude. Figure 5.23d shows the coherent nonlinear image with the compensation, from which it can be seen that the background level in the nonlinear metric drops from approximately 0.2 to 0.02.

On the basis of the aforementioned compensation method, crack growth was monitored from 0 to 100,000 fatigue cycles, where the loading step was chosen to be 10,000 fatigue cycles to allow the early-stage damage to be monitored and detection performance limits to be investigated. As references, the sequential images from 10,000 to 100,000 fatigue cycles are respectively shown in Fig. 5.24a–j. In the early stages up to 50,000 fatigue cycles, cracks were not detected because of their

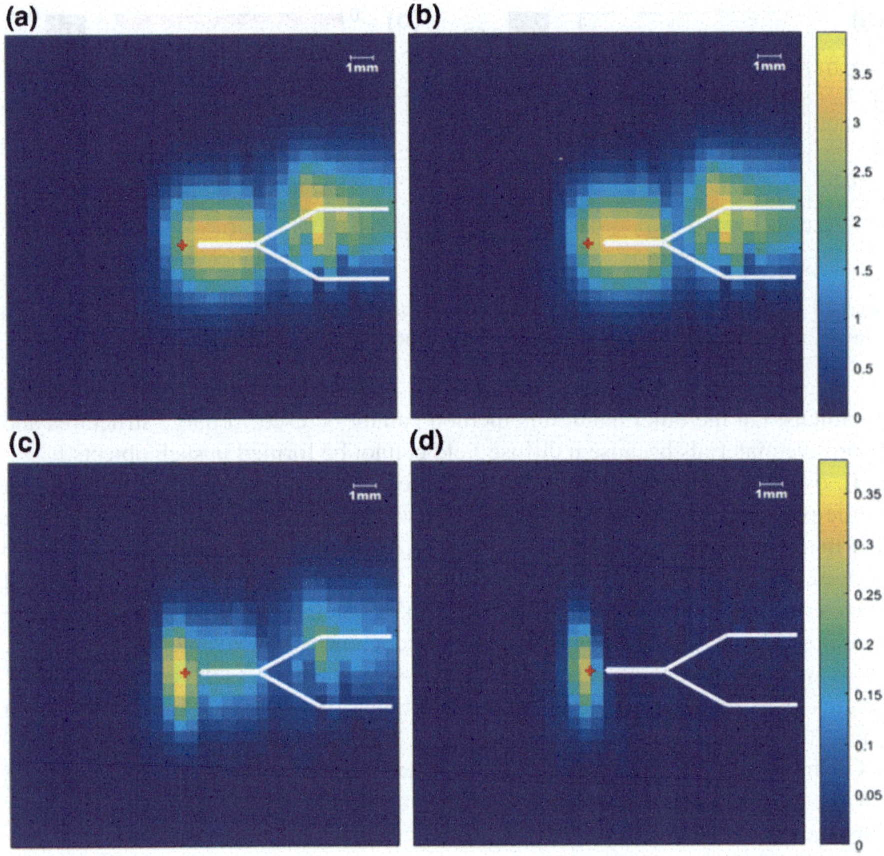

Fig. 5.23 Imaging results of the fatigue crack at 30,000 fatigue cycles. **a** Sequential image (I_S), **b** parallel image (I_P), **c** coherent nonlinear image, and **d** coherent nonlinear image with compensation applied. The red cross indicates the position of the micrographically measured crack tip. The white lines denote the geometry of the starting notch [77] (Copyright (2018) IOP Publishing Ltd.)

small size (i.e., maximum 1.3 mm). Beyond 50,000 fatigue cycles, the cracks were visualized and their lengths can be measured from these sequential images.

The coherent nonlinear images from 10,000 to 100,000 fatigue cycles are shown in Fig. 5.25. In contrast to the linear images (Fig. 5.24), the coherent nonlinear images exhibit excellent selectivity to nonlinear features. The crack was first distinguished at 20,000 fatigue cycles, corresponding to a crack length of 0.35 mm. Note that the selectivity to the nonlinear features in the early stages was significantly improved by the compensation [77]. Subsequently, the nonlinear metric increased until 60,000 fatigue cycles and then started decreasing gradually. It can be seen that the position of the maximum nonlinear metric at each stage is close to the micrographically measured crack tip. Furthermore, the results suggest that the crack became progressively more open from 60,000 fatigue cycles, which is consistent with the increase in the linear

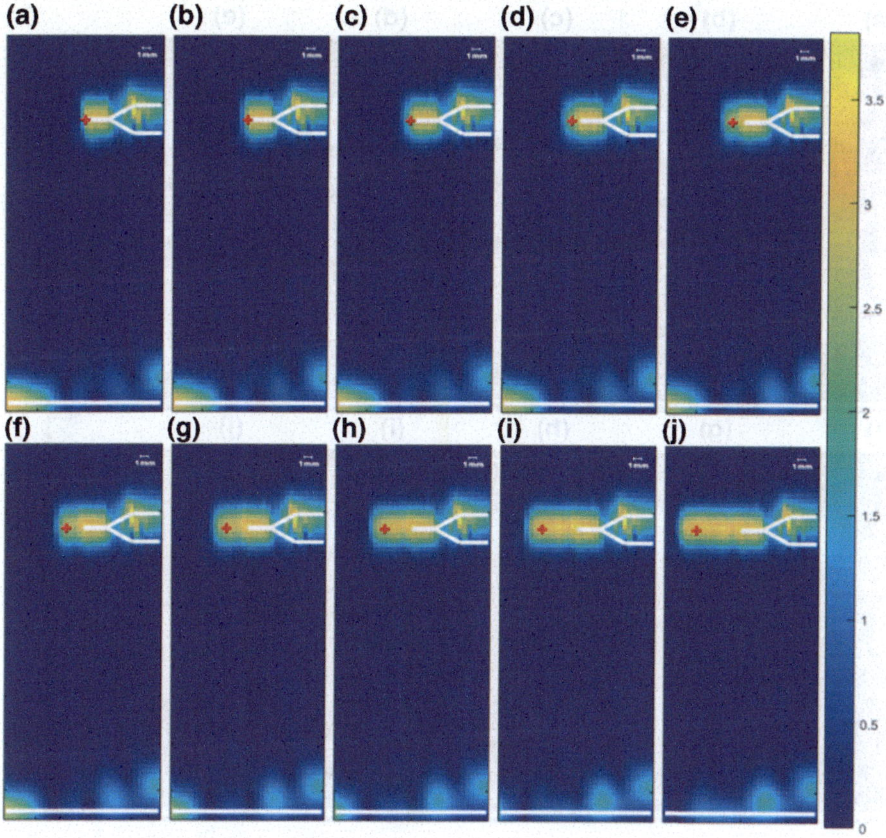

Fig. 5.24 Sequential images in arbitrary units taken at **a** 10,000, **b** 20,000, **c** 30,000, **d** 40,000, **e** 50,000, **f** 60,000, **g** 70,000, **h** 80,000, **i** 90,000, and **j** 100,000 fatigue cycles. The red cross denotes the position of the micrographically measured crack tip [77] (Copyright (2018) IOP Publishing Ltd.)

response (Fig. 5.25b–j). Thus, the efficacy of the coherent nonlinear image for parallel and sequential transmission was also demonstrated in the crack growth monitoring.

Finally, the important points of this section and points to be considered in the application of the discussed methods are summarized as follows:

- All nonlinear components can be indirectly measured through the incident-wave-amplitude dependence of fundamental responses. This enables us to make use of the center frequency of an array transducer.
- Excellent selectivity to closed cracks can be achieved because of the very high amplitude ratio between parallel and sequential transmission.
- Diffuse-field measurement can achieve very high sensitivity, although the spatial resolution may not be high, which will be determined by the transmission sound field.

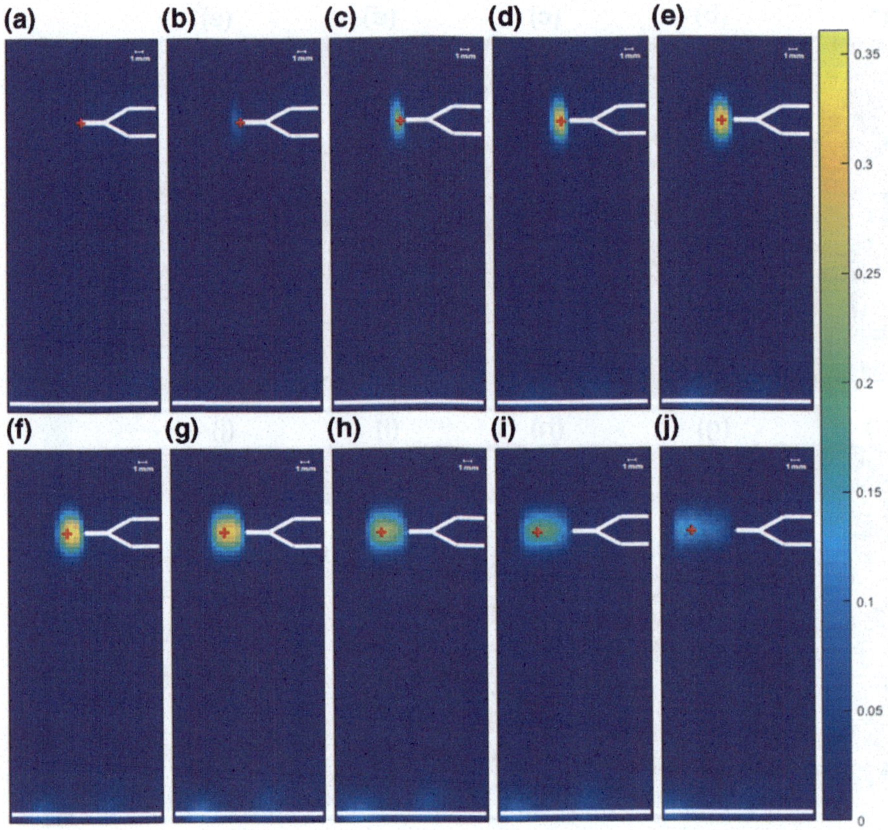

Fig. 5.25 Coherent nonlinear images after compensation taken at **a** 10,000, **b** 20,000, **c** 30,000, **d** 40,000, **e** 50,000, **f** 60,000, **g** 70,000, **h** 80,000, **i** 90,000, and **j** 100,000 cycles. The red cross denotes the position of the micrographically measured crack tip [77] (Copyright (2018) IOP Publishing Ltd.)

- In reality, an ideal diffuse field cannot be achieved: hence, a compromise between maximizing the amplitude and the diffuse-field convergence needs to be found for each specimen. The optimization of the reception delay and the time window for the diffuse-field measurement may be tricky for manual operation.
- The implementation of diffuse-field measurement for parallel and sequential transmission may be acceptable only for flexible PA hardware (e.g., for research).
- Large structures and absorptive materials cannot be inspected by diffuse-field measurement because an observable diffuse field cannot be formed in such objects. Coherent-field measurement is suitable for the inspection of such objects.
- The compensation method used in coherent-field measurement will be practically useful not only for this method but also for the other nonlinear imaging methods.

- Coherently scattered waves for sequential transmission will be much smaller than those for parallel transmission. To receive all signals without clipping, an ADC with a sufficient dynamic range will be required.
- Comparison between linear and nonlinear images is quite important to clarify the necessity of nonlinear ultrasonic PA and to gain physical insight.
- The bandwidth of both the incident wave and the diffuse field should be checked to choose an appropriate bandwidth for the BPF used to effectively extract the fundamental component.

5.4 All-Elements, Odd-Elements, and Even-Elements Transmission

5.4.1 Principles

After describing SPACE that measures a specific nonlinear component, i.e., a subharmonic wave, in Sect. 5.2, we introduced two methods based on diffuse- and coherent-field measurement for parallel and sequential transmission in Sect. 5.3. The concept of the imaging method introduced in this section is similar to the coherent-field measurement for parallel and sequential transmission in terms of the use of the fundamental wave. The main difference is the transmission mode. Instead of parallel and sequential transmission, three transmission modes, i.e., all-elements, odd-elements, and even-elements transmission [45, 79–81], are implemented here. This method is referred to as the fixed-voltage fundamental wave amplitude difference (fixed-voltage FAD) [79–81]. All-elements transmission is regarded as having the largest amplitude among the three transmission modes. This is equivalent to parallel transmission. Odd-elements and even-elements transmission are regarded as having a smaller amplitude, i.e., half of the amplitude for all-elements transmission. The amplitude ratio between all-elements transmission and odd-elements and even-elements transmission is two, which is smaller than that between parallel and sequential transmission. This may lead to lower sensitivity to closed cracks than the method described in Sect. 5.3. On the other hand, the difference in the magnitudes of received waves for all-elements, odd-elements, and even-elements transmission will not be so large, in contrast to the coherent-field method described in Sect. 5.3. ADCs installed in typical commercial PA hardware will be sufficient for this method; therefore, this method may be easily implemented into typical commercial PA hardware. As with the methods introduced in Sect. 5.3, this method is based on the measurement of the loss of the fundamental wave with increasing incident wave amplitude. Thus, this method has many common points with the method described in Sect. 5.3. In this section, we describe this method in a different way, which will hopefully enrich readers' understanding of the concept of using the fundamental wave.

In fixed-voltage FAD (Fig. 5.26), all-elements, odd-elements, and even-elements

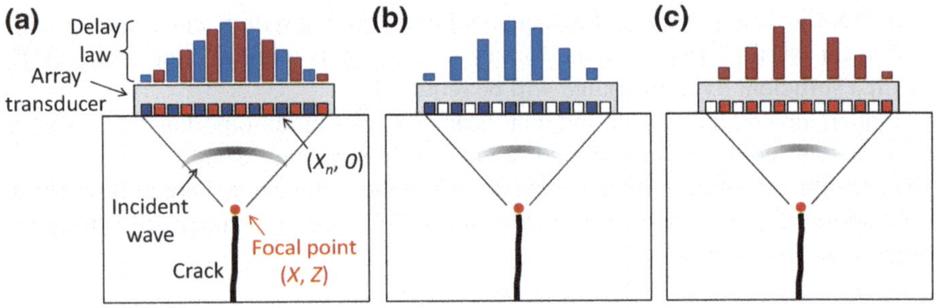

Fig. 5.26 Schematic illustration of fixed-voltage FAD with all-elements, odd-elements, and even-elements transmission: **a** All-elements transmission, T_{All}, **b** odd-elements transmission, T_{Odd}, and **c** even-elements transmission, T_{Even}

transmission are employed to change the incident wave amplitude at a fixed excitation voltage and are referred to as T_{All}, T_{Odd}, and T_{Even}, respectively. In T_{All}, all elements within the transmitting aperture of an array transducer are employed for transmission. T_{All} is regarded as having the largest-amplitude incidence among the three transmission modes. In T_{Odd} and T_{Even}, odd-numbered and even-numbered elements within the transmitting aperture of the array transducer are employed for transmission, respectively. T_{Odd} and T_{Even} are regarded as having a smaller-amplitude incidence than T_{All}. The incident wave amplitude at the focal point for T_{Odd} and T_{Even} will be approximately half of that for T_{All}.

To describe the concept of fixed-voltage FAD, the responses to a slit and a closed crack in the frequency domain are schematically illustrated in Figs. 5.27 and 5.28, respectively. The first example is the imaging of a slit by fixed-voltage FAD (Fig. 5.27). One of the measures indicating the performance of a nonlinear ultrasonic PA is the capability of suppressing the artifacts due to linear defects. Although the slit does not generate any nonlinear components regardless of the incident wave amplitude, higher harmonics can be generated because of the nonlinearities originating from the piezoelectric elements and liquid couplant. These harmonics will be superimposed on the incident wave. Note that the amplitude of the higher harmonics generated for T_{Odd} and T_{Even} will be half of that for T_{All} because it is proportional to the number of elements employed for transmission at a fixed voltage of excitation. Consequently, the loss of the fundamental amplitude will become equal between the response of T_{All} (Fig. 5.27a) and the summation of the responses of T_{Odd} (Fig. 5.27b) and T_{Even} (Fig. 5.27c). Thus, as illustrated in Fig. 5.27d, the effect of nonlinearities due to the piezoelectric elements and liquid couplant can be canceled by subtracting the response for T_{All} from the sum of the responses for T_{Odd} and T_{Even}. This means that, in principle, no linear scatterers appear in nonlinear images, providing a high SNR for nonlinear defects.

The second example is the imaging of a closed crack (Fig. 5.28). The nonlinearity of the closed crack is approximated as follows. Although analytical solutions that treat nonlinear scattering in the region of quadratic and/or cubic nonlinearities have been derived recently [82–84], a simple model is adopted here because the exact

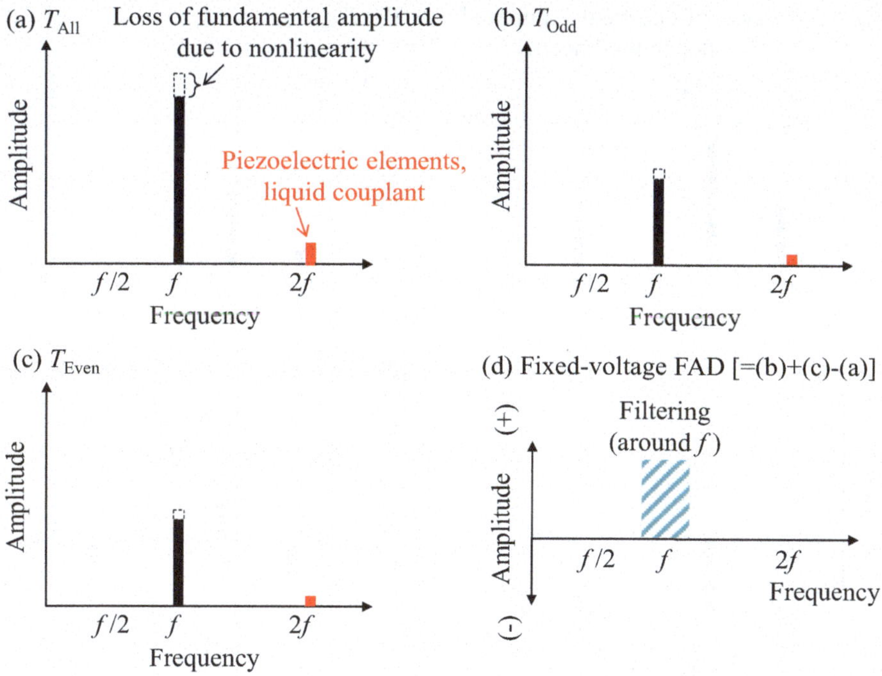

Fig. 5.27 Imaging of a slit by fixed-voltage FAD. Amplitude spectra of the response at a slit for **a** all-elements transmission (T_{All}), **b** odd-elements transmission (T_{Odd}), and **c** even-elements transmission (T_{Even}). **d** Amplitude spectrum of the subtracted response based on fixed-voltage FAD [=(**b**) + (**c**) − (**a**)]

modeling of the nonlinearity of closed fatigue cracks is not easy because of the complicated microscopic asperities of crack faces [85, 86], the distributions of the crack opening displacement, and the compressive residual stress [87, 88], as well as the oxide debris generated between crack faces [17]. In T_{All} (Fig. 5.28a), the generation of higher harmonics and/or subharmonics owing to the contact vibration of crack faces may occur. Note that additional higher harmonic generation occurs because of the nonlinearities originating from the piezoelectric elements and liquid couplant. In T_{Odd} (Fig. 5.28b) and T_{Even} (Fig. 5.28c), the higher harmonics and/or subharmonics generated at a closed crack will be smaller than those in T_{All} (Fig. 5.28a). Assuming that the response of the closed crack behaves as a quadratic nonlinearity, which is the simplest approximation, and noise is negligible, the amplitude of the nonlinear components generated for T_{Odd} and T_{Even} will be approximately one-quarter of that for T_{All}. Although it is possible that the generated harmonics will interact and mix back into the fundamental wave owing to cubic nonlinearity, only quadratic nonlinearity is considered here for simplicity. On the other hand, the amplitude of the higher harmonics generated by the nonlinearity of the piezoelectric elements and liquid couplant will be half of that for T_{All}. Thus, as illustrated in Fig. 5.28d, the effect of nonlinearities due to the piezoelectric elements and liquid couplant can be

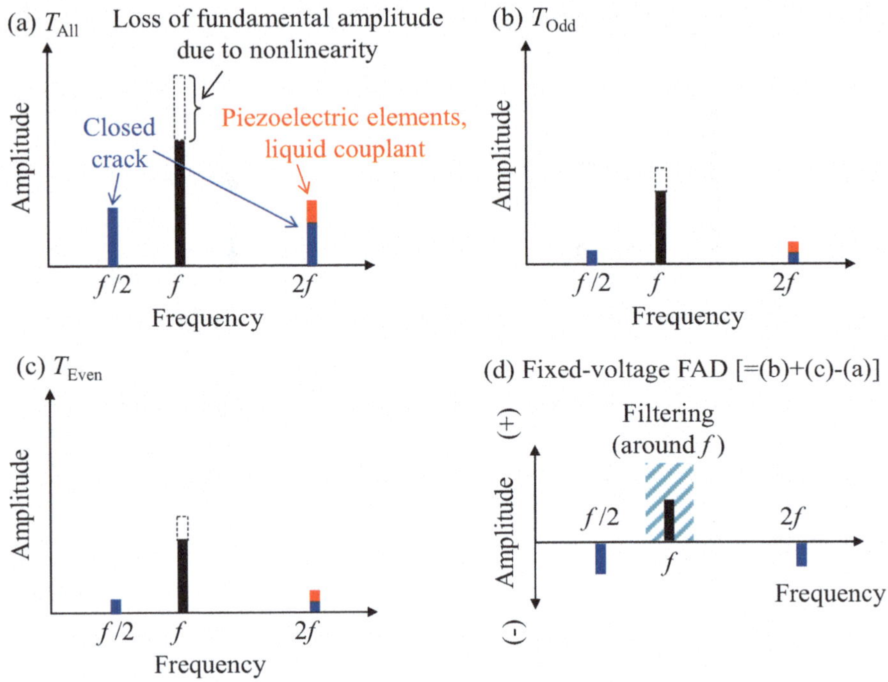

Fig. 5.28 Imaging of a closed crack by fixed-voltage FAD. Amplitude spectra of the response at a closed crack for **a** all-elements transmission (T_{All}), **b** odd-elements transmission (T_{Odd}), and **c** even-elements transmission (T_{Even}). **d** Amplitude spectrum of the subtracted response based on fixed-voltage FAD [=(**b**) + (**c**) − (**a**)]. Modified from [80] (Copyright (2019) Acoustical Society of America)

canceled by fixed-voltage FAD by subtracting the response (Fig. 5.28a) for T_{All} from the sum of the responses (Fig. 5.28a, b) for T_{Odd} and T_{Even}. Hence, the nonlinear response originating only from the closed crack can be effectively extracted. Importantly, the energies of nonlinear components generated at closed cracks are supplied from that of the fundamental wave. Therefore, by measuring only the fundamental amplitude in the subtracted response with frequency filtering around *f*, all the nonlinear components originating only from the closed crack can be indirectly measured. It is expected that this approach of measuring all the nonlinear components can achieve higher sensitivity than the method of measuring a specific nonlinear component.

The schematic illustration shown in Fig. 5.26 is also used to formulate the imaging algorithm of fixed-voltage FAD. Here, the origin of the *x*–*z* Cartesian coordinate system is at the center of the array transducer on the *x*-axis and at the surface of the specimen on the *z*-axis. The specimen to be inspected is assumed to be isotropic and homogeneous. Under this assumption, the propagation time and delay law are identical to those for single-array SPACE (Sect. 5.2). The propagation time $t_{\mathrm{T}}(X, Z, n)$ and delay law $\Delta t_{\mathrm{T}}(X, Z, n)$ for focusing ultrasound at $F(X, Z)$ on transmission are expressed as Eqs. (5.13) and (5.14), respectively. Likewise, the propagation time

$t_R(X, Z, n)$ and delay law $\Delta t_R(X, Z, n)$ for focusing on reception are expressed as Eqs. (5.15) and (5.16), respectively. Assuming that the signals received at the nth element after filtering around the fundamental frequency for transmission focusing are $u_{n,\mathrm{All}}(t)$, $u_{n,\mathrm{Odd}}(t)$, and $u_{n,\mathrm{Even}}(t)$ for a transmitting aperture using all elements (Fig. 5.26a), odd elements (Fig. 5.26b), and even elements (Fig. 5.26c), respectively, the nonlinear signal can be obtained as

$$u_{n,NL}(t) = u_{n,\mathrm{Odd}}(t) + u_{n,\mathrm{Even}}(t) - u_{n,\mathrm{All}}(t). \tag{5.36}$$

When $u_{n,\mathrm{All}}(t)$ is also used to produce a linear image, the linear and nonlinear DAS signals, $U_L(X, Z, t)$ and $U_{NL}(X, Z, t)$, can be expressed by

$$U_L(X, Z, t) = \frac{1}{N}\sum_{n=1}^{N} u_{n,All}(t - \Delta t_R(X, Z, n)), \tag{5.37}$$

$$U_{NL}(X, Z, t) = \frac{1}{N}\sum_{n=1}^{N} u_{n,NL}(t - \Delta t_R(X, Z, n)), \tag{5.38}$$

respectively, where an N-element array transducer is used. The corresponding scattering intensities, $I_L(X, Z)$ and $I_{NL}(X, Z)$, are calculated as the RMS values of $U_L(X, Z, t)$ and $U_{NL}(X, Z, t)$, respectively, as

$$I_L(X, Z) = \left(\frac{1}{\Delta\tau} \int_{t_R(X,Z,1)}^{t_R(X,Z,1)+\Delta\tau} U_L^2(X, Z, t)dt \right)^{1/2}, \tag{5.39}$$

$$I_{NL}(X, Z) = \left(\frac{1}{\Delta\tau} \int_{t_R(X,Z,1)}^{t_R(X,Z,1)+\Delta\tau} U_{NL}^2(X, Z, t)dt \right)^{1/2}. \tag{5.40}$$

By plotting $I_L(X, Z)$ and $I_{NL}(X, Z)$ at each pixel over an imaging area, the linear and nonlinear images are produced.

5.4.2 Experimental Conditions

To demonstrate the imaging capability of fixed-voltage FAD, two different experiments were performed under the following experimental conditions as examples. First, to experimentally demonstrate the importance of large-amplitude incidence in fixed-voltage FAD and its efficacy in terms of high-SNR imaging, a fatigue crack was grown in a CT specimen made of A7075 aluminum alloy ($V_L = 6230$ m/s), as shown in Fig. 5.29. The shape of the CT specimen was in accordance with ASTM-E399.

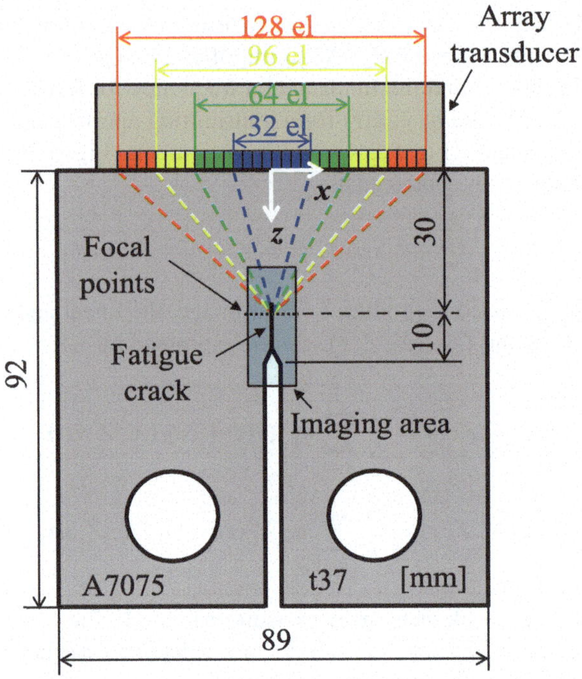

Fig. 5.29 Experimental configurations of fixed-voltage FAD with all-elements, odd-elements and even-elements transmission. The transmitting apertures for the three transmission modes were varied from 32 to 128 elements with a step of 32 elements. Modified from [80] (Copyright (2019) Acoustical Society of America)

The distance between the tip of the starting notch and the top surface was 37 mm. Under the fatigue conditions of $K_{max} = 9.0$ MPa m$^{1/2}$ and $K_{min} = 0.6$ MPa m$^{1/2}$, a crack was grown to a depth of approximately 10 mm on the side surface after 76,000 fatigue cycles [15, 46, 80].

Figure 5.29 shows the experimental configuration. A 128-element array transducer (5 MHz, 0.5 mm pitch) manufactured by Imasonic (Voray-sur-l'Ognon, France) was placed on the top surface and positioned centrally over the crack. In this experiment, to increase the incident wave amplitude at the focal point, the number of elements of the transmitting aperture was increased from 32 to 128 in steps of 32, while the number of elements of the receiving aperture was fixed to 128 to investigate only the effect of the incident wave amplitude. PA hardware (OEM-PA 128/128) produced by Advanced OEM Solutions (AOS) Inc. (Cincinnati, USA) was employed [45, 80]. The positions of the transmission focal points were set to $x = -5$ to 5 mm with 0.5 mm steps at $z = 30$ mm, and the positions of the receiving focal points for post-processing were set over the imaging area (10 mm × 25 mm) with 0.1 mm steps, where the origin of the x–z Cartesian coordinate system was at the center of the array transducer on the x-axis and at the surface of the specimen on the z-axis. Each element was excited by a three-cycle burst with a frequency of 5 MHz and a voltage of 145 Vp-p. An FFT BPF with a bandwidth from 2.5 to 7.5 MHz was employed to extract the fundamental components from the received waves.

To demonstrate the importance of large-amplitude incidence, the absolute displacement of the incident wave amplitude at the crack tip should be investigated for the various transmitting apertures with 32, 64, 96, and 128 elements before the imaging experiments. To this end, a crack-free specimen made of the same material (A7075) as the fatigue-crack specimen was prepared, as shown in Fig. 5.30. The thickness of the sample was 30 mm, which corresponds to the distance between the top surface and the position of the crack tip optically observed on the side surface. The array transducer positioned on the top surface was excited following the delay law formulated as Eq. (5.14) to focus ultrasound on the back wall at $x = 0$ and $z = 30$ mm. The absolute displacements of the incident wave for the transmitting apertures with 32, 64, 96, and 128 elements were measured at the focal point on the back wall of the sample by a laser Doppler vibrometer (OFV505) produced by Polytec (Waldbronn, Germany). Figure 5.31 shows the peak-to-peak incident wave amplitude plotted against the number of elements employed for transmission, from which the relationship between the number of elements used for transmission and the incident wave amplitude can be seen to be nonlinear. This is because the elements

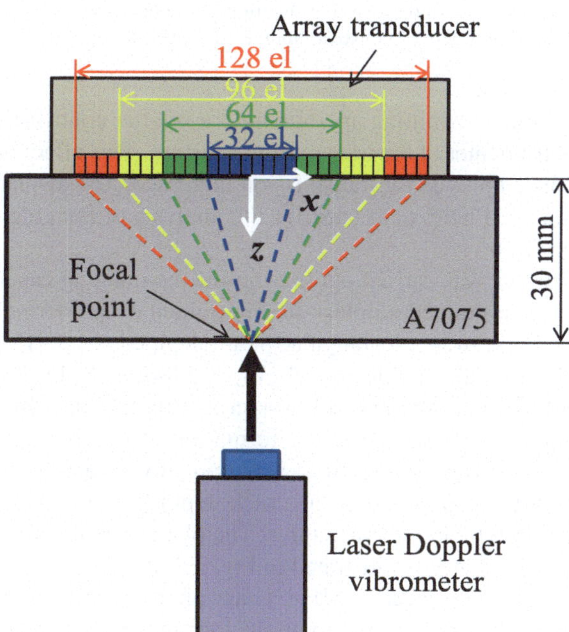

Fig. 5.30 Measurement of the absolute displacement of incident wave amplitude for the transmitting apertures of 32, 64, 96, and 128 elements in a crack-free sample by a laser Doppler vibrometer. The transmission focal point was set at $x = 0$ and $z = 30$ mm, corresponding to the back wall of the sample and the crack position on the side surface [80] (Copyright (2019) Acoustical Society of America)

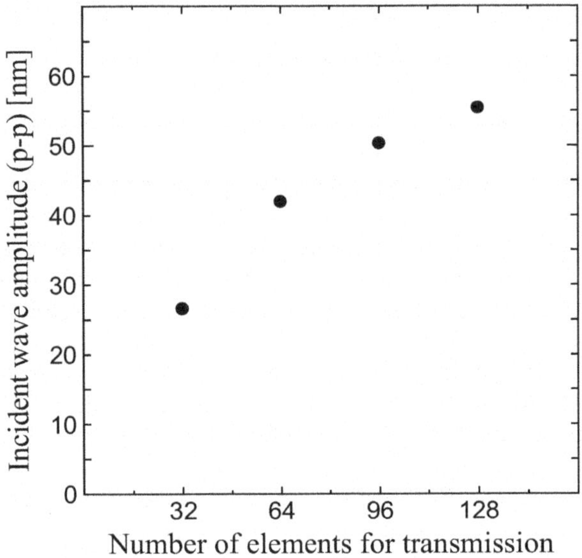

Fig. 5.31 Displacements of incident wave amplitude measured at $x = 0$ and $z = 30$ mm (shown in Fig. 5.30) using a laser Doppler vibrometer as a function of the number of elements employed for transmission [80] (Copyright (2019) Acoustical Society of America)

around the edge of the transmitting aperture have a smaller contribution to focusing than those around the center of the transmitting aperture. This effect increased as the number of elements employed for transmission increased. The displacements of the incident wave measured here were used for the analysis of imaging results, which will be introduced later.

The next experiment was carried out to examine the imaging capability of fixed-voltage FAD in a coarse-grained stainless-steel specimen with a thermal fatigue crack. The specimen used here was a rectangular parallelepiped ($61 \times 150 \times 100$ mm^3) made of AISI304 austenitic stainless steel ($V_L = 5700$ m/s) [45, 79], as shown in Fig. 5.32. A small thermal fatigue crack was grown by thermal fatigue loading by Trueflaw Ltd. (Espoo, Finland). After the formation of the thermal fatigue crack, the crack length (i.e., in the y-direction) was optically measured to be 24.2 mm by dye penetrant inspection, whereas the crack depth (i.e., in the z-direction) was estimated to be 5.9 mm by the manufacturer. The plane in which the crack exists is approximately the y–z plane, as illustrated in Fig. 5.32.

A 64-element array transducer (5 MHz, 1 mm pitch) manufactured by Imasonic (Voray-sur-l'Ognon, France) was placed on the top surface and was operated by PA hardware (OEM-PA) produced by AOS (Cincinnati, USA). Each element was excited by a pulse of 145 Vp-p. As illustrated in Fig. 5.32, mechanical scans were carried out in two directions by manually translating the array transducer: (1) along the crack (i.e., translation in the y-direction in steps of $\Delta y = 1$ mm; the array transducer was positioned centrally over the crack, perpendicular to the plane in which the crack

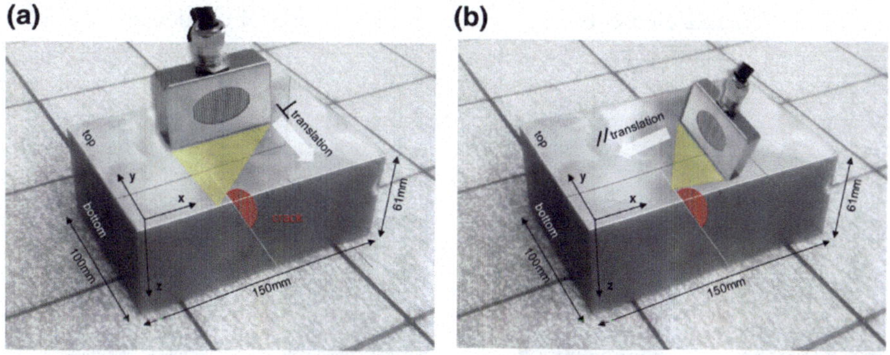

Fig. 5.32 Schematic of AISI304 stainless steel specimen with the position of the array transducer and the two configurations: a perpendicular scan (i.e., translation along the y-direction with the focal points set in the x–z plane) and a parallel scan (i.e., translation along the x-direction with the focal points set in the y–z plane). The plane in which the crack exists was the y–z plane. Modified from [79] with permission from Elsevier

exists) and (2) across the crack (i.e., translation in the x-direction in steps of $\Delta x =$ 1 mm; the array transducer was in the same plane (y–z) as the crack plane). For the electronic scan, the focal points were set at $z = 57$ mm, corresponding to the position of the crack tip, and from −15 to 15 mm in steps of 250 μm in the x–z plane for the perpendicular scan or in the y–z plane for the parallel scan. Sixteen acquisitions were performed at each position and averaged to achieve a high SNR.

5.4.3 Imaging Results

First, the experiment for demonstrating the importance of large-amplitude incidence in fixed-voltage FAD was carried out in the experimental configuration shown in Fig. 5.29. Figure 5.33 shows linear (T_{All}) and nonlinear ($= T_{\mathrm{Odd}} + T_{\mathrm{Even}} - T_{\mathrm{All}}$) images obtained for the transmitting apertures with 32, 64, 96, and 128 elements and schematic illustrations.

When the transmitting aperture had 32 elements (Fig. 5.33a), corresponding to an incident wave amplitude of 26.5 nm at the crack tip, the fatigue crack and the starting notch were visualized in the linear image (Fig. 5.33b). The intensity of the crack was even weaker than that of the notch N_T. Although the crack is visible, accurate crack depth measurement was difficult because of the low SNR. In the nonlinear image (Fig. 5.33c), the notch, which was a strong linear scatterer, was successfully canceled. This shows the fundamental performance of fixed-voltage FAD in terms of the capability of suppressing the artifacts due to strong linear scatterers. However, the crack response at C_2 can be seen to be very weak. This may be because the incident wave amplitude (i.e., 26.5 nm) for the transmitting aperture with 32 elements was insufficient to cause the contact vibration of the crack.

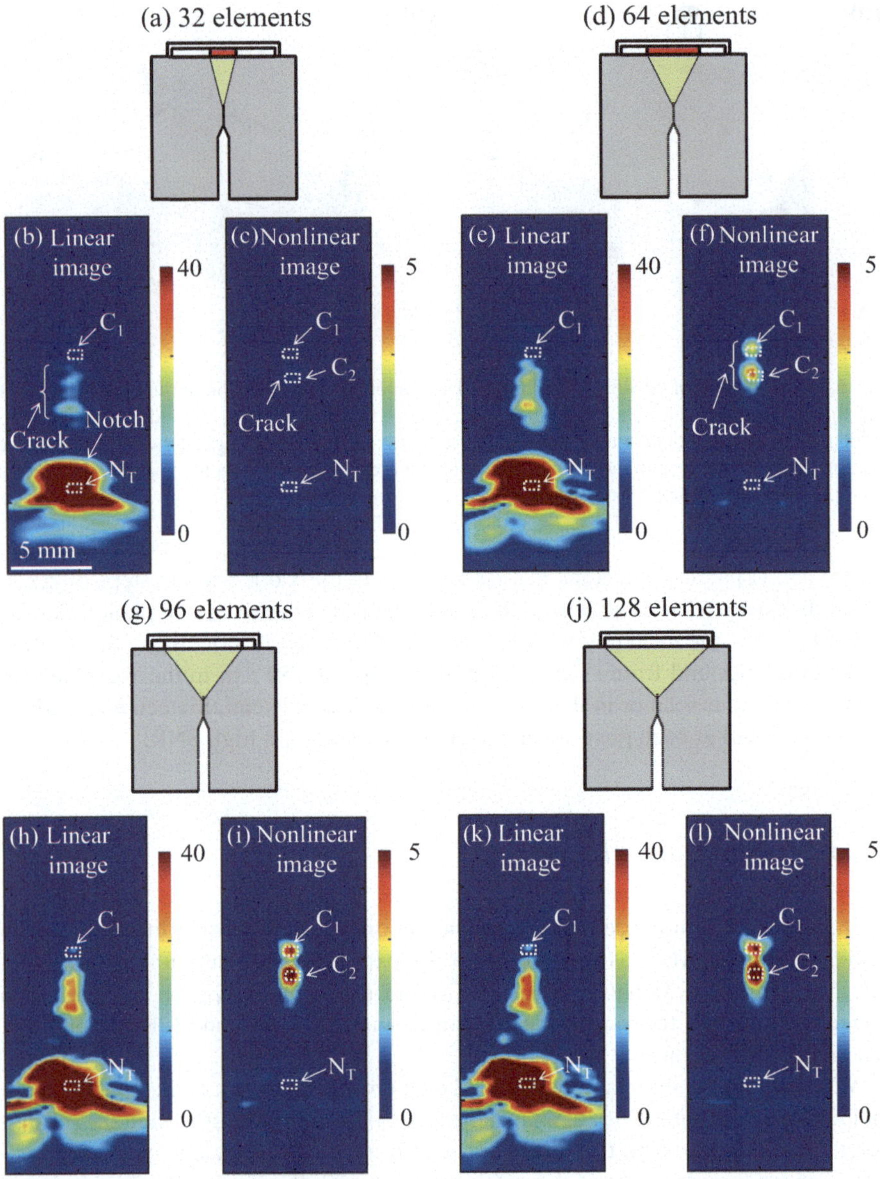

Fig. 5.33 Images of the closed fatigue crack obtained by linear PA and fixed-voltage FAD. **a** Schematic illustration and **b** linear and **c** nonlinear images for the transmitting aperture with 32 elements. **d** Schematic illustration and **e** linear and **f** nonlinear images for the transmitting aperture with 64 elements. **g** Schematic illustration and **h** linear and **i** nonlinear images for the transmitting aperture with 96 elements. **j** Schematic illustration and **k** linear and **l** nonlinear images for the transmitting aperture with 128 elements. Modified from [80] (Copyright (2019) Acoustical Society of America)

When the transmitting aperture had 64 elements (Fig. 5.33d), corresponding to an incident wave amplitude of 41.9 nm at the crack tip, the linear image (Fig. 5.33e) was very similar to that (Fig. 5.33b) for the transmitting aperture with 32 elements. On the other hand, in the nonlinear image (Fig. 5.33f), two strong crack responses appeared, in contrast to Fig. 5.33c. This suggests that the thresholds of the incident wave amplitudes at C_1 and C_2 were between the incident wave amplitudes for the transmitting apertures with 32 and 64 elements (i.e., 26.5 and 41.9 nm, respectively). Note that the response at N_T was successfully canceled, as with Fig. 5.33c.

When the transmitting apertures had 96 and 128 elements, corresponding to incident wave amplitudes of 50.3 and 55.4 nm, respectively, the linear images (Fig. 5.33h, k) did not change from those (Fig. 5.33b, e) for the transmitting apertures with 32 and 64 elements. This suggests that the linear images were insufficient to measure the closed crack depth even for a large-amplitude incidence. In the nonlinear images (Fig. 5.33i, l), the responses at C_1 and C_2 increased further, while the responses at N_T were still successfully canceled. For quantitative evaluation of the SNR achieved in the above images, the SNRs were examined using the equation

$$S = \frac{I_{C1}}{I_N}, \tag{5.41}$$

where I_{C1} and I_N are the intensities at C_1 and N_T, respectively, examined as the mean intensity in the area surrounded by a white dotted rectangle in Fig. 5.33. It was found that S was less than 0.1 for the linear images, irrespective of the incident wave amplitude. This shows that C_1 was invisible in the linear images (Fig. 5.33b, e, h, k). In contrast, in the nonlinear image (Fig. 5.33l) obtained with the transmission aperture of 128 elements, S was 5.3, which was more than 30 dB higher than those in the linear images. Thus, the above experiment and analysis explicitly reveal that large-amplitude incidence is one of the most important parameters for achieving the high-SNR imaging of closed cracks in fixed-voltage FAD.

In the previous experiment, the imaging capability of fixed-voltage FAD with large-amplitude incidence was demonstrated in the mechanical-fatigue-crack specimen. Here, the applicability of fixed-voltage FAD to an AISI304 coarse-grained specimen with a thermal fatigue crack was explored, where mechanical scans of the array transducer were carried out in two directions.

Figure 5.34 shows the linear (upper) and nonlinear (lower) images at the position where the highest SNR was obtained by the mechanical scan (i.e., perpendicular scan) in y-direction. Here, both images are plotted in dB, where 0 dB is defined as the maximum amplitude of the signal in the linear image, i.e., the echo from the back wall, for the comparison of both results. In the linear image (obtained by the transmission mode: T_{All}), the crack appeared as a consequence of weak linear scattering. The SNR, however, was not high because of the structural noise due to grains over the imaging area. In the nonlinear image, the crack was clearly visualized with a higher SNR than that in the linear image. In contrast to the noise like speckle observed in the linear image, the noise pattern appeared as sparse spikes, which can be attributed to fast low thermal and/or electronic fluctuations. Note that strong residue

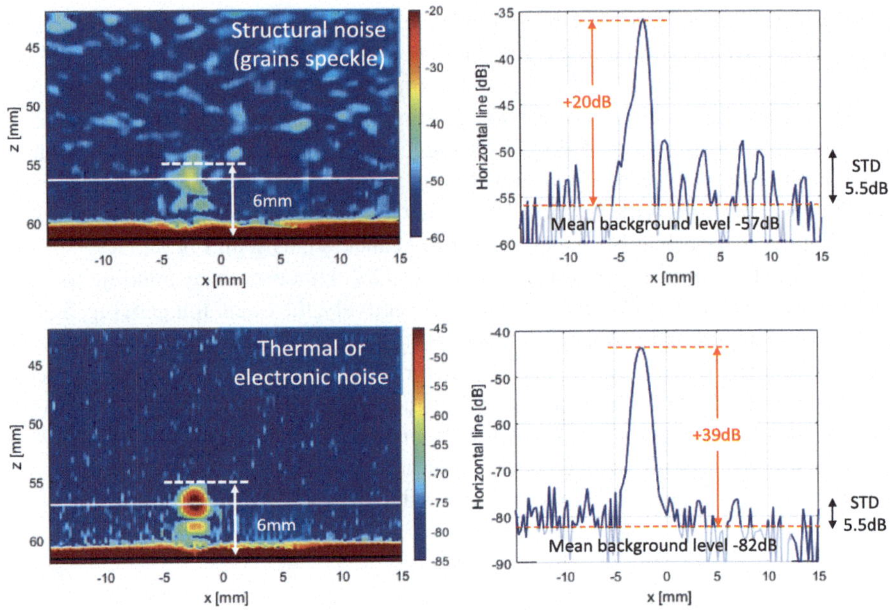

Fig. 5.34 Comparison of crack response and background noise between the linear image (upper) and the nonlinear image (lower) obtained by fixed-voltage FAD performed with the mechanical scan across the crack direction (i.e., perpendicular scan). Modified from [79] with permission from Elsevier

can be seen at the back wall in the nonlinear image, although the intensity of the back wall in the nonlinear image was reduced by 36 dB in comparison with that in the linear image. This shows that the fixed-voltage FAD could not perfectly cancel the strong linear reflector, presumably because of the nonlinearity of the instrumentation as also discussed in Sect. 5.3. Although this issue can be mitigated by adopting the compensation method described in Sect. 5.3.3, the present experimental setup achieved a higher contrast in the nonlinear image (i.e., 39 dB) than that in the linear image (i.e., 20 dB), with the background noise reduced from −57 dB in the linear image to −82 dB in the nonlinear image.

Figure 5.35 shows the linear (upper) and nonlinear (lower) images at the position where the highest SNR was obtained by the mechanical scan (i.e., parallel scan) in the *x*-direction. In the linear image (obtained by the transmission mode: T_{All}), the crack was visualized; however, the estimation of the crack shape was difficult because of the low SNR. In the nonlinear image, the crack was clearly visualized with a higher SNR than that in the linear image. Interestingly, a semielliptical crack profile was revealed, as outlined by the white curve. This also suggests that an electronic scan in the *y*-direction is suitable for visualizing an entire crack. Thus, the efficacy of fixed-voltage FAD was demonstrated for the coarse-grained stainless steel with a thermal fatigue crack.

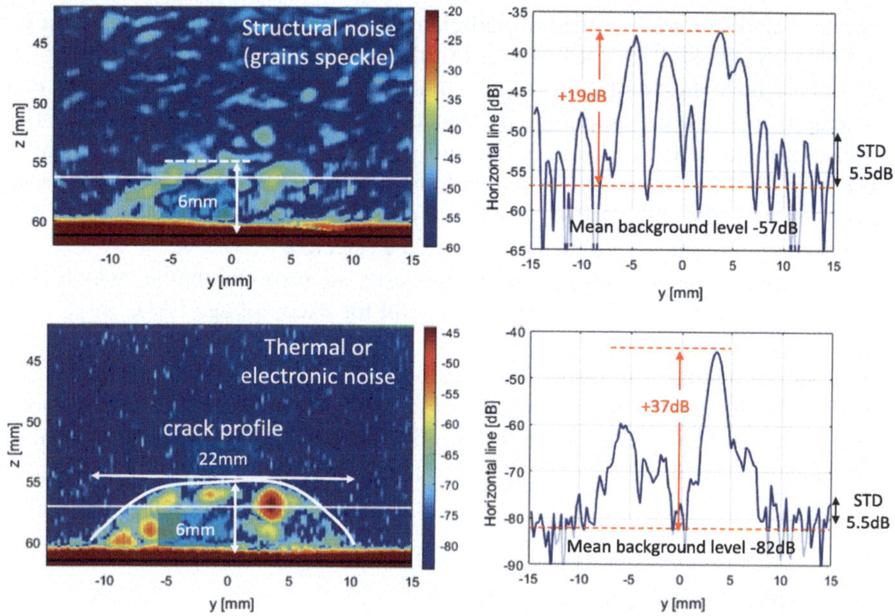

Fig. 5.35 Comparison of crack response and background noise between the linear image (upper) and the nonlinear image (lower) obtained by fixed-voltage FAD performed with the mechanical scan along the crack direction (i.e., parallel scan). To compare both results, images are plotted in dB, where 0 dB is defined as the maximum amplitude of the signal in the linear image, i.e., the echo from the back wall. Modified from [79] with permission from Elsevier

Finally, the important points of this section and points to be considered in the application of the discussed methods are summarized as follows:

- All nonlinear components can be indirectly measured through the incident-wave-amplitude dependence of the fundamental responses. This enables us to make use of the center frequency of an array transducer.
- The concept of fixed-voltage FAD is essentially similar to that of coherent-field measurement for parallel and sequential transmission, except for the transmission modes for small-amplitude incidence.
- The implementability of fixed-voltage FAD into commercial PA hardware may be higher than that of the methods introduced in Sect. 5.3, since the transmission modes employed are similar to those in typical linear PA, and the ADCs installed in typical commercial PA hardware will be sufficient.
- This technique can be applied to large structures and absorptive materials because directly scattered waves (i.e., in a coherent field) are used similarly to typical PA.
- Fixed-voltage FAD with an amplitude ratio of 2 was introduced in this section. On the other hand, the amplitude ratio can be increased by using a narrower transmitting aperture (e.g., one-third aperture, one-quarter aperture, and one-eighth

aperture, and so forth.) instead of odd-elements and even-elements transmission, and thereby, the SNR can be enhanced further [79].

- To obtain the highest amplitude ratio, a transmitting aperture with an element can be employed for small-amplitude incidence, which is equivalent to the sequential transmission described in Sect. 5.3. In this case, fixed-voltage FAD is identical to coherent field measurement for parallel and sequential transmission. It has also been reported that the SNR saturates with increasing amplitude ratio [79], and thus, an appropriate amplitude ratio should be selected.
- The compensation method for the nonlinearity of instrumentation, which was described in Sect. 5.3 [77], will be very useful for fixed-voltage FAD.
- Comparison between linear and nonlinear images is quite important to show the necessity of nonlinear ultrasonic PA and to gain physical insight.
- The bandwidth of both the incident and received waves should be checked to choose an appropriate bandwidth for the BPF used to effectively extract the fundamental component.

5.5 Utilization of Thermal Stress

5.5.1 Principles

Sections 5.2–5.4 involve the nonlinear ultrasonic PAs categorized into method A, in which a large-amplitude probe wave is used in a MHz-order frequency range. Such a large-amplitude probe wave, however, cannot induce the contact vibration of crack faces if cracks are tightly closed with a compressive residual stress larger than the tensile stress of the large-amplitude probe wave. Furthermore, some hardware requirements (e.g., the fabrication of a LN transmitter, the synchronization of two pieces of hardware when using LN SPACE, modification of the PA hardware to realize the transmission modes and post-processing when using the methods described in Sects. 5.3 and 5.4) must be satisfied to implement method A. In this section, we describe a nonlinear ultrasonic PA categorized into method B, which involves combining a commercial PA with a cooling spray, which is used to temporarily apply a tensile thermal stress on a closed crack [46]. This is one of the methods combining a small-amplitude probe wave with a large-displacement pump wave, and can be very easily implemented without any hardware modification. Further, it has excellent imaging capability for very tightly closed cracks that cannot be visualized by the other nonlinear ultrasonic PA methods.

A brief historical context of the combination of PA with low-frequency excitation is summarized as follows. In 2011, the combination of PA with low-frequency loading by a servohydraulic testing machine was proposed [15], which produced the imaging results shown in Fig. 5.3. This will be categorized into method B when the low-frequency loading is regarded as pump excitation. The hydraulic testing machine used for low-frequency loading, however, can only be used for lab experiments.

For the pump excitation, low-frequency vibration by a mechanical shaker, such as a piezoelectric disk, a piezostack, a giant magnetostrictive vibrator, or an ultrasonic welder [36, 42, 70–72, 89–93], has been used with a monolithic transducer (i.e., not PA). This approach is very effective for thin objects because the vibration energy of the pump wave is confined in small-volume objects, whereas it may not be easy to focus the vibration energy at a closed crack when objects are thick. As an alternative means, the application of thermal stress induced by a cooling spray [94, 95], laser irradiation [96–101], or an ice cylinder [102] has been proposed for enhancing the detectability of cracks in UT. Among them, a cooling spray will be the most practical approach since it is low-cost and easy to use. (A cooling spray is essentially the same as an air duster.) On the other hand, as described later, it was found that the use of only a cooling spray was sometimes insufficient to open a very tightly closed crack because the applicable thermal stress is restricted because of the cooling temperature, determined by the cooling medium in the spray. To solve this problem, global preheating and local cooling (GPLC) was proposed as a means of applying an arbitrary thermal stress [46, 103–106], which can be used in combination with PA to monitor the change in the crack state. GPLC can be regarded as a pump excitation with a very low frequency; hence, the combination of PA with GPLC is categorized into method B.

The principle of GPLC is illustrated with that of load difference phased array (LDPA) in Fig. 5.36. The following description is for GPLC. Suppose that a specimen has a crack with a closed tip and that it is visualized by PA. After the global preheating (GP) of the specimen, an open crack is imaged by PA (Fig. 5.36a), whereas a closed crack tip is not imaged because ultrasound penetrates through the closed part. Subsequently, by cooling the top surface of the specimen instantaneously via the injection of a cooling spray, only the vicinity of the top surface thermally contracts, and thereby, tensile thermal stress is applied to the closed crack by a principle similar to that of a three-point bending test. It should be noted that the use of only local cooling (LC) can also apply tensile thermal stress to the crack. However, it may not open tightly closed cracks because of the limited cooling temperature, determined by the cooling medium. In contrast, the thermal tensile stress can be increased by carrying out GP before LC since the temperature difference between the top surface and the crack region increases [46]. Therefore, by selecting an appropriate GP temperature, the closed crack tip can be opened. As a result, the crack tip can be imaged by PA (Fig. 5.36b).

On the other hand, the PA image shown in Fig. 5.36b includes not only the closed crack but also other linear scatterers such as coarse grains. In principle, this will make it difficult to achieve a high SNR in closed-crack imaging. To extract only closed cracks, LDPA [15] with thermal stress [46, 103–106] will be useful. LDPA is a method of taking the difference between PA images at different loads (or stresses). When GPLC is applied, the response of the closed crack tip appears, whereas the responses of the other linear scatterers including coarse grains are not changed since they are independent of tensile stress. Therefore, by subtracting the PA images obtained before the onset of LC from those after the onset, only the crack

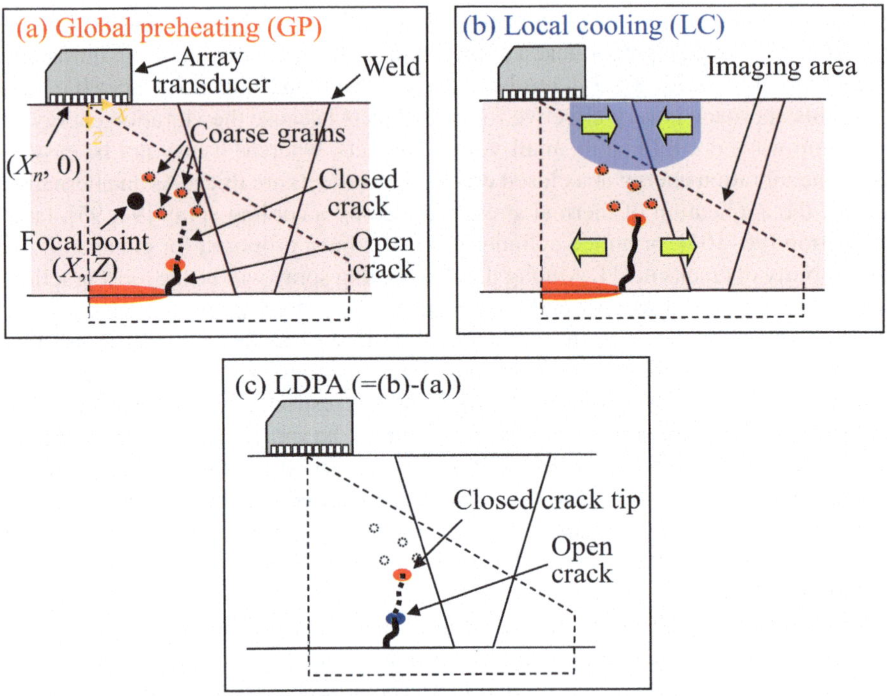

Fig. 5.36 Principles of GPLC and LDPA. PA images **a** before and **b** after LC. **c** LDPA image obtained by subtracting (**a**) from (**b**). The area surrounded by the dashed line denotes the area imaged by PA. The yellow arrows denote the direction of thermal stress. The origin of the coordinates is the center of the array transducer. Modified from [106], with permission from Elsevier

can be selectively extracted while canceling the other linear scatterers, as illustrated in Fig. 5.36c.

The schematic illustration shown in Fig. 5.36 is also used to formulate the imaging algorithm. Here, the origin of the x–z Cartesian coordinate system is at the center of the array transducer on the x-axis and at the surface of the specimen on the z-axis. The specimen to be inspected is assumed to be isotropic and homogeneous. Under this assumption, the propagation time and delay law are identical to those for single-array SPACE (Sect. 5.2). The propagation time $t_{\mathrm{T}}(X, Z, n)$ and delay law $\Delta t_{\mathrm{T}}(X, Z, n)$ for focusing ultrasound at $F(X, Z)$ on transmission are expressed as Eqs. (5.13) and (5.14), respectively. Likewise, the propagation time $t_{\mathrm{R}}(X, Z, n)$ and delay law $\Delta t_{\mathrm{R}}(X, Z, n)$ for focusing on reception are expressed as Eqs. (5.15) and (5.16), respectively. Assuming that the signals received at the nth element for focusing on transmission are $u_{n,0}(t)$ and $u_{n,\mathrm{LC}}(t)$ before and after the onset of LC, the corresponding DAS signals, $U_0(X, Z, t)$ and $U_{LC}(X, Z, t)$, can be expressed by

$$U_0(X, Z, t) = \frac{1}{N}\sum_{n=1}^{N} u_{n,0}(t - \Delta t_{\mathrm{R}}(X, Z, n)), \tag{5.42}$$

$$U_{LC}(X, Z, t) = \frac{1}{N}\sum_{n=1}^{N} u_{n,LC}(t - \Delta t_R(X, Z, n)), \tag{5.43}$$

respectively. The scattering intensities $I_0(X, Z)$ and $I_{LC}(X, Z)$ are calculated as the RMS values of $U_0(X, Z, t)$ and $U_{LC}(X, Z, t)$, respectively, as

$$I_0(X, Z) = \left(\frac{1}{\Delta\tau} \int_{t_R(X,Z,1)}^{t_R(X,Z,1)+\Delta\tau} U_0^2(X, Z, t)dt \right)^{1/2}, \tag{5.44}$$

$$I_{LC}(X, Z) = \left(\frac{1}{\Delta\tau} \int_{t_R(X,Z,1)}^{t_R(X,Z,1)+\Delta\tau} U_{LC}^2(X, Z, t)dt \right)^{1/2}. \tag{5.45}$$

The LDPA image can be obtained by subtracting Eq. (5.44) from Eq. (5.45):

$$I_{LDPA}(X, Z) = I_{LC}(X, Z) - I_0(X, Z). \tag{5.46}$$

By plotting $I_0(X, Z)$ and $I_{LDPA}(X, Z)$ at each pixel over an imaging area, the linear and LDPA images are produced, where the LDPA image corresponds to a nonlinear image.

5.5.2 Experimental Conditions

To demonstrate the principle and efficacy of GPLC, the conventional method of using only LC and the proposed method, GPLC, were applied to the A7075 aluminum alloy CT specimen ($V_L = 6230$ m/s) with a fatigue crack [15, 46, 80] used to demonstrate fixed-voltage FAD in Sect. 5.4. Figure 5.37 shows the experimental configuration. A 32-element array transducer (5 MHz, pitch 0.5 mm) manufactured by Imasonic (Voray-sur-l'Ognon, France) was operated by PA hardware (PAL) produced by KJTD (Tokyo, Japan). Each element was excited by a pulse of 96 V. The sampling rate was 50 MS/s. The DAS processing was carried out in 1 mm steps in depth and 1° steps in angle, following the delay law formulated as Eq. (5.16). For the conventional method of using only a cooling spray (i.e., LC) [94, 95], the vicinity of the top surface was locally cooled from the area of the top surface confined by the acrylic lid at room temperature by two cooling sprays (HFC-125a) for 10 s. During cooling, the crack was monitored by PA in real time. Subsequently, an experiment involving GPLC was carried out. The tensile thermal stress applied to a crack depends on the temperature difference between the top surface and the crack area. Here, the GP temperature was selected to be 323 K and the specimen was heated by a hot plate. Subsequently, the

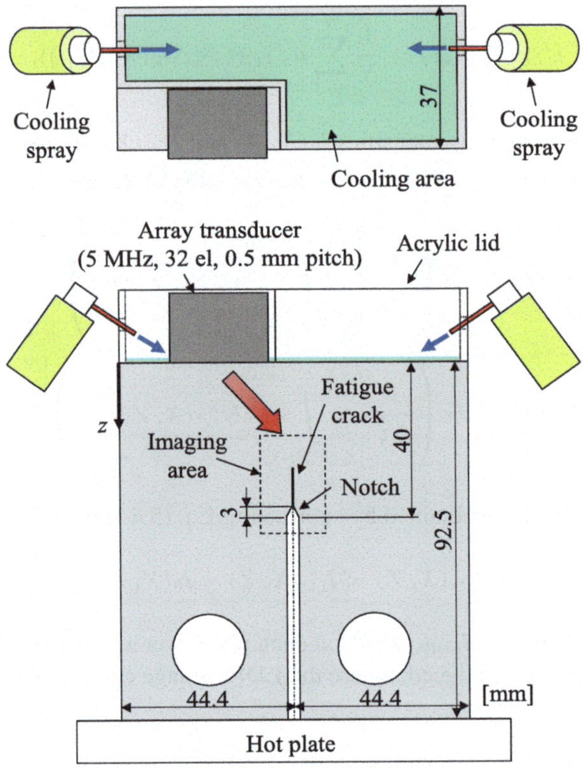

Fig. 5.37 Experimental configuration for PA with GPLC. The CT specimen had a closed crack extending from a notch. The specimen was placed on a hot plate for GP. The hot plate was not used for the experiment using only LC. The top surface confined by an acrylic lid was cooled by two cooling sprays (HFC-132a) from the left and right sides. Here, the two commercially available cooling sprays were used for rapid cooling to 218 K over the confined area. The area surrounded by the dashed line was imaged by the PA in real time while applying thermal stress. Taken from [106], with permission from Elsevier

same area (Fig. 5.37) as that in the experiment using only LC was cooled by two cooling sprays for 10 s.

As a preliminary experiment on the A7075 CT specimen for reference, the closed crack was imaged by PA before and after applying a mechanical load by a servo-hydraulic testing machine. When no load was applied, as shown in Fig. 5.38a, the notch was imaged, whereas the crack was not imaged. This shows that the crack was closed. When a stress intensity factor K of 6.8 MPa m$^{1/2}$ was mechanically applied to the crack, the crack became open, as visualized in the PA image (Fig. 5.38b), from which the crack depth was measured to be 11.3 mm. Obtaining this crack depth as a value for comparison was the goal in the measurement using the thermal stress.

After the experiment on the CT specimen (A7075), GPLC and LDPA were applied to the CT specimen (coarse-grained SUS316L austenitic stainless steel, $V_L =$

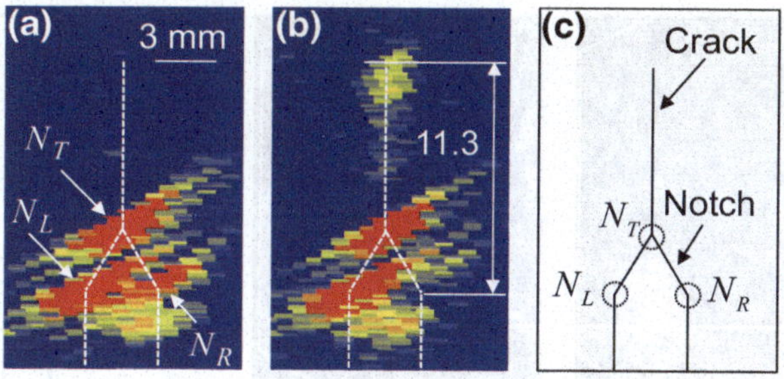

Fig. 5.38 PA images of the closed crack in the A7075 CT specimen, in which a mechanical load was applied with the servohydraulic testing machine, for the measurement of the actual crack depth and schematic illustration. **a** PA images for $K = 0$ and **b** $K = 6.8$ MPa m$^{1/2}$. **c** Schematic of imaging area. The white dotted lines in (**a**) and (**b**) denote the outlines of the notch and crack. N_T, N_L, and N_R denote the top, left side, and right side of the notch, respectively. Note that the response at N_L was stronger than that at N_R. Here, the array was positioned on the left side of the top surface, as shown in Fig. 5.37. Given the geometric relationship between the array and the notch, it is conceivable that the scattered wave fields from N_L and N_R mainly exist on the left and right sides of the crack, respectively. Therefore, the array can receive stronger scattered waves from N_L than those from N_R. Taken from [46] (Copyright 2013 © American Institute of Physics)

5900 m/s) with a fatigue crack [15, 46, 80] to demonstrate the efficacy of combining GPLC and LDPA. The shape of the CT specimen was identical to that of the previous CT specimen, as shown in Fig. 5.37. To introduce a closed fatigue crack, a stepwise-decremental method was employed [106]; K_{max} was gradually decreased in a stepwise manner with the crack propagation, while R was maintained at less than 0.1 during the fatigue test. K_{max} and K_{min} were initially set to 18.6 and 0.6 MPa m$^{1/2}$, respectively; $\Delta K = 18.0$ MPa m$^{1/2}$. Each time the crack propagated by 1 mm, K_{max} was decreased by 1 MPa m$^{1/2}$. During the fatigue test, K_{max} was automatically controlled on the basis of the crack depth measured by a clip gauge. Finally, the fatigue-crack growth (FCG) rate became extremely low when K_{max} became 8.6 MPa m$^{1/2}$. Here, $\Delta K = 8.0$ MPa m$^{1/2}$, which is in good agreement with the FCG threshold reported in the literature [107]. The advantage of this method is that it reduces the formation time of closed cracks because the FCG rate before entering the near-threshold regime was much higher than that in the near-threshold regime. In addition, in the fatigue test with a constant ΔK, it is difficult to select a near-threshold ΔK because of the variability of the FCG threshold, whereas in this method, the FCG automatically stops in the near-threshold regime; a tightly closed crack can be formed without selecting a near-threshold ΔK even if the near-threshold ΔK varies. Thus, a tightly closed fatigue crack was formed in the SUS316L CT specimen. Note that this specimen was a coarse-grained material; such materials are frequently used in actual structures and components. Thus, this specimen is suitable for demonstrating the imaging capability of GPLC for practical applications. The experimental conditions for GPLC were the same as those for the A7075 CT specimen, except for the cooling time, which was selected to be 20 s. This is longer than that for the A7075 CT specimen, because of the difference in the thermal conductivity between A7075 and SUS316L.

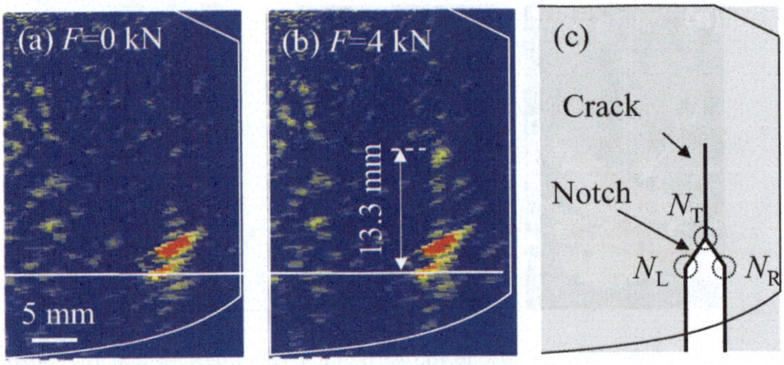

Fig. 5.39 PA images of the closed crack in the SUS316L CT specimen, in which a mechanical load was applied with the servohydraulic testing machine, for the measurement of the actual crack depth and schematic illustration. **a** PA images for $F = 0$ and **b** $F = 4$ kN. **c** Schematic of imaging area. N_T, N_L, and N_R denote the top, left side, and right side of the notch, respectively. Note that the response at N_L was stronger than that at N_R. Here, the array was positioned on the left side of the top surface, as shown in Fig. 5.37. Given the geometric relationship between the array and the notch, it is conceivable that the scattered wave fields from N_L and N_R mainly exist on the left and right sides of the crack, respectively. Therefore, the array can receive stronger scattered waves from N_L than those from N_R. In contrast to Fig. 5.38, speckle noise appeared over the imaging area because of multiple linear scattering at coarse grain boundaries. Modified from [106], with permission from Elsevier

As a preliminary experiment on the SUS316L CT specimen for reference, the closed crack was imaged by PA before and after applying a mechanical load by a servohydraulic testing machine. When no load was applied, as shown in Fig. 5.39a, the notch was imaged, whereas the crack was not imaged. This shows that the crack was originally closed. When a load of 4 kN was mechanically applied to the crack, the crack became open, as shown in the PA image (Fig. 5.39b), from which the crack depth was tentatively measured to be 13.3 mm. Note that the crack depth measured here may not be accurate because of the low SNR due to coarse grains and therefore is a tentative goal in the measurement using the thermal stress.

5.5.3 Imaging Results

To demonstrate the efficacy of GPLC, an experiment was first carried out on the basis of the conventional method [94, 95] of using only a cooling spray (i.e., LC). As illustrated in Fig. 5.37, the vicinity of the top surface was locally cooled from the area of the top surface confined by the acrylic lid at room temperature by two cooling sprays (HFC-125a) for 10 s. During cooling, the crack was monitored by PA in real time, as shown in Fig. 5.40. Before cooling, the crack was not imaged (Fig. 5.40a). In contrast, after the onset of LC, the crack was clearly imaged (Fig. 5.40c, d). The maximum crack depth was measured to be 7.9 mm in Fig. 5.40c at 4 s after the onset of LC. The measured crack depth, however, was smaller than the actual value

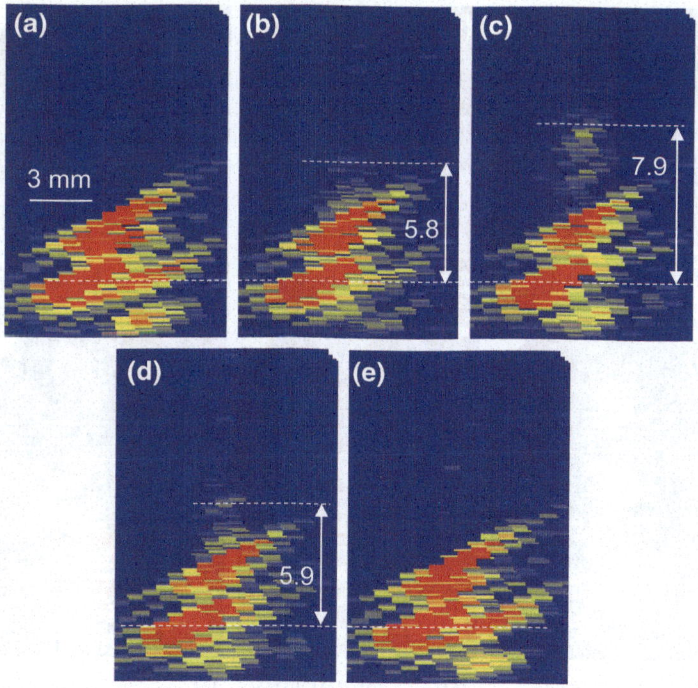

Fig. 5.40 Snapshots of PA images with local cooling (LC). **a** PA image before LC and **b–e** PA images taken at 2, 4, 10, and 15 s after the onset of LC, respectively. Modified from [46] (Copyright 2013 © American Institute of Physics)

of 11.3 mm, which was confirmed in the preliminary experiment (Fig. 5.38). This reveals that the crack tip was still closed and the tensile thermal stress was insufficient to open the tightly closed crack by the conventional method of using LC.

To overcome this problem, GPLC was applied to the same specimen. In this experiment, the specimen was globally preheated to 323 K by GP. Snapshots of the PA images obtained before and after the onset of LC are shown in Fig. 5.41a–f. Before LC, the crack was not imaged (Fig. 5.41a), similarly to the case shown in Fig. 5.40a. After the onset of LC, the crack was clearly imaged (Fig. 5.41b–d). Note that the maximum crack depth was measured to be 11.3 mm in Fig. 5.41d at 4 s, in contrast to the case in Fig. 5.40c. Importantly, the measured crack depth is the same as the actual value of 11.3 mm (Fig. 5.38). This reveals that GPLC is effective for opening a tightly closed crack. Then, as the cooling time increased from 4 s, the crack appeared in a shallower part, as shown in Fig. 5.41d. Finally, it diminished at 15 s (Fig. 5.41e). This is because the thermal stress decreased since the temperature distribution within the specimen gradually became uniform. Figure 5.42 shows the change in the crack depth d measured in the PA images with increasing time t after the onset of LC. This can be understood from the schematic illustrations of the crack state (Fig. 5.41g–l). At $t \leq 2$ s, $d \leq 8.7$ mm. This shows that the crack was partially

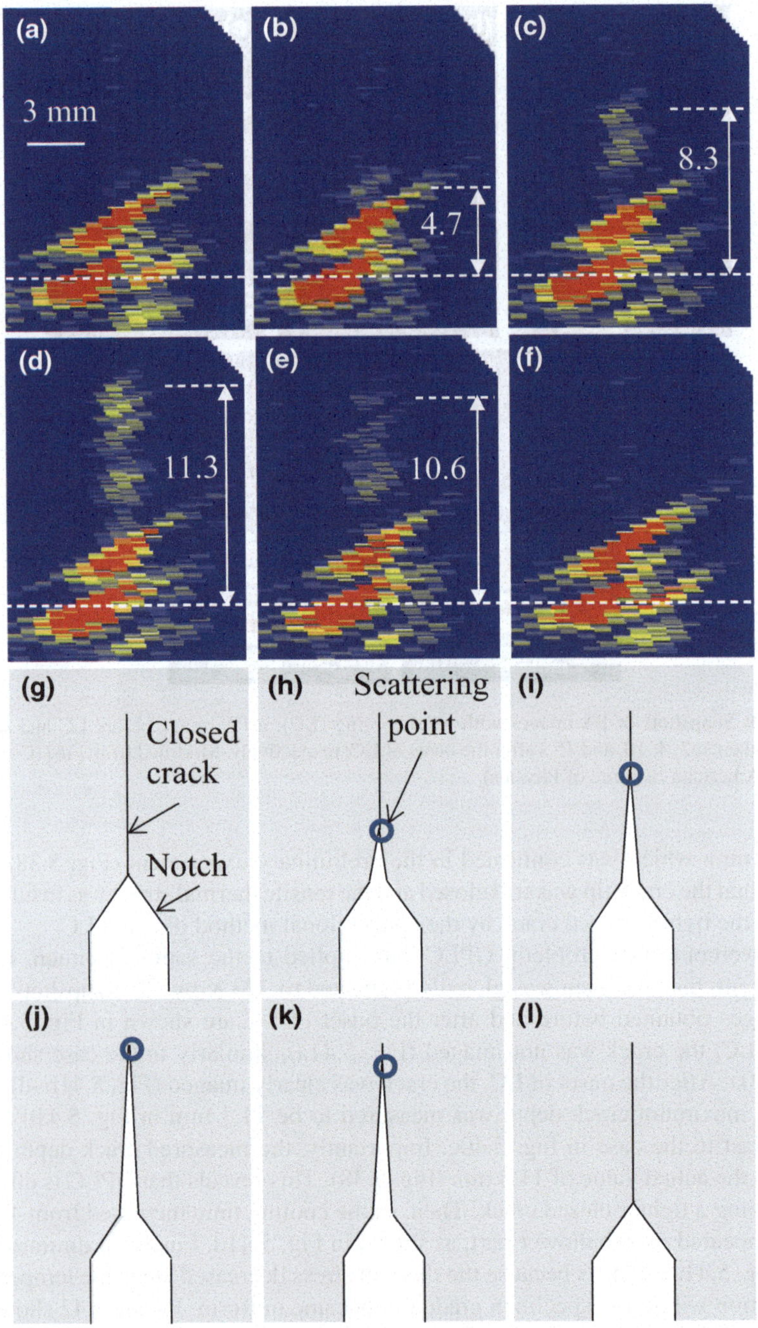

Fig. 5.41 Snapshots of PA images with GP (323 K) and LC: **a** PA image before LC and **b**–**f** PA images at $t = 1, 2, 4, 10$, and 15 s, respectively, after the onset of LC, respectively. **g**–**l** Schematics of crack states for (**a**–**f**), respectively. In **h**–**k**, the circle represents the scattering point of the incident wave. Taken from [104] (Copyright (2014) The Japan Society of Applied Physics)

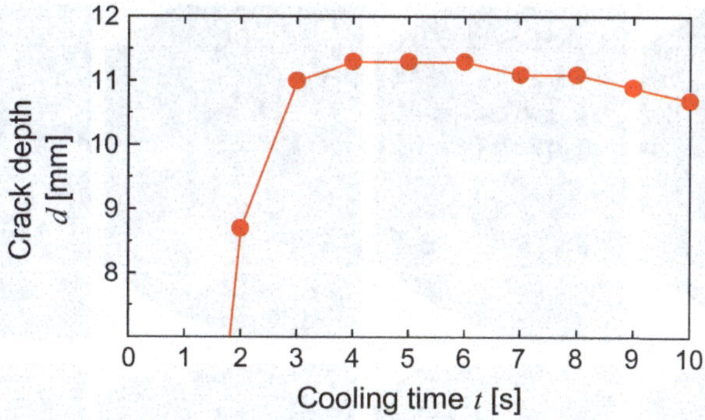

Fig. 5.42 Crack depth as a function of cooling time. The red circles denote the crack depths measured in the PA images (Fig. 5.41b–e) obtained by applying GPLC. Modified from [46] (Copyright 2013 © American Institute of Physics)

open since the thermal stress was lower than the crack closure stress. At t = 4 s, d reached a maximum value of 11.3 mm. This suggests that the thermal stress at t = 4 s was higher than the crack closure stress. From the practical viewpoint, this analysis suggests that the short cooling time of approximately 4 s was sufficient for the accurate measurement of the crack depth in this specimen.

After confirming the superior performance of GPLC over LC, the combination of GPLC and LDPA was applied to the coarse-grained SUS316L CT specimen with the closed crack. Snapshots of the PA images obtained while applying GPLC are shown in Fig. 5.43. At t = 0 and 1 s (Fig. 5.43a, b, respectively), the crack was not imaged, although the top and left side of the notch were imaged. Note that speckle noise appeared over the imaging area, which was due to multiple linear scattering at the coarse grain boundaries. At t = 2 s (Fig. 5.43c), the crack gradually appeared in the PA image. At $t > 3$ s (Fig. 5.43d–f), the intensity of the response of the crack at a depth of 13.3 mm gradually increased with increasing t. At $t \geq 10$ s (Fig. 5.43g, h), the intensity was sufficiently higher than the speckle noise due to the coarse grains. The results show that the closed crack was opened by the thermal tensile stress induced by GPLC. It is, however, still difficult to identify the position of the crack tip because of the low SNR, although the crack depth of 13.3 mm measured in Fig. 5.43d–h was equal to that measured in Fig. 5.39. It is possible that the response of the crack tip may be obscured by the strong linear scattering at the coarse grain boundaries.

To selectively extract the change in the response of the crack because of the application of the thermal stress, LDPA was applied to the PA images (Fig. 5.43). In LDPA, the PA image before LC (Fig. 5.43a) was subtracted from the PA images after the onset of LC, as shown in Fig. 5.44. The detailed changes in the crack depth measured in the PA and LDPA images are shown in Fig. 5.45. Note that the LDPA images correspond to nonlinear images. At t = 0.2 s, the linear scatterings at the coarse grains and the notch were successfully canceled because they were independent of the thermal stress. It was also confirmed that there was no response, suggesting that the crack was still closed. At t = 1 s, an increase in the response

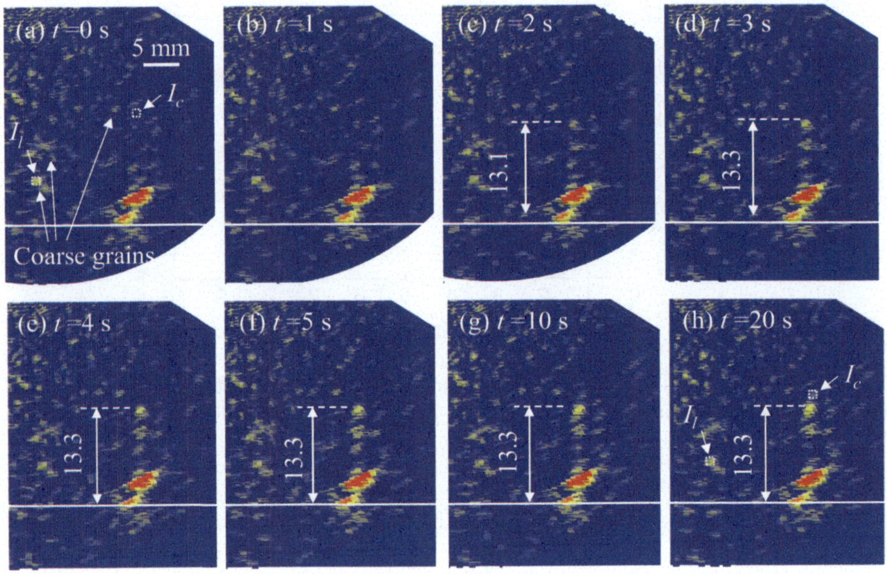

Fig. 5.43 Snapshots of PA images obtained while applying GPLC. **a** PA image before LC, **b**–**h** PA images between $t = 1$ s and $t = 20$ s after the onset of LC. Taken from [106], with permission from Elsevier

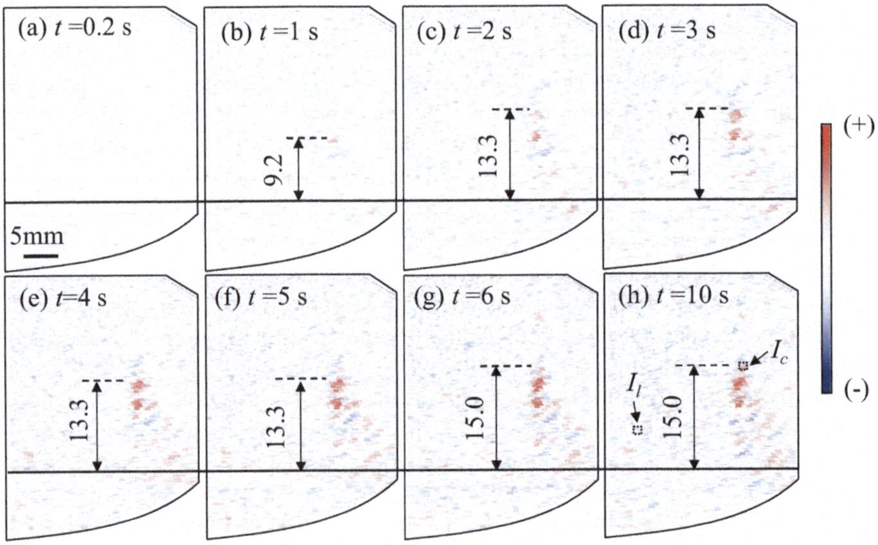

Fig. 5.44 Snapshots of LDPA images obtained by subtracting the PA image before LC (Fig. 5.43a) from that at t between 0.2 and 10 s. Modified from [106], with permission from Elsevier

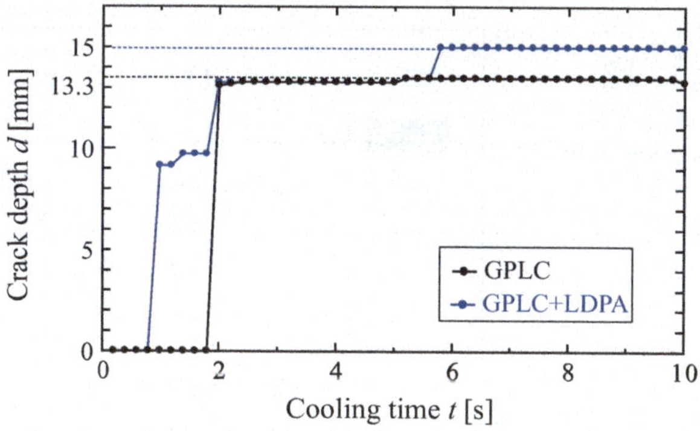

Fig. 5.45 Crack depth as a function of cooling time. The black circles denote the crack depths measured in the PA images (Fig. 5.43) obtained by applying GPLC. The blue circles denote the crack depths measured in the LDPA images (Fig. 5.44). Modified from [106], with permission from Elsevier

at the crack appeared at a depth of 9.2 mm, which was obscured by the responses due to coarse grains in Fig. 5.43b. Note that because the intensity was low, the crack was not observed in the PA image (Fig. 5.43b). Nevertheless, the change in the crack response was successfully observed because the linear scattering was eliminated by LDPA. Between $t = 2$ and 5 s (Fig. 5.44c–f), the intensities of the crack response at 9.2 and 13.3 mm increased with increasing t. The greatest depth of the crack in Fig. 5.44c–f was the same as that observed in the PA images (Fig. 5.43d–h). At $t \geq 6$ s (Fig. 5.44g, h), an increase in the intensity of the response of the crack was observed at a depth of 15.0 mm, which could not be identified in the PA images (Fig. 5.43) because of the speckle noise.

Finally, the selectivity of cracks for coarse grains was quantitatively examined in Figs. 5.43a, h and 5.44h. As a measure of the selectivity, the intensity ratio of the cracks to the linear scatterers (coarse grains) is defined as

$$S = \frac{I_c}{I_l}, \tag{5.47}$$

where I_c is the intensity of the response at the closed crack tip and I_l is the intensity of the response at the coarse grains. These were calculated as mean intensity values in the region surrounded by the dotted square in Figs. 5.43a, h and 5.44h. As shown in Fig. 5.46, S was improved by applying GPLC, although it was still lower than unity. In contrast, S was markedly enhanced by 25.8 dB by applying GPLC and LDPA. Thus, it was demonstrated that the combination of GPLC and LDPA is useful for imaging a tightly closed fatigue crack in a coarse-grained specimen, in spite of the very simple means of using only commercial PA and cooling sprays.

Finally, the important points of this section and points to be considered in the application of the discussed methods are summarized as follows:

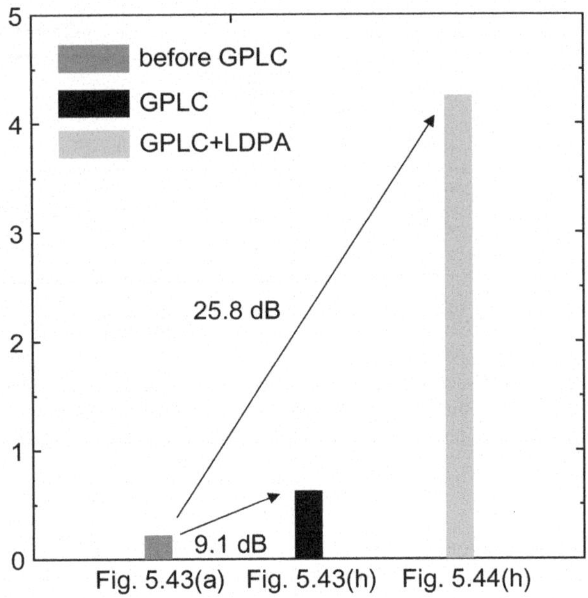

Fig. 5.46 Selectivity of the closed crack against coarse grains in the SUS316L CT specimen. Modified from [105], with permission from Elsevier

- The implementability of GPLC is excellent because it is a very simple method of using only commercial PA and cooling sprays, where the latter are essentially equivalent to air dusters, making the method very cheap to employ. No modification of commercially available PA hardware is required. The cost of this method will be the lowest among (I)–(IV) summarized in Sect. 4.1.
- The tensile thermal stress that can be induced by cooling sprays is much larger than the stress of the large-amplitude incidence in (I)-(III). At present, only this technique has the potential to visualize a very tightly closed crack in practical applications.
- The thermal stress can be arbitrarily changed by varying the temperature of GP.
- LDPA is useful in canceling linear scatterers in PA images. LDPA images can be regarded as nonlinear images.
- A GP temperature sufficiently below the Curie temperature of an array transducer can be used to apply a large tensile thermal stress.
- The thermal stress induced by GPLC can be roughly calculated analytically [103–105]. The crack closure stress can be also estimated by comparing the analytical results with experimental results.
- GP was carried out using a hot plate to prove the concept, whereas more practical heating units including flash and halogen lamps, fan heaters, belt heater, and so forth, can be readily employed.
- The repetition of GPLC is time-consuming because one has to wait until the temperature of the specimen has returned to its original value (e.g., room temperature). The inspection of a large region by, for example, the mechanical scan of an array transducer, while applying GPLC may not be realistic. On the other hand, the

detection of a closed crack may be easy by measuring the echo from the root of the crack, because the root of a crack is typically open. It will be a smart way that GPLC and LDPA is applied only to areas with cracks detected by, for example, real-time PA to achieve accurate and reliable crack depth measurement.

5.6 Conclusions

In this chapter, the nonlinear ultrasonic PA methods based on PA with nonlinear ultrasonics for the measurement of closed-crack depth were comprehensively reviewed. First, various nonlinear ultrasonic PA methods were categorized into four groups: (I) subharmonics, (II) parallel and sequential transmission, (III) all-elements, odd-elements, and even-elements transmission, and (IV) utilization of thermal stress. A common point among these methods is the utilization of the contact vibration of crack faces. In (I)–(III), a large-amplitude incident (probe) wave with a MHz-order frequency range is employed to induce the contact vibration. In (IV), a thermal stress is applied as a large-displacement pump excitation while the effect of the contact vibration is examined by a small-amplitude probe wave in a MHz-order frequency range. Specifically, (I) is based on the measurement of a specific nonlinear component, i.e., subharmonics, generated by large-amplitude ultrasonic incidence. Both (II) and (III) are based on the measurement of the fundamental component to indirectly measure all nonlinear components, although they utilize different transmission modes and post-processing. (IV) involves the combination of a pump wave (i.e., thermal stress induced by cooling sprays) and a probe wave (i.e., PA). The efficacy of each technique was demonstrated using specific closed-crack specimens, and each technique has its own merits and demerits as summarized in the last part of each section. Thus, after understanding the key features of each method and the points to be considered in its use, an appropriate technique should be selected depending on the objective. On the other hand, nonlinear ultrasonics and PA have been evolving, which may lead to more powerful nonlinear ultrasonic PA methods. Current and future nonlinear ultrasonic PA methods will be available for actual industrial applications as key techniques to solve serious problems that cannot be resolved by conventional (i.e., linear) UT.

Acknowledgements It is our great pleasure to thank all those who have collaborated with us regarding nonlinear ultrasonic PA. Financial support by Japan Society for the Promotion of Science (JSPS) KAKENHI and other various projects for part of the work described in this chapter is gratefully acknowledged.

References

1. J. Blitz, G. Simpson, *Ultrasonic Methods of Non-Destructive Testing* (Chapman & Hall, London, 1996)
2. L.W. Schmerr, *Fundamentals of Ultrasonic Nondestructive Evaluation* (Plenum, New York, 1998)
3. J.D. Achenbach, Quantitative nondestructive evaluation. Int. J. Solids Struct. **37**, 13–27 (2000)
4. B.W. Drinkwater, P.D. Wilcox, Ultrasonic arrays for non-destructive evaluation: a review. NDT&E Int. **39**, 525–541 (2006)
5. L.W. Schmerr, *Fundamentals of Ultrasonic Phased Array* (Springer, Cham, 2015)
6. T.L. Szabo, *Diagnostic Ultrasound Imaging: Inside Out* (Academic, New York, 2004)
7. S.-C. Wooh, Y. Shi, Optimum beam steering of linear phased arrays. Wave Motion **29**, 245–265 (1999)
8. B. Puel, D. Lesselier, S. Chatillon, P. Calmon, Optimization of ultrasonic arrays design and setting using a differential evalution. NDT&E Int. **44**, 797–803 (2011)
9. D.H. Johnson, D.E. Dudgeon, *Array Signal Processing, Concepts and Techniques* (Prentis Hall, Upper Saddle River, 1993)
10. C. Holmes, B.W. Drinkwater, P.D. Wilcox, Post-Processing of the full matrix of ultrasonic transmit-receive array data for non-destructive evaluation. NDT&E Int. **38**, 701–711 (2005)
11. M.-L. Zhu, F.-Z. Xuan, S.-T. Tu, Effect of load ratio on fatigue crack growth in the near-threshold regime: a literature review, and a combined crack closure and driving force approach. Eng. Fract. Mech. **141**, 57–77 (2015)
12. S. Horinouchi, M. Ikeuchi, Y. Shintaku, Y. Ohara, K. Yamanaka, Evaluation of closed stress corrosion cracks in Ni-based alloy weld metal using subharmonic phased array. Jpn. J. Appl. Phys. **51**, 07GB15-1-5 (2012)
13. J.D. Frandsen, R.V. Inman, O. Buck, A comparison of acoustic and strain gauge techniques for crack closure. Int. J. Fract. **11**, 345–348 (1975)
14. T. Mihara, S. Nomura, M. Akino, K. Yamanaka, Relationship between crack opening behavior and crack tips scattering and diffraction of longitudinal waves. Mater. Eval. **62**, 943–947 (2004)
15. Y. Ohara, T. Mihara, K. Yamanaka, Nonlinear ultrasonic imaging method for closed cracks using subtraction of responses at different external loads. Ultrasonics **51**, 661–666 (2011)
16. W. Elber, Fatigue crack closure under cyclic tension. Eng. Fract. Mech. **2**, 37–45 (1970)
17. A.T. Stewart, The influence of environment and stress ratio on fatigue crack growth at near threshold stress intensities in low-alloy steels. Eng. Fract. Mech. **13**, 463–478 (1980)
18. R.O. Ritchie, S. Suresh, C.M. Moss, Near-threshold fatigue crack growth in 2 1/4Cr-1Mo pressure vessel steel in air and hydrogen. J. Eng. Mater. Tech. **102**, 293–299 (1980)
19. K. Minakawa, A.J. McEvily, On crack closure in the near-threshold region. Scr. Metall. **15**, 633–636 (1981)
20. Y. Zheng, R.G. Maev, I.Y. Solodov, Nonlinear acoustic applications for material characterization: a review. Can. J. Phys. **77**, 927–967 (1999)
21. K.-Y. Jhang, Nonlinear ultrasonic techniques for nondestructive assessment of micro damage in material: a review. Int. J. Precis. Eng. Manuf. **10**(1), 123–135 (2009)
22. K.H. Matlack, J.-Y. Kim, L.J. Jacobs, J. Qu, Review of second harmonic generation measurement techniques for material state determination in metals. J. Nondestruct. Eval. **34**, 273-1-23 (2015)
23. T. Kundu (ed.), *Nonlinear Ultrasonic and Vibro-Acoustical Techniques for Nondestructive Evaluation* (Springer, New York, 2018)
24. M.A. Breazeale, D.O. Thompson, Finite-amplitude ultrasonic waves in aluminum. Appl. Phys. Lett. **3**(5), 77–78 (1963)
25. O. Buck, W.L. Morris, J.M. Richardson, Acoustic harmonic generation at unbonded interfaces and fatigue cracks. Appl. Phys. Lett. **33**(5), 371–373 (1978)
26. I.Y. Solodov, Ultrasonics of non-linear contacts: propagation, reflection and NDE-applications. Ultrasonics **36**, 383–390 (1998)

27. I.Y. Solodov, N. Krohn, G. Busse, CAN: an example of nonclassical acoustic nonlinearity in solids. Ultrasonics **40**, 621–625 (2002)
28. R.A. Guyer, P.A. Johnson, Nonlinear mesoscopic elasticity: evidence for a new class of materials. Phys. Today **52**, 30–36 (1999)
29. R.A. Guyer, P.A. Johnson, *Nonlinear Mesoscopic Elasticity: The Complex Behaviour of Rocks, Soil, Concrete* (Wiley, New York, 2009)
30. M.C. Remillieux, T.J. Ulrich, H.E. Goodman, J.A. Ten Cate, Propagation of a finite-amplitude elastic pulse in a bar of Berea sandstone: a detailed look at the mechanisms of classical nonlinearity, hysteresis, and nonequilibrium dynamics. J. Geophys. Res. Sol. Earth **122**(11), 8892–8909 (2017)
31. H. Ogi, M. Hirao, S. Aoki, Noncontact monitoring of surface-wave nonlinearity for predicting the remaining life of fatigued steels. J. Appl. Phys. **90**(1), 438–442 (2001)
32. Y. Ohara, K. Kawashima, Detection of internal micro defects by nonlinear resonant ultrasonic method using water immersion. Jpn. J. Appl. Phys. **43**(5B), 3119–3120 (2004)
33. S. Biwa, S. Hiraiwa, E. Matsumoto, Pressure-dependent stiffnesses and nonlinear ultrasonic response of contacting surfaces. J. Sol. Mech. Mater. Eng. **3**(1), 10–21 (2009)
34. I.Y. Solodov, C.A. Vu, Popping nonlinearity and chaos in vibrations of a contact interface between solids. Acoust. Phys. **39**, 476–479 (1993)
35. B.A. Korshak, I.Y. Solodov, E.M. Ballad, DC effects, sub-harmonics, stochasticity and "memory" for contact acoustic non-linearity. Ultrasonics **40**, 707–713 (2002)
36. I. Solodov, J. Wackerl, K. Pfleiderer, G. Busse, Nonlinear self-modulation and subharmonic acoustic spectroscopy for damage detection and location. Appl. Phys. Lett. **84**, 5386–5388 (2004)
37. K. Yamanaka, T. Mihara, T. Tsuji, Evaluation of closed cracks by model analysis of subharmonic ultrasound. Jpn. J. Appl. Phys. **43**, 3082–3087 (2004)
38. Y. Ohara, T. Mihara, K. Yamanaka, Effect of adhesion force between crack planes on subharmonic and DC responses in nonlinear ultrasound. Ultrasonics **44**, 194–199 (2006)
39. J.G. Sessler, V. Weiss, Crack Detection Apparatus and Method. US Patent, 38667836 (1975)
40. K.E.-A. Van Den Abeele, P.A. Johnson, A. Sutin, Nonlinear elastic wave spectroscopy (NEWS) techniques to discern material damage, part I: nonlinear wave modulation spectroscopy (NWMS). Res. Nondestr. Eval. **12**, 17–30 (2000)
41. D. Donskoy, A. Sutin, A. Ekimov, Nonlinear acoustic interaction on contact interfaces and its use for nondestructive testing. NDT&E Int. **34**, 231–238 (2001)
42. V.V. Kazakov, A. Sutin, P.A. Johnson, Sensitive imaging of an elastic nonlinear wave-scattering source in a solid. Appl. Phys. Lett. **81**(4), 646–648 (2002)
43. Y. Ohara, T. Mihara, R. Sasaki, T. Ogata, S. Yamamoto, Y. Kishimoto, K. Yamanaka, Imaging of closed cracks using nonlinear response of elastic waves at subharmonic frequency. Appl. Phys. Lett. **90**, 011802-1-3 (2007)
44. J.N. Potter, A.J. Croxford, P.D. Wilcox, Nonlinear ultrasonic phased array imaging. Phys. Rev. Lett. **113**, 144031-1-5 (2014)
45. S. Haupert, G. Renaud, A. Schnumm, Ultrasonic imaging of nonlinear scatterers buried in a medium. NDT&E Int. **87**, 1–6 (2017)
46. Y. Ohara, K. Takahashi, S. Murai, K. Yamanaka, High-selectivity imaging of closed cracks using elastic waves with thermal stress induced by global preheating and local cooling. Appl. Phys. Lett. **103**, 031917-1-5 (2013)
47. Y. Ohara, S. Yamamoto, T. Mihara, K. Yamanaka, Ultrasonic evaluation of closed cracks using subharmonic phased array. Jpn. J. Appl. Phys. **47**(5), 3908–3915 (2008)
48. S. Yamamoto, Y. Ohara, T. Mihara, K. Yamanaka, Application of laser interferometer to subharmonic phased array for crack evaluation (SPACE). J. Jpn. Soc. Nondestr. Insp. **57**(4), 198–203 (2008)
49. Y. Ohara, H. Endo, T. Mihara, K. Yamanaka, Ultrasonic measurement of closed stress corrosion crack depth using subharmonic phased array. Jpn. J. Appl. Phys. **48**(7), 07GD01-1-6 (2009)

50. Y. Ohara, Y. Shintaku, S. Horinouchi, M. Hashimoto, Y. Yamaguchi, M. Tagami, K. Yamanaka, Ultrasonic imaging of stress corrosion crack formed in high temperature pressurized water using subharmonic phased array. Proc. Mtgs. Acoust. **10**, 045007-1-8 (2010)
51. Y. Ohara, H. Endo, M. Hashimoto, K. Yamanaka, Monitoring growth of closed fatigue crack using subharmonic phased array. AIP Conf. Proc. **1211**, 903–909 (2010)
52. Y. Ohara, S. Horinouchi, Y. Shintaku, R. Shibasaki, Y. Yamaguchi, M. Tagami, K. Yamanaka, High-selectivity imaging of closed cracks in weld part of stainless steel using subharmonic phased array with a single array transducer. J. Jpn. Soc. Nondestr. Insp. **60**(11), 658–664 (2011)
53. K. Yamanaka, Y. Ohara, M. Oguma, Y. Shintaku, Two-dimensional analyses of subharmonic generation at closed cracks in nonlinear ultrasonics. Appl. Phys. Express **4**, 076601-1-3 (2011)
54. Y. Ohara, Y. Shintaku, S. Horinouchi, M. Ikeuchi, K. Yamanaka, Enhancement of selectivity in nonlinear ultrasonic imaging of closed cracks using amplitude difference phased array. Jpn. J. Appl. Phys. **51**, 07GB18-1-6 (2012)
55. K. Jinno, A. Sugawara, Y. Ohara, K. Yamanaka, Analysis on nonlinear images of vertical closed cracks by damped double node model. Mater. Trans. **55**(7), 1017–1023 (2014)
56. T. Mihara, H. Ishida, Improvement in the identification of a crack tip echo in ultrasonic inspection using large displacement ultrasound transmission. J. Phys. Conf. Ser. **520**, 012010-1-6 (2014)
57. A. Ouchi, A. Sugawara, Y. Ohara, K. Yamanaka, Subharmonic phased array for crack evaluation using surface acoustic wave. Jpn. J. Appl. Phys. **54**, 07HC05-1-6 (2015)
58. A. Sugawara, K. Jinno, Y. Ohara, K. Yamanaka, Closed-crack imaging and scattering behavior analysis using confocal subharmonic phased array. Jpn. J. Appl. Phys. **54**, 07HC08-1-8 (2015)
59. C.-S. Park, J.-W. Kim, S. Cho, D.-C. S, A high resolution approach for nonlinear sub-harmonic imaging. NDT&E Int. **79**, 114–122 (2016)
60. Y. Ohara, J. Potter, S. Haupert, H. Nakajima, T. Tsuji, T. Mihara, Multi-mode nonlinear ultrasonic phased array for closed crack imaging. Proc. Mtgs. Acoust. **34**, 055001-1-5 (2018)
61. Y. Ohara, J. Potter, H. Nakajima, T. Tsuji, T. Mihara, Multi-mode nonlinear ultrasonic phased array for imaging closed cracks. Jpn. J. Appl. Phys. **58**, SGGB06-1-7 (2019)
62. I.Y. Solodov, N. Krohn, G. Busse, *Nonlinear Ultrasonic NDT for Early Defect Recognition and Imaging*. Proceedings of 10th European Conference on Non-Destructive Testing (2010)
63. R. Koda, T. Mihara, K. Inoue, G. Konishi, Y. Udagawa, Transmission of larger amplitude ultrasound with SiC transistor pulser for subharmonic signal measurements at closed cracks. Phys. Proc. **70**, 528–531 (2015)
64. M. Scalerandi, A.S. Gliozzi, C.L.E. Bruno, D. Masera, P. Bocca, A scaling method to enhance detection of a nonlinear elastic response. Appl. Phys. Lett. **92**, 101912-1-3 (2008)
65. C.L.E. Bruno, A.S. Gliozzi, M. Scalerandi, P. Antonaci, Analysis of elastic nonlinearity using the scaling subtraction method. Phys. Rev. B **79**, 0641108-1-13 (2009)
66. M. Scalerandi, M. Griffa, P. Antonaci, M. Wyrzykowski, P. Lura, Nonlinear elastic response of thermally damaged consolidated granular media. J. Appl. Phys. **113**, 154902-1-9 (2013)
67. P. Antonaci, C.L.E. Bruno, M. Scalerandi, F. Tondolo, Effects of corrosion on linear and nonlinear elastic properties of reinforced concrete. Cem. Concr. Res. **51**, 96–103 (2013)
68. M. Ikeuchi, K. Jinno, Y. Ohara, K. Yamanaka, Improvement of closed crack selectivity in nonlinear ultrasonic imaging using fundamental wave amplitude difference. Jpn. J. Appl. Phys. **52**, 07HC08-1-5 (2013)
69. Y. Ohara, K. Yamanaka, Japan Patent, 6,025,049 (2016)
70. X. Han, W. Li, Z. Zeng, L.D. Favro, R.L. Thomas, Acsoutic chaos and sonic infrared imaging. Appl. Phys. Lett. **81**, 3188–3190 (2002)
71. F. Mabrouki, M. Thomas, M. Genest, A. Fahr, Frictional heating model for efficient use of vibrothermography. NDT&E Int. **42**, 345–352 (2009)
72. L. Pieczonka, F. Aymerich, G. Brozek, M. Szwedo, W.J. Staszewski, T. Uhl, Nonlinear vibroacoustic wave modulations for structural damage detection: an overview. Struct. Control Health Monit. **20**, 626–638 (2013)

73. I. Solodov, G. Busse, Resonance ultrasonic thermography: highly efficient contact and air-coupled remote modes. Appl. Phys. Lett. **102**, 061905-1-3 (2013)
74. K. Truyaert, V. Aleshin, K.V.D. Abeele, S. Delrue, Theoretical calculation of the instantaneous friction-induced energy losses in arbitrarily excited axisymmetric mechanical contact systems. Int. J. Solids Struct. **158**, 268–276 (2019)
75. J.N. Potter, J. Chen, A.J. Croxford, B.W. Drinkwater, Ultrasonic phased array imaging of contact acoustic nonlinearity. Proc. Mtgs. Acoust. **29**, 045002-1-6 (2016)
76. J. Cheng, J.N. Potter, A.J. Croxford, B.W. Drinkwater, Monitoring fatigue crack growth using nonlinear ultrasonic phased array imaging. Smart Mater. Struct., **26**, 05506-1-10 (2017)
77. J. Cheng, J.N. Potter, B.W. Drinkwater, The parallel-sequential field subtraction technique for coherent nonlinear ultrasonic imaging. Smart Mater. Struct. **27**, 065002-1-10 (2018)
78. J. Potter, A.J. Croxford, Characterization of nonlinear ultrasonic diffuse energy imaging. IEEE Trans. Ultrason. Ferroelectr. Freq. Control **65**(5), 870–880 (2018)
79. S. Haupert, Y. Ohara, E. Carcreff, G. Renaud, Fundamental wave amplitude difference imaging for detection and characterization of embedded cracks. Ultrasonics **96**, 132–139 (2019)
80. Y. Ohara, H. Nakajima, S. Haupert, T. Tsuji, T. Mihara, Nonlinear ultrasonic phased array with fixed-voltage fundamental wave amplitude difference for high-selectivity imaging of closed cracks. J. Acoust. Soc. Am. **146**(1), 266–277 (2019)
81. Y. Ohara, H. Nakajima, T. Tsuji, T. Mihara, Nonlinear surface-acoustic-wave phased array with fixed-voltage fundamental wave amplitude difference for imaging closed cracks. NDT&E Int. **108**, 102170–1–10 (2019)
82. G. Tang, L.J. Jacobs, J. Qu, Scattering of time-harmonic elastic waves by an elastic inclusion with quadratic nonlinearity. J. Acoust. Soc. Am. **131**, 2570–2578 (2012)
83. C.M. Kube, Scattering of harmonic waves from a nonlinear elastic inclusion. J. Acoust. Soc. Am. **141**, 4756–4767 (2017)
84. Y. Wang, J.D. Achenbach, Reflection of ultrasound from a region of cubic material nonlinearity due to harmonic generation. Acta Mech. **229**, 763–778 (2018)
85. N. Walker, C.J. Beevers, A fatigue crack closure mechanism in titanium. Fatigue Eng. Mater. Struct. **1**, 135–148 (1979)
86. J. Jin, J. Rivière, Y. Ohara, P. Shokouhi, Dynamic acoustic-elastic response of single fatigue cracks with different microstructural features: an experimental investigation. J. Appl. Phys. **124**, 075303-1-14 (2018)
87. A. Steuwer, M. Rahman, A. Shterenlikht, M.E. Fitzpatrick, L. Edwards, P.J. Withers, The evolution of crack-tip stresses during a fatigue overload event. Acta Mater. **58**, 4039–4052 (2010)
88. J.D. Carroll, W. Abuzaid, J. Lambros, H. Sehitoglu, High resolution digital image correlation measurements of strain accumulation in fatigue crack growth. Int. J. Fatigue **57**, 140–150 (2013)
89. I.Y. Solodov, B.A. Korshak, Instability, chaos, and 'memory' in acoustic-wave-crack interaction. Phys. Rev. Lett. **88**(1), 014303-1-3 (2001)
90. A. Moussatov, V. Gusev, B. Castagnede, Self-induced hysteresis for nonlinear acoustic waves in cracked material. Phys. Rev. Lett. **90**(12), 124301-1-4 (2003)
91. R.B. Mignogna, R.E. Green Jr., J.C. Duke, E.G. Henneke, K.L. Reifsnifer, Thermographic investigation of high-power ultrasonic heating in materials. Ultrasonics **19**, 159–163 (1981)
92. I. Solodov, G. Busse, Nonlinear air-coupled emission: the signature to reveal and image microdamage in solid materials. Appl. Phys. Lett. **91**, 251910-1-3 (2007)
93. I. Solodov, J. Bai, S. Bekgulyan, G. Busse, A local defect resonance to enhance acoustic wave-defect interaction in ultrasonic nondestructive evaluation. Appl. Phys. Lett. **99**, 211911-1-3 (2011)
94. P.B. Nagy, G. Blaho, Identification of distributed fatigue cracking by dynamic crack-closure. Rev. Prog. Quant. Nondestr. Eval. **14**, 1979–1986 (1995)
95. S.R. Ahmed, M. Saka, Y. Matsuura, D. Kobayashi, Y. Miyachi, Y. Kagiya, An effective method of local thermal treatment for sensitive NDE of closed surface cracks. Res. Nondestruct. Eval. **21**, 51–70 (2009)

96. H. Xiao, P.B. Nagy, Enhanced ultrasonic detection of fatigue cracks by laser-induced crack closure. J. Appl. Phys. **83**(12), 7453–7460 (1998)
97. Z. Yan, P.B. Nagy, Thermo-optical modulation of ultrasonic surface waves for NDE. Ultrasonics **40**, 689–696 (2002)
98. C.-Y. Ni, N. Chigarev, V. Tournat, N. Delorme, Z.-H. Shen, V.E. Gusev, Probing of laser-induced crack modulation by laser-monitored surface waves and surface skimming bulk waves. JASA Express Lett. **131**(3), EL250–EL255 (2012)
99. C. Ni, N. Chigarev, V. Tournat, N. Delorme, Z. Shen, V.E. Gusev, Probing of laser-induced crack closure by pulsed laser-generated acoustic waves. J. Appl. Phys. **113**, 014906-1-8 (2013)
100. S. Mezil, N. Chigarev, V. Tournat, V. Gusev, Two dimensional nonlinear frequency-mixing photo-acoustic imaging of a crack and observation of crack phantoms. J. Appl. Phys. **114**, 174901-1-17 (2013)
101. S. Mezil, N. Chigarev, V. Tournat, V. Gusev, Evaluation of crack parameters by a nonlinear frequency-mixing laser ultrasonics method. Ultrasonics **69**, 225–235 (2016)
102. H. Tohmyoh, M. Saka, Y. Kondo, Thermal opening technique for nondestructive evaluation of closed cracks. J. Pressure Vessel Technol. **129**, 103–108 (2007)
103. Y. Ohara, K. Takahashi, K. Jinno, K. Yamanaka, High-selectivity ultrasonic imaging of closed cracks using global preheating and local cooling. Mater. Trans. **55**(7), 1003–1010 (2014)
104. K. Takahashi, K. Jinno, Y. Ohara, K. Yamanaka, Evaluation of crack closure stress by analyses of ultrasonic phased array images during global preheating and local cooling. Jpn. J. Appl. Phys. **53**, 07KC20-1-7 (2014)
105. K. Tkahashi, K. Ohmachi, Y. Ohara, K. Yamanaka, Estimation of saturated duration in phased array imaging of closed cracks by global preheating and local cooling. J. Jpn. Soc. Nondestr. Inspect. **65**(10), 513–520 (2016)
106. Y. Ohara, K. Takahashi, Y. Ino, K. Yamanaka, T. Tsuji, T. Mihara, High-selectivity imaging of closed cracks in a coarse-grained stainless steel by nonlinear ultrasonic phased array. NDT&E Int. **91**, 139–147 (2017)
107. N. Noraphaiphipaksa, T. Putta, A. Manonukul, C. Kanchanomai, Interaction of plastic zone, pores, and stress ratio with fatigue crack growth of sintered stainless steel. Int. J. Fract. **176**, 25–38 (2012)

Chapter 6
Nonlinear Frequency-Mixing Photoacoustic Characterisation of a Crack

Sylvain Mezil, Nikolay Chigarev, Vincent Tournat and Vitalyi Gusev

Abstract A one and two dimensional imaging of a crack by a novel nonlinear frequency-mixing photoacoustic method is presented. Acoustic waves are initiated by a pair of laser beams intensity-modulated at two different frequencies. The first laser beam, intensity modulated at a low frequency f_L, generates a thermoelastic wave which modulates the local crack rigidity up to complete closing/opening of the crack, corresponding to crack breathing. The second laser beam, intensity modulated at much higher frequency f_H, generates an acoustic wave incident on the breathing crack. The detection of acoustic waves at mixed frequencies $f_H \pm n f_L$ ($n = 1, 2, \ldots$), absent in the excitation frequency spectrum, provides detection of the crack, which can be achieved all-optically. The theory attributes the generation of the frequency-mixed spectral components to the modulation of the acoustic waves reflected/transmitted by the time-varying nonlinear rigidity of the crack. The crack rigidity is modified due to stationary and oscillating components from the laser-induced thermoelastic stresses. The amplitudes of the spectral sidelobes are non-monotonous functions of the increasing thermoelastic loading. Fitting such experimental evolutions with theoretical ones leads to estimating various local parameters of the crack, including its width and rigidity.

S. Mezil (✉) · N. Chigarev · V. Tournat · V. Gusev
LAUM, Le Mans Université, 72085 Le Mans, France
e-mail: sylvain.mezil@eng.hokudai.ac.jp

N. Chigarev
e-mail: nikolay.chigarev@univ-lemans.fr

V. Tournat
e-mail: vincent.tournat@univ-lemans.fr

V. Gusev
e-mail: vitali.goussev@univ-lemans.fr

K.-Y. Jhang et al. (eds.), *Measurement of Nonlinear Ultrasonic Characteristics*, Springer Series in Measurement Science and Technology,
https://doi.org/10.1007/978-981-15-1461-6_6

6.1 Introduction to Nonlinear Photoacoustics

Photoacoustics and laser ultrasonics are largely overlapping experimental techniques for monitoring acoustic waves with light and lasers [1–3]. Their essential common feature is the generation of acoustic waves by optical radiation, while the methods of acoustic waves detection could be various and multiple and, in principle, non-optical. Both techniques are widely used for material evaluation, non-destructive testing and imaging via generation and detection of linear acoustic waves. The application of light in acoustics provides well-known opportunities to conduct material evaluation without contact and from a distance, in a hostile environment, to evaluate large surface areas fast and with a high spatial resolution by using scanning focused laser beams. Lasers provide an opportunity to realise an efficient synchronous generation of the coherent acoustic waves [2, 4]. The application of ultrafast lasers provides the opportunity to generate and detect coherent acoustic waves up to THz frequencies (down to single digit nanometers wavelengths) which are not accessible by piezoelectric transducers [5–7]. Depth spatial resolution of imaging with laser-based acoustic waves is now already better than 100 nm in some applications [8–11], while lateral spatial resolution could be better than light diffraction limited by application of the near-field optical techniques [12, 13]. This provides an opportunity to apply laser-based acoustic waves for the evaluation of nanomaterials and nanostructures [11, 14–18]. There are many physical mechanisms leading to the transformation of the optical radiation into acoustic waves [2, 19, 20], the most known of them is thermoelastic expansion of the light absorbing materials and electrostriction of the transparent materials. There are also several opportunities to detect the acoustic waves, either when they arrive at the material surface by monitoring optically the motion of the material surface (via interferometry or beam deflection techniques) [1–3], or both in the bulk and near the surface based on the acousto-optic effect [11, 21, 22]. Thus, the optical techniques for the generation and detection of the acoustic waves are very universal. It would be highly advantageous for a variety of applications to apply these profitable features provided by optics to the monitoring of nonlinear acoustic waves as well [23–26].

In comparison with linear acoustic waves, nonlinear waves, i.e., finite amplitude acoustic waves, are modifying their frequency spectrum when propagating in materials. These modifications are due to the deviations of the sound propagation laws from the linear ones. Therefore, monitoring of the nonlinear acoustic waves modifications could provide fundamental knowledge on the nonlinear elasticity/inelasticity of the materials such as the information on interatomic potentials in perfect crystals [23, 27]. At the same time, strongly increased acoustic nonlinearity of the materials could be caused by the presence of defects, dislocations, cracks, actually of many possible types of material damage, caused by variety of reasons starting from penetrating radiations and finishing with different types of fatigue loading [25, 26, 28–30]. High sensitivity of several nonlinear acoustic phenomena to the presence of cracks or contacting interfaces in the material is well documented. Among them are the frequency-mixing (parametric modulation) [29, 31, 32], harmonics generation [33–35], subharmonics generation [35, 36], demodulation [36, 37], self-modulation [38, 39], modulation transfer [40, 41] and acoustoelasticity [42]. These different nonlinear acoustic phenomena are rather frequently applied for the evaluation of the materials and, particularly for the detection of the cracks [29, 43–45]. However, until now the advantages of the nonlinear acoustics methods are rarely combined with optical techniques for the generation and/or the detection of nonlinear acoustic waves, or even for the modulation of the acoustic waves.

6.1.1 An Overview of a Few Non-destructive Methods Combining Laser Optics with Nonlinear Acoustics

In nonlinear acoustics pulsed lasers have been more frequently applied for the generation of the coherent nonlinear acoustic pulses than continuous wave lasers. The formation of the weak shock fronts in laser generated bulk longitudinal acoustic pulses were first studied in liquids [2, 46, 47] and later in solids [48]. Nonlinear laser-generated acoustic pulses were applied to study materials exhibiting non-classical hysteretic nonlinearities such as the micro-inhomogeneous and thermally aged materials [49–51]. Recently, through the application of ultrafast lasers, experiments with nonlinear longitudinal pulses were extended to the GHz frequency range [52–54]. Ultrafast (femtosecond) lasers were also applied to excite and study acoustic solitons in crystals [55, 56]. The application of intense nanosecond laser pulses provided an opportunity to generate nonlinear pulsed surface acoustic waves and to study the phenomena of shock fronts and solitary waves formation in surface acoustic waves [57–61]. Very recently laser ultrasonics was extended to the evaluation of the nonlinear edge waves [62]. By creating a periodic optical pattern on the material surface (via slits mask) it was possible to generate surface acoustic wave packets of sufficient amplitude for the observation of harmonics generation [63, 64]. Quasi-monochromatic laser-generated surface acoustic wave packets were applied to study the phenomenon of acoustoelasticity caused by dynamic and static (mechanical and acoustical) loading of materials [65, 66].

Loading of a crack by laser-induced thermoelastic stresses caused by laser radiation absorption in the vicinity of the crack was first suggested to modulate reflection/transmission by the crack of surface acoustic waves generated by piezoelectrical transducers [67–69]. Later, thermoelastically-induced acoustoelasticity of the crack was studied with laser-generated surface Rayleigh and bulk skimming acoustic waves [70–72]. Lasers were, in general, more frequently used for the detection of the nonlinear acoustic waves than for their generation [73–75].

This chapter is devoted to the description of the nonlinear frequency-mixing photoacoustic experimental technique where the laser radiation is applied both for the generation of the monochromatic acoustic wave and periodic thermoelastic loading of the crack. This is accomplished at two different frequencies and the new spectral components excited by the frequency-mixing processes caused by the breathing crack nonlinearity are detected [76–84]. The technique was already applied to the imaging of cracks [77, 80, 81], to quantitative evaluation of the local parameters of a crack [83] and can be accompanied by the optical detection of the nonlinear acoustic waves to become all-optical [80]. More recently the nonlinear frequency mixing processes have been studied with laser generated narrowband surface Rayleigh waves [85] and laser-excited acoustic waves and vibrations of the structures [86, 87]. In all experimental realisations of the photoacoustic frequency-mixing technique reported until now, the acoustic frequencies were rather low and laser-induced generation of the acoustic waves was due to thermoelastic mechanism of optoacoustic conversion.

6.1.2 Generation of Thermoelastic Stresses and Acoustic Waves by Modulation of Continuous Wave Laser Radiation

Laser ultrasonics techniques can generate and detect acoustic waves in a solid through several physical mechanisms, depending on the laser type and power. Lasers can be divided into two main categories: continuous wave (cw) ones and pulsed ones. A cw laser continuously emits an electromagnetic wave of a constant amplitude at a constant optical frequency while a pulsed laser emits its optical power in pulses (quasi-monochromatic wave packets) of a certain duration at a given repetition rate.

Although in general the acoustic wave can be generated by lasers even in transparent materials (by electrostriction for example), the most common of the physical mechanisms of optoacoustic conversion relies on light absorption. This is possible when the solid is opaque to the laser wavelength (at least partially) in order to absorb some of the incident energy. In general, the amplitude of the photo-excited acoustic waves grows nearly proportionally to the laser intensity up to the material optical ablation threshold, where the non-destructive mechanisms of the optoacoustic transformation become destructive ones. This threshold depends on the sample properties (light absorption coefficient and reflectivity at the laser wavelength, material heat capacity, thermal conductivity, etc.). Below the ablation threshold, the generation most commonly takes place due to the thermoelastic effect, without damaging the material. Above the ablation threshold, the optical energy absorbed in the material is sufficient for its melting and evaporating in the vicinity of the surface. From momentum conservation, laser-induced emission of the material from the surface results in the recoil pressure on the material surface, which generates the acoustic waves. In the so-called thermoelastic regime, the motion of the material is induced by the bulk forces which are due to the spatial gradients in the thermoelastic stresses, and which are proportional to material elastic modulus, thermal expansion coefficient, and temperature. Thus, to induce the motion of the material by the laser, in particular in the form of the acoustic waves, the temperature gradients should be induced by the laser. This naturally takes place when the laser radiation, due to its absorption, penetrates inside the materials just to a certain depth, creating a temperature distribution which, even under the action of thermal conduction processes, is inhomogeneous with temperature maximum at the surface. However, the laser-induced temperature gradients is a necessary but not a sufficient condition for the generation of the acoustic waves. This can be understood by considering the cw laser action on the material, where the material inhomogeneous stationary heating results in the inhomogeneous stationary thermal expansion in the material. To generate the acoustic waves, temporal variations in temperature distribution are required in addition to its spatial variations. In the case of pulsed lasers applications, the temporal modulation of heating appears naturally, while in the case of cw lasers, the laser intensity needs to be additionally modulated in time. This can be achieved, for example, by application of an acousto-optic modulator (AOM) at a chosen frequency f. This modulation of the electromagnetic wave intensity will subsequently modulate the heat deposition inside the sample and, as a consequence, will modulate the temperature variations, the thermoelastic stresses, and the acoustic waves will be generated by the latter at the frequency f. In the general case of a linear response of a material to its laser excitation, only acoustic frequencies present in the spectrum of the laser intensity temporal variations related either to pulsed-periodic emission of laser radiation or to temporal modulation of the cw laser radiation can be thermoelastically generated by the lasers [2]. In the following, we are only considering the generation with cw lasers in the thermoelastic regime (in order to remain non-destructive).

The temperature field T induced by an intensity modulated cw laser can be determined by solving the heat diffusion equation. For the case of a laser radiation with a normal incidence on the surface (as depicted in Fig. 6.1a), this equation can be written, in cylindrical coordinates (r, ϕ, z), as:

$$\frac{\partial T}{\partial t} = \chi \Delta T + \frac{I}{\ell \rho c_p} g(t) \Psi(r, \phi) e^{-z/\ell}, \tag{6.1}$$

where χ, ρ, and c_p are the thermal diffusivity, the density and the specific heat of the sample, respectively, I is the absorbed part of the laser light intensity, ℓ is the penetration depth of the laser intensity imposing spatial distribution of heating source along the z axis, while g and Ψ are the temporal and lateral spatial distribution of the laser intensity, respectively. Assuming a gaussian beam ($\Psi(r, \phi) = e^{-(r/a)^2}$, with a the beam radius), a 100% sinusoidal modulation ($g(t) = H(t) \cdot [1 + \cos(\omega t)]/2$ with $H(t)$ the Heaviside step function and $\omega = 2\pi f$ the cyclic frequency) and with the initial and boundary conditions ($T(t = 0) = 0$, $\partial_z T|_{z=0} = 0$, $\partial_\phi T|_{\phi=0[\pi]} = 0$), it is possible to evaluate the temperature rise associated with the laser heating at any given point in space and time [88]. In the following, the intensity modulation is always assumed to be of 100%; therefore, the laser intensity is equal to 0 at a certain moment of its period.

The choice of the modulation frequency f has an important impact on the resulting thermal field. Without modulation, the temperature continuously increases towards an asymptotic temperature limit T_∞ (see the purple curve corresponding to $f = 0$ Hz in Fig. 6.1b). In such a case, no acoustic wave is generated after the stabilisation of

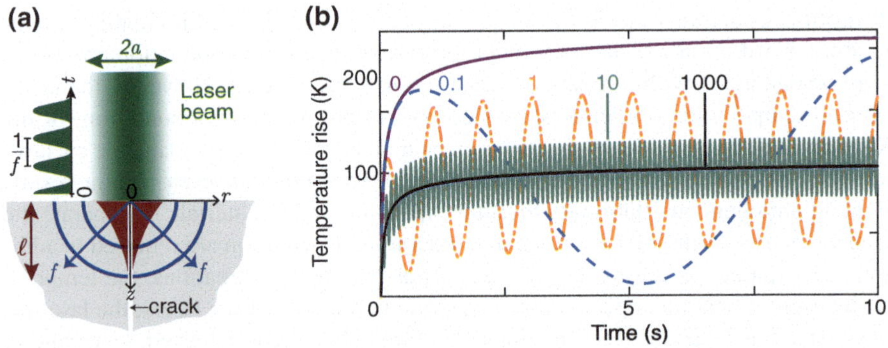

Fig. 6.1 **a** Schematic representation in cylindrical coordinates (r, ϕ, z) of an intensity modulated laser beam over time t heating a sample surface with a normal incidence. The sample contains a crack along z centred with the laser beam in $r = 0$. The laser beam is invariant over angle ϕ. ℓ is the laser intensity penetration depth, f is the modulation frequency, a is the beam radius. **b** Temperature rise evolution as a function of time in the centre of a gaussian beam ($\lambda = 532$ nm, $r = 0$, $z = 0$, $a = 100\,\mu$m, 100 mW power) in the considered absorbing glass (see Table 6.1) and for the modulation frequencies $f = 0, 0.1, 1, 10$ and 1000 Hz

the temperature distribution. For a very low frequency f, the characteristic time of heat diffusion in the sample across the region of light absorption is much shorter than the modulation period and the temperature field oscillates, at frequency f, between (almost) 0 and (almost) T_∞ (see the blue curve associated with $f = 0.1$ Hz in Fig. 6.1b), in parallel with the laser intensity variations; the system is quasistatic. This generates an acoustic wave at the frequency f. For a very high frequency f, an acoustic wave is also generated at the frequency f. However, the laser intensity variations are too fast compared to the heat diffusion time, resulting in a sample temperature field evolution similar to one without modulation but towards $T_\infty/2$ (as the heat release is on average twice smaller due to the 100% sinusoidal intensity modulation, see the black curve in Fig. 6.1b associated with $f = 1$ kHz). In this regime, the oscillating temperature component is much smaller than the average temperature. In-between these two extrema, the sample temperature field oscillates with the laser intensity modulations but without being able to reach neither $T = 0$ nor T_∞. In other words, the temperature field oscillates between T_1 and T_2 (with $0 \leqslant T_1 \leqslant T_\infty/2 \leqslant T_2 \leqslant T_\infty$ and $(T_1 + T_2) = T_\infty$, see the orange and green curves associated with $f = 1$ Hz and $f = 10$ Hz, respectively, in Fig. 6.1b). For a modulated beam ($f > 0$), the temperature rise inducing the laser ultrasonics generation can be decomposed into constant and oscillating parts, which ratio depends on the chosen frequency and the material thermal characteristics. The constant part is also referred, here and after, as average part or stationary part. It corresponds to the stationary temperature rise and is equal to $T_\infty/2$ in the steady regime. It does not depend on the frequency f. The oscillating part, at the opposite, corresponds to the temporal temperature rise range $(T_2 - T_1)$ and its magnitude is inversely proportional to the frequency: the lower the frequency f is, the lower the temperature T_1 and the higher

Table 6.1 Some physical parameters of the experimental glass sample

Parameter	Symbol	Value	Parameter	Symbol	Value
Density	ρ	2616 $kg\,m^{-3}$	Light penetration length	ℓ_{532}	0.31 mm
Specific heat	c_p	720J $kg^{-1}\,K^{-1}$		ℓ_{800}	0.22 mm
Thermal diffusivity	χ	0.547 $\mu\,m^2\,s^{-1}$	Light reflection coefficient	r_{532}	0.11
Bulk modulus	K	38.9 GPa		r_{800}	0.14
Poisson ratio	ν	0.22	Linear thermal expansion coefficient	α	5.5×10^{-6} K
Sound celerity	c	5750 $m\,s^{-1}$			

Subscript 532 (800) indicates the considered laser wavelength in nm

the temperature T_2 are. With the increasing modulation frequency, the amplitude of the modulated temperature field becomes progressively smaller and smaller ($T_{1,2} \rightarrow T_\infty/2$, and thus $T_2 - T_1 \rightarrow 0$). In Fig. 6.1b, for example, the oscillating part ($T_2 - T_1$) is equal to 183, 126, 47 and 0.7 K for the frequencies $f = 0.1$, 1, 10 and 1000 Hz, respectively, while the constant part is always equal to ~104 K. Therefore, for a very high frequency, the oscillating part is much smaller than the average part. The temperature rise is proportional to the laser intensity and is influenced by the laser beam radius a and light penetration ℓ (Eq. 6.1). Thus, both constant and oscillating parts are affected by changing such parameters.

The temperatures reached in the beam centre in Fig. 6.1b are calculated with parameters similar to the experimental conditions presented in the following. For the highest temperature elevation of ~200 K, calculated at the surface, there is no risk of glass melting (which is present for $T > 500$ K) but dilatation occurs. This can be measured at the sample boundaries. In particular, if a surface breaking crack is present (see Fig. 6.1a), the dilatation induced by the laser heating will modify the crack width. The (one dimensional) dilatation of the glass along r can be estimated, at a given depth z by the formula $\Delta(z) = \alpha \int_0^\infty T(r, z)dr$, where α is the linear thermal coefficient of the glass (see Table 6.1) and $T(r, z)$ the temperature rise reached in the point (r, z). This formula assumes that the modulation frequency is low enough to induce a quasistatic motion of the crack faces. With parameters similar as the ones used in the experiments, the maximum dilatation, reached at the surface (in $z = 0$), is estimated to be of a few hundreds of nanometers.

6.1.3 *Influence of Stationary Laser Heating on a Crack*

To generate an acoustic wave with a cw laser, the latter needs to be intensity modulated at the desired frequency, as explained earlier. In addition to the acoustic wave generation, this also induces heating that can produce (or cause) local stationary thermoelastic stress and dilatation within the laser spot generation area. Provided there is a crack within this generation spot, the distance between the crack faces (or crack thickness) will be affected. If the crack thickness is smaller than the sum of laser-induced displacements of its surfaces caused by thermal dilatation, the crack, under such thermal stress, will evolve from an open crack (the crack faces are not in contact) to a closed crack (the crack faces are in contact).

With a very low frequency f_L (e.g. $f_L = 1$ Hz), and in the considered glass sample, the oscillating part associated with the heating is large (see Fig. 6.1b), meaning that the distance between the crack faces can be changed significantly in this slow dynamic regime. Providing the laser power is large enough, these oscillating temperature changes offer the possibility to make the crack evolve dynamically from an open to a closed state and vice versa. This is then referred to as 'crack breathing'. If this low frequency thermal wave induces a motion of the crack faces but without creating contact between them, the closed state is not reached and this is not considered as 'crack breathing' here. Similarly, if the constant heating is so high that the contact between the crack faces is varying but is never lost, not even when the thermal loading is at its minimum (the crack remains closed all the time), this is not considered as crack breathing either. Furthermore, this crack breathing can also be referred as either 'tapping' -when it starts with increasing modulated heating from an open crack- or 'clapping' -when it starts from the closed state [81]. In the considered situation of a crack breathing caused by the thermoelastic loading modulated at frequency f_L, 'clapping' ('tapping') will correspond to a crack spending more time in the open (closed) state over the breathing period $T = 1/f_L$.

An acoustic wave incident on the crack is reflected by an open crack; while it is transmitted by a closed crack (see Fig. 6.2). Therefore, in the presence of a crack breathing, this reflection (or transmission) coefficient evolves temporally between 0 and 1 depending on the crack state. This dynamic evolution could be used to discriminate cracks from other possible defects that would not be affected by the low frequency thermal field in such way. The main idea of the photoacoustic frequency-mixing technique presented here is to generate an acoustic wave that interacts with a

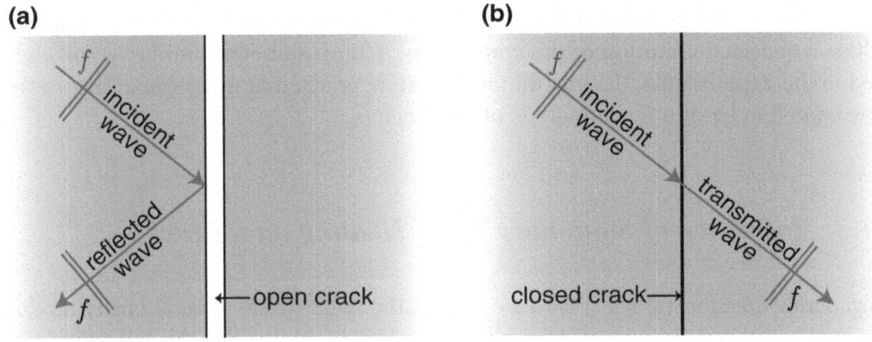

Fig. 6.2 Schematic representation of an acoustic wave being **a** reflected by an open crack and **b** transmitted by a closed crack

crack, which breathing is induced by a low frequency thermal heating. This acoustic wave is chosen on a resonant frequency of the sample and is of a few tens of kHz in the experiments. Note that stationary heating accompanying the generation of the acoustic wave by the modulated cw laser radiation can reduce the crack width due to material dilatation. However, the oscillating heating leading to the high frequency acoustic wave generation as well as the amplitude of this wave are negligibly small in comparison with the amplitude of the low frequency thermoelastic wave, so only the latter can enable the crack tapping/clapping. Prior to studying crack breathing, one should ensure that it is possible to locally modify the distance between the crack faces up to a closed state with a laser power below the ablation threshold.

In the first experiment, the effect of a cw laser heating on a crack was studied [77, 80]. The sample was a $50 \times 25 \times 3\ \text{mm}^3$ plate of light absorbing glass containing a single surface breaking crack. The glass physical properties are given in Table 6.1. The crack had been artificially created with a thermal shock via local heating with a flame followed by a fast cooling process. The crack had a length of several centimetres and crossed the plate thickness. The distance between the crack faces h depends on the position along the crack. Application of the atomic force microscopy (AFM) revealed, in similarly prepared samples, cracks of up to several hundred nanometers in width [71]. Thus, such cracks can theoretically be closed in the conditions of the considered experiments (see Sect. 6.1.2). The experimental setup is presented in Fig. 6.3a. A cw diode laser ('probe' beam in Fig. 6.3a) is modulated in intensity at high frequency $f_H = 16$ kHz, by modulation of the current in its power supply. This laser is referred to as 'probe laser' because it is applied for the generation of the acoustic wave which probes the crack state. Without modifying probe laser parameters such as intensity, beam radius, and optical wavelength, the changes in the amplitude of the detected probe acoustic wave are related to the modifications in the local crack state. The modulation frequency $f_H = 16$ kHz corresponds to one of the acoustic resonances of the plate and is chosen to maximise the amplitude of the detected signal. It has been estimated that this frequency corresponds to one of the standing wave resonances of the asymmetric Lamb modes propagating parallel to the long side of the plate. The acoustic resonances due to the symmetric Lamb modes are expected at frequencies exceeding 45 kHz. The dominant component of the plate surface motion, induced by the flexural waves, is an out-of-plane surface displacement which favours its detection by a vibrometer. A cw solid-state laser ('heating' beam in Fig. 6.3a), co-focused with the probe beam on the crack down to a $\sim 100\,\mu\text{m}$ spot, is locally heating the crack.[1] The absorption of the optical energy of the cw laser induces temperature rise that causes a thermoelastic expansion in the laser-irradiated region. As a result, the local separation between the crack faces diminishes as the cw laser power increases. A laser vibrometer ('detection' beam in Fig. 6.3a), focused a few centimetres away from the crack, detects the out-of-plane displacement resulting from the acoustic signal at f_H.

[1] The beam sizes given, here and after, correspond to the 1/e level in intensity.

Figure 6.3b presents the amplitude of the acoustic signal at the fundamental frequency f_H as a function of the power of the heating laser. Experimental data indicate a transition from low to high efficiency of the optoacoustic conversion, i.e., of the probe acoustic wave photo-generation, at this frequency when the power of the cw heating laser increases from 40 to 120 mW. The strongest dependence of the optoacoustic conversion efficiency on the power of the heating laser is found at approximately 80 mW. These observations are attributed to the heating-induced transition of the crack from an open to a closed state [78]. From a physics point of view, thermoelastic generation of sound near the faces of an open crack is similar to the one near a mechanically free surface and can be very inefficient [2]. Thermoelastic expansion of the locally heated region could first lead to the creation of a small number of contacts between the crack faces and then to complete local closing of the gap between them. This process is accompanied by increasing mechanical loading of one face of the crack by another, which can be viewed as a process of increasing the crack rigidity [78], with a consequent increase in the optoacoustic conversion efficiency. The results, presented in Fig. 6.3b, indicate that the rigidity of the crack is the most sensitive to external action (the crack is the most nonlinear acoustically) when the power of the heating laser is about 80 mW. For powers below 40 mW and higher than 120 mW, the influence of the variations in the heating power on the acoustic wave at frequency f_H is insignificant, indicating that the crack is nearly not modified by these variations. In other words, this indicates that below 40 mW, the crack is in the open state and no contact between the faces are made. Above 120 mW, the crack is closed, both faces are in contact, and an additional increase of the power does not change the crack state anymore.

This result demonstrates that the crack local nonlinearity, both in the open and closed states, is weak in comparison with the one in the state that is transitional from the open to the closed one. This transitional state is characterised by an incomplete local contact between the crack faces. In the open state, where there is no contact

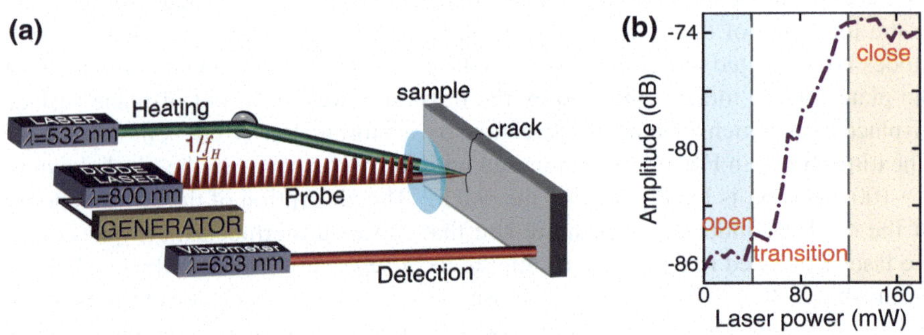

Fig. 6.3 **a** Experimental setup to probe the crack state. **b** Amplitude of the acoustic signal at f_H as a function of the heating laser power [80]

between the asperities located at opposite faces of the crack, and in the closed state, where the contact between the crack faces is nearly perfect, crack rigidity weakly depends on elastic loading. Weak nonlinearity of the open state is due to the absence of the highly-nonlinear weakly-loaded contacts between the asperities located at opposite faces of the crack. In the closed state, the contacts between the asperities are strongly loaded, weakly nonlinear, and the material behaves as nearly intact.

This first experiment demonstrated the possibility to achieve crack closing with a laser heating. This effect can be advantageously used to detect cracks. For example, in Ref. [70, 71], Ni et al. study the propagation of an acoustic pulse between two points on the sample surface separated by a crack with and without an extra heating of the crack vicinity (provided by a cw-laser) and evaluate possible amplitude difference and/or mode conversion due to the crack state evolution. A similar idea has also been proven efficient by Ohara et al. although the method relies on opening a closed crack (by the use of cooling spray) to affect the crack width between two measurements [89] (see Chap. 5 for more details). These experiments are denoted as acoustoelasticity because the nonlinear mechanical parameters of damaged (cracked) solids are evaluated via the measurement of the variations in the linear acoustic field (at the fundamental frequency) following the stationary external action on the sample (stress, temperature).

In this chapter, we describe crack detection via frequency-mixing nonlinear acoustic phenomena (instead of acoustoelasticity). In frequency-mixing nonlinear acoustics, the nonlinear elastic parameters of the materials could be assessed by measuring the nonlinear acoustic field (new spectral components of the signal) appearing due to the interaction of the probe acoustic field with the elastic field of different frequency launched by an additional periodic action on the sample. In the case of a cracked sample, the interaction of acoustic waves takes place mostly at the cracks, which are the most nonlinear elements of the sample. When the modulation frequency of the heating laser is far from the acoustic resonances of the sample while the heating laser is focused in the vicinity of the crack, the probe acoustic field interacts at the crack predominantly not with the acoustic waves emitted by the heating laser but with the thermoelastic strain field which periodically modulates the crack parameters via the modulation of the crack width. The aim is therefore to achieve crack breathing so that the crack oscillates between open and closed states. This is motivated by the observation from the above-described experiment demonstrating that the signal is particularly sensitive to crack modulation in the transition zone (see Fig. 6.3b). Therefore, crack breathing can be expected to have a dramatic influence over the acoustic signal generated in the crack area. In the frequency domain, this crack breathing between open and closed states will manifest by the appearance in the acoustic spectrum of sidelobes around the fundamental frequency, separated from it by integer numbers of the inverse of the heating period (corresponding to the modulation frequency of the heating laser). The process of the sidelobes excitation can be identified as a frequency-mixing parametric process (see Part III and Ref. [78]) and is the core of the presented nonlinear photoacoustic technique.

6.2 Nonlinear Frequency-Mixing Photoacoustic Method for Crack Detection

In this section we first describe the setup used in the experiments, followed by the principle of the method to perform one dimensional (1D) and two dimensional (2D) scans over a crack. The spatial resolution of the method is then discussed.

6.2.1 *Experimental Setup*

During this study, various experimental setups, based on the same physical principle, have been used. Two different excitation setups, with the same detection system, corresponding to the two most used setups, are represented in Fig. 6.4. The excitation can be divided into two parts: the probe beam and the pump beam. Here, the probe beam refers to the beam generating acoustic waves, at frequency f_H, corresponding to a resonant frequency of the sample (as introduced in the preliminary experiment presented in Sect. 6.1.3) near which the acoustic waves are detected. The chosen frequency varies with the experiment but is always of a few tens of kHz. The other beam, denoted as pump beam, is modulated at much lower frequency f_L; typically $f_L \sim 1$ Hz, and is used to generate the thermoelastic wave. These beams and their properties are denoted by a subscript H and L where H (L) stands for high (low) frequency.

In Fig. 6.4a, pump and probe beams are generated with two different lasers: the probe beam originates from a 1 W diode laser ($\lambda = 800$ nm), which current is modulated by an external generator at the chosen frequency f_H and the pump beam is generated by another laser (Coherent, Inc. Verdi, 2W, $\lambda = 532$ nm), 100% intensity modulated by an acousto-optic modulator (AA Opto-Electronics, Inc., Model MQ180). Both beams are then co-focused on the sample surface, that can be moved over x, y axes (see Fig. 6.4a). In the second configuration (Fig. 6.4b), the pump and the probe waves are generated with a single laser (Coherent, Inc. Verdi, 2W, $\lambda = 532$ nm). In this setup, the laser beam is split into two beams that are 100% independently in-

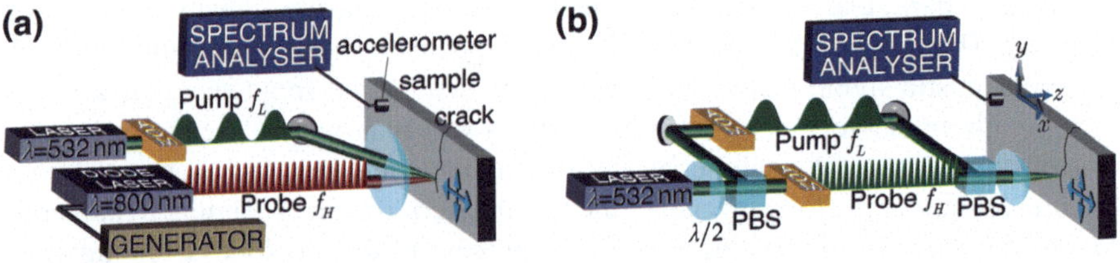

Fig. 6.4 Illustration of the two main experimental setups with **a** two different lasers **b** a single laser. Detection is represented with an accelerometer. AOM: Acousto-optic modulator. PBS: Polarised beam splitter

tensity modulated by two distinct acousto-optic modulators (AA Opto-Electronics, Inc., Model MQ180). The relative intensity of the pump and the probe beams can be modified by tuning the $\lambda/2$ plate prior to the beam split. The two beams are then recombined and focused on the sample surface. In both configurations, the two beams are independently modulated to avoid any possible cross-talk, that could lead to the appearance of the mixed-frequency components of f_L and f_H in the spectrum of the total laser intensity modulation. In general, there is not any significant advantage of using one or the other configuration. This excitation principle has also been proven efficient with two identical diode lasers [77] or a single diode laser with a beam doubly modulated at f_L and f_H. In the case of a double modulation of a single laser beam, the negligible presence of the cross-talk (frequency-mixing) in the modulation process should be independently verified. In such experiments, conducted in [80], no cross-talk was observed when the beam was focused away from the crack.

Regarding the detection, several systems have been used [80]. In both setups presented in Fig. 6.4, it is accomplished with accelerometers, which is the most frequently used detection sensor here. The accelerometer is placed a few centimetres from the crack so that the thermoelastic wave cannot reach it. It was reported in Ref. [80] that such experiment could also be realised with optical detection, either with a vibrometer (as in Fig. 6.3a) or via optical deflectometry [90, 91], adapted for the considered frequency range. In the case of all-optical setups, using the configuration with a single laser (Fig. 6.4b) can be advantageous to have the other laser used for detection at a different optical wavelength.

The sample is similar as the one introduced in the previous section. It consists of light absorbing glass plate, which properties are displayed in Table 6.1. Its surface is of a several squared centimetres, its thickness is of 3 mm, and it contains a single surface breaking crack. The crack, created by a thermal shock, has a dimension of few centimetres long and hundreds of nanometers width (corresponding to the distance between the crack faces). Several samples with different surface dimensions and different cracks have been studied. Therefore, from one experiment to another, the resonant frequency f_H, and the crack properties differ. The beam radii $a_{H,L}$ and powers $P_{H,L}$ are also usually modified.

6.2.2 Principle of the Method

As introduced earlier, the method relies on the interaction of an acoustic wave generated at f_H with a crack breathing at f_L. In acoustics, the interaction of two waves at different frequencies caused by the nonlinearity of the materials leads, in first approximation, to the simplest frequency-mixing, corresponding to the generation of the sum and difference frequencies ($f_H \pm f_L$). This process could also be understood as a parametric process, i.e., as a result of the modulation of the material parameters by the waves. The first wave modulates the material parameters at its frequency for the propagation of the second wave and vice versa. In the nonlinear frequency-mixing photoacoustic phenomenon of interest here, these two waves are both generated by intensity modulated laser radiation. The parametric process can be observed through the apparition of new spectral components in the signal at the mixed-frequencies $mf_H \pm nf_L$ (with $m, n = 1, 2, \ldots$). If the width of the crack is modulated at frequency f_L, from an open to a closed state of the crack by an external loading, it leads to a nonlinear process. The amplitude of the reflected and transmitted acoustic waves at f_H are modulated at frequency f_L. It is usual to say that the wave interacting with the crack is parametrically modulated as it is modulated by the variation of the crack parameters. The modulation of the monochromatic wave amplitude in the time domain is equivalent to frequency-mixing in the spectral domain. The frequency spectrum of each of the waves (the reflected and the transmitted ones) is influenced by its interaction with the crack, breathing at frequency f_L. It results in the nonlinear generation of the mixed-frequencies both in reflection and in transmission. The nonlinear frequency-mixing processes in the interaction of an acoustic wave with a breathing crack were already theoretically described [78, 92–96] as well as experimentally observed with acoustic experimental setups [33, 35, 97] and, more recently, with photoacoustic ones [79]. The first experimental evidence of this nonlinear frequency-mixing photoacoustic phenomenon based on the crack breathing was presented by Chigarev et al. [77], then followed by Mezil et al. [80, 81]. In these last experiments, the setups correspond to the one presented in Fig. 6.4a. The experimental setup in Ref. [77] is similar but realised with two different diode lasers.

The pump beam frequency f_L is chosen low enough, with a power P_L high enough, to ensure sufficient amplitude of the temperature oscillations in the generation spot (see Sect. I.2), and therefore large crack faces displacement to provide the crack tapping/clapping (assuming a crack is present in the heated region). The local separation between the crack faces in the laser-irradiated region diminishes (increases) due to thermoelastic expansion (contraction) when the temperature in the vicinity of the crack increases (diminishes). The temperature rise is caused by the absorption of optical energy from the intensity modulated cw laser, while the fall in temperature is due to heat conduction. The interaction of the acoustic wave, generated by the probe beam at f_H, with the breathing crack induces the generation of the nonlinear mixed-frequencies. The mixed acoustic frequencies $f_{m\pm n} = mf_H \pm nf_L$ are necessarily generated by nonlinear acoustic processes in the material as they cannot be present in the modulation spectrum of the cw lasers because the lasers are independently modulated and at different frequencies. This is an advantage in comparison to methods relying on sub- and superharmonic excitation, where harmonics and subharmonics could be potentially present in the spectrum of the cw laser modulation [73, 74, 98]. The nonlinearities due to the tapping/clapping of a crack are several orders of magnitude higher than those possible due to the intrinsic sample properties [77]. Outside the crack, the mixed acoustic frequencies are weakly generated or not generated at all. Therefore, the detection and localisation of a crack can be done through the analysis of the signal spectrum in order to detect the possible generation of the mixed acoustic frequencies at $f_{m\pm n} = mf_H \pm nf_L$. Two examples of the resulting spectrum outside and on a crack, obtained with the setup presented in Fig. 6.3a, are presented in Fig. 6.5a and b, respectively. On the crack (Fig. 6.5b),

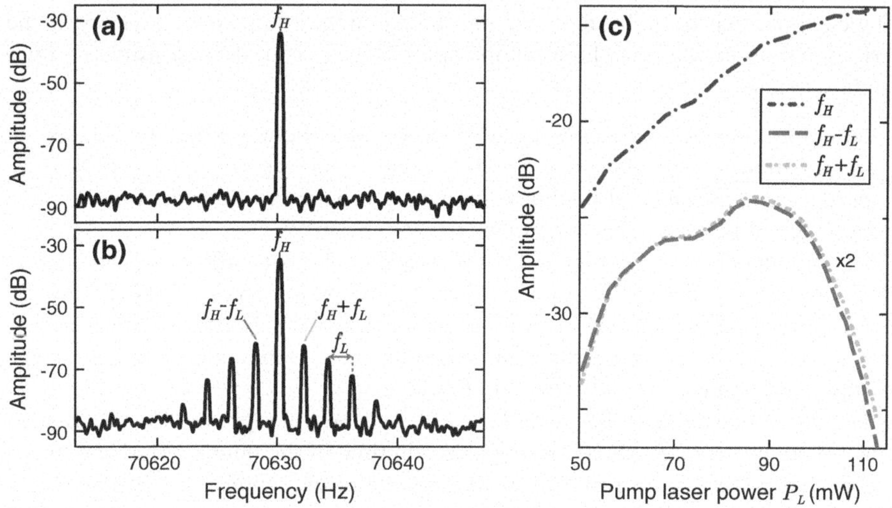

Fig. 6.5 Detected acoustic spectrum for the frequency window $f_H \pm 9 f_L$ when the power and the probe beams are cofocused **a** on an intact sample surface few mm away from the crack and **b** on the crack. **c** Amplitude of the acoustic signal at f_H, $f_H - f_L$, and $f_H + f_L$ as a function of the average power of the pump laser, which power is modulated at f_L

one can observe the generation of the mixed acoustic frequencies and the symmetry between left ($f_H - nf_L$) and right ($f_H + nf_L$) sidelobes.

To demonstrate the effect of the crack breathing on the resulting frequency spectrum, the experiment introduced in Sect. 6.1.3 (to study the stationary heating) is repeated but the heating beam is now modulated at $f_L = 1$ Hz and the signal detection is achieved by an accelerometer. The experimental setup thus corresponds to the one presented in Fig. 6.3a. The corresponding spectrum is recorded for the different pump powers $P_L \in [50; 110]$ mW at 29 different magnitudes. As the other experimental parameters remain identical with the one introduced in Fig. 6.3a, the influence of the pump power can be evaluated. Results for the linear spectral component at f_H and the nonlinear ones at $f_H \pm f_L$ are presented in Fig. 6.5c. For heating power lower than 50 mW and for heating power higher than 110 mW, corresponding to the open and closed states of the crack, respectively (see Fig. 6.3a), the amplitude of the mixed-frequencies is small. This confirms that the crack local nonlinearity is low in such states. In-between these two values, the nonlinear mixed frequencies are clearly observed. The maximum signal at mixed frequencies is detected at approximately 85 mW average power of the heating laser, where the nonlinear behaviour of the crack is the highest (see Figs. 6.3a and 6.5c). This transitional state is characterised by the incomplete local contact between the crack faces. As this maximum occurs roughly in the intermediate value between the two limiting ones (50 and 110 mW), this could correspond to the tapping/clapping transition, when the crack spends the same amount of time in the open and closed states. It will be theoretically confirmed in Sect. 6.3 that the amplitude maximum of $f_H \pm f_L$ is indeed reached at the tapping/clapping transition. In summary, depending on the pump beam intensity on the surface and on the distance between the crack faces at rest, three possible regimes can be expected:

- the pump intensity is too low to generate contacts between the crack faces. In this situation, the system is linear and the crack remains permanently open. Thus, there is no appearance of nonlinear frequency-mixing components. In the results presented in Fig. 6.5c, this corresponds to $P \lesssim 50$ mW;
- the pump intensity is too high to let the sample cool down sufficiently, preventing the crack to be opened at some moment of the pump period. This situation is also linear and as the crack remains permanently closed when the pump laser is switched on, there is no apparition of nonlinear frequency-mixing components. This corresponds to $P \gtrsim 110$ mW in Fig. 6.5c;
- the pump intensity is in-between these two laser power regimes. The crack faces are in contact (closed crack) during a certain time of the pump period and out of contact (open crack) the rest of the time. This nonlinear regime is responsible for the generation of the nonlinear frequency-mixing components, that can be observed when analysing the spectrum around f_H (Fig. 6.5b). The powers corresponding to the transition from open to breathing regime, and from breathing to closed regime, are denoted, here and after, as P_o and P_{cc}, respectively. In the experimental configuration realised in [77], leading to the results partially illustrated in Fig. 6.5c, it follows $P_o \simeq 50$ mW and $P_{cc} \simeq 110$ mW. While crack breathing is necessary to generate the nonlinear components, it can be anticipated that, in order to enhance their amplitude, the crack needs to spend a reasonable part of the period in both open and closed states.

6.2.3 One Dimensional Imaging for Crack Localisation

The results in Fig. 6.5a, b demonstrate that, provided the pump power is correctly adapted to ensure crack breathing (see Sect. 6.2.2), this technique can be used to discriminate the possible presence of a crack in the common focus spot of the pump and probe beams. By moving the sample relatively to the co-focused pump and probe laser beams, it is possible to monitor the amplitude of the nonlinear frequency-mixing components point by point and therefore to map the presence of the crack in order to localize it. The sample is progressively moved step by step in a direction perpendicular to the crack faces (along x axis in Fig. 6.4b). The step length, corresponding to the distance between the consecutive image points, is similar to or smaller than the diameter of the smallest beam in order to fully scan the line. At each step, the signal is acquired and a spectrum analyser evaluates the signals in the spectral window $f_H \pm 10 f_L$ (similarly as in Fig. 6.5a, b).

In Fig. 6.6 the evolution of the amplitude of some of the spectral components for two different cracks is presented. In the first experiment (Fig. 6.6a), the excitation is realised with only one diode laser doubly-modulated (with $f_L = 2$ Hz and $f_H =$

18.33 kHz, and $\lambda = 800$ nm) while the other laser (at $\lambda = 532$ nm) is used for the detection by adapting it for the deflectometry technique. The use of different laser wavelengths for excitation and detections ensures to be sensitive only to the detection beam light by filtering out the light at $\lambda = 800$ nm. Deflection technique is based on the dependence of the probe beam reflection from the surface on the spatial orientation of the latter (inclination relative to the probe beam incidence direction). Therefore, when an acoustic wave propagating at the sample surface introduces spatially inhomogeneous normal displacement of the surface, and thus modifies locally its inclination, the wave is detected by recording the modifications in the scattered light direction [90, 91]. It is important to adapt the distance from the probed surface to the photodetector to be sensitive to the targeted frequencies, which are here of ~18 kHz. In this experiment, this distance is set to 10 m and the scan step is equal to 50 μm (see Ref. [80] for details). The evolutions of the amplitude of the main peak f_H and of the nonlinear frequency-mixing sidelobes at $f_H \pm 2f_L$ are presented in Fig. 6.6a as a function of the probed position on the surface and the crack. In other words, for each point, the spectrum is registered and the amplitudes corresponding to the spectral peaks are measured from the acquired spectrum. The evolution of the main peak amplitude does not contain useful information, but the peaks in the nonlinear sidelobes clearly demonstrate the presence of the crack and can be used to localize it. This experiment demonstrates, not only the potential of this new technique but also the possibility to realise such detection all-optically. Furthermore, the symmetry between $f_H - nf_L$ and $f_H + nf_L$ (also referred to as left and right sidelobes, respectively) is clear on the images (Figs. 6.3b, and 6.6a), as predicted by the theory [78]. Thus, analysing only left (or right) sidelobes is sufficient for crack detection.

Another crack is then analysed with the setup presented in Fig. 6.4a and with a different set of parameters ($f_H = 70.6$ kHz, $f_L = 2$ Hz, $P_L = 90$ mW, $P_H = 120$ mW), with the scan step again equal to 50 μm (about half of each beam radius as $a_H = 106$ μm and $a_L = 95$ μm) and with a detection by accelerometer. The evolution of the amplitudes of the main peak (at f_H) and of the first three left nonlinear mixed frequencies (at $f_H - f_L$, $f_H - 2f_L$, and $f_H - 3f_L$, respectively) is presented in Fig. 6.6b.[2] The right nonlinear frequency-mixed spectral components are not presented due to the revealed symmetry (Fig. 6.6a). Again, the main peak amplitude evolution does not allow for the crack localisation, but each of the three nonlinear sidelobes clearly demonstrates the presence of the crack and can be used to localize it. The nonlinear sidelobes are consistent on the crack position. The peak widths, defined as a 6 dB decrease from the amplitude maximum, are also similar and read 235, 240 and 210 μm for the first, second and third sidelobe, respectively, which roughly match the beam diameters.

These two experiments demonstrate the possibility to detect and localize a crack thanks to the different nonlinearly mixed frequencies that are generated only when the two beams are focused on the breathing crack. Similar evolutions are observed

[2]The spectra in Fig. 6.5a,b were extracted from this 1D scan when the co-focused beams are localised 500 and 0 μm away from the crack, respectively.

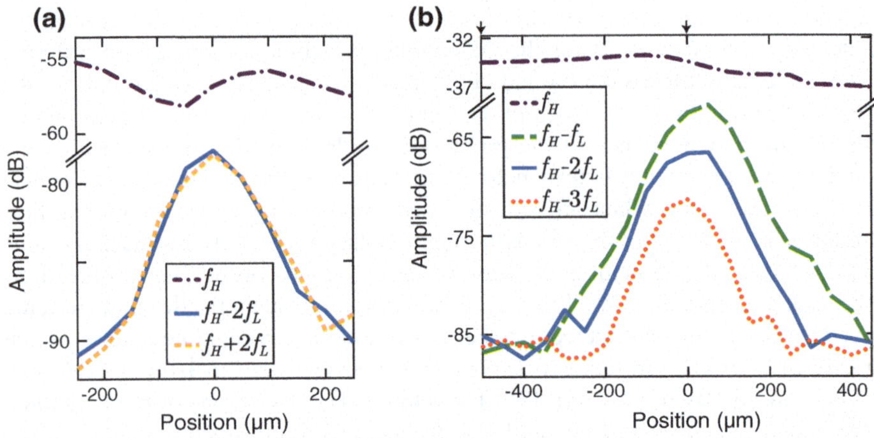

Fig. 6.6 Dependence of the acoustic amplitudes on the relative position between the common focus spot of the heating and the sound generating laser beams and the crack at the frequencies **a** f_H and $f_H \pm 2f_L$, **b** $f_H - nf_L$ with $n = 0, 1, 2$ and 3. The detection is realised by **a** deflectometry technique and **b** by accelerometer. Arrows indicate the position of the spectrum presented in Fig. 6.5a, b

for the sidelobes symmetric relative to the main peak ($f_H + nf_L$ and $f_H - nf_L$). The main peak, corresponding to the generated acoustic wave at f_H, is however useless for such precise detection as its amplitude does not vary significantly with the laser focus position relative to the crack. This crack detection can be achieved all-optically as it was demonstrated with the deflectometry technique. Vibrometry technique, as used in the initial experiment (Sect. 6.1.3), is also a possibility [80]. The experiments discussed in the following, however, are realised with an accelerometer detection for simplicity.

6.2.4 Two Dimensional Imaging of a Crack

Two 1D scans have been presented before (see Fig. 6.6), and the next step is to perform two dimensional (2D) imaging of a crack. While it could seem obvious to transform a 1D technique into a 2D one, it is actually not straightforward. This is because the crack properties evolve along the crack: the distance between its faces and its rigidity are not constant and it is not possible to adapt the beam power at each location for imaging with maximal sensitivity without a drastic increase in the imaging time. In addition to that, the other sources of nonlinearities are also not constant over the sample surface and could interfere with the crack nonlinearity which could cause false detections. For this 2D scan, realised on another crack, the same apparatus is used (Fig. 6.4a) with the notable difference that the sample is now moved in both x and y directions. (Other parameters are $f_L = 1$ Hz and $f_H = 25$ kHz, $P_L = 76.3$ mW and

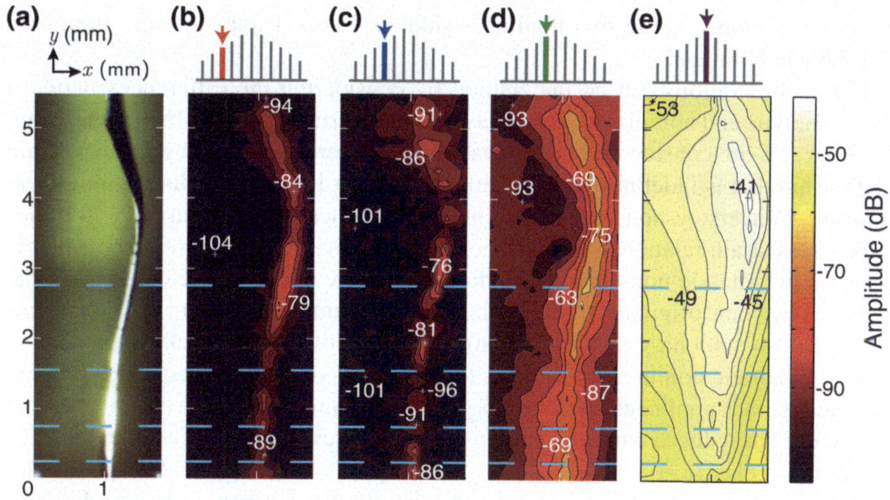

Fig. 6.7 **a** Optical image of the scanned area with the crack. **b–e** Top: Schematic representation of the analysed spectrum component. Bottom: Two dimensional scans of the crack achieved by detection of different sidelobes at $f_H + nf_L$. From left to right, the first three nonlinear left sidelobes: **b** $n = -3$, **c** $n = -2$, **d** $n = -1$, and **e** the main peak, at f_H ($n = 0$). All scans are represented with the same amplitude scale. Amplitude difference between two isolines is of 2 dB for $n = 0$, 6 dB for $n = -1$, and 5 dB for $n = -2, -3$. Dashed lines: Position of the 1D scans presented in Fig. 6.8

$P_H = 46.5$ mW, $a_L = 108\,\mu$m and $a_H = 328\,\mu$m). Here, the choice of the frequency f_H, near one of the sample resonances, is carefully selected so that the variation of the signal amplitude at this frequency over the whole 2D surface is small: only a ~3dB variation is observed with the probe beam alone.

Results of the crack imaging, initially published in Ref. [81], are presented in Fig. 6.7 for a 5.5×1.8 mm^2 area. Each image corresponds to the amplitude mapping of a frequency component of the spectrum. From left to right are the results for the frequencies $f_H + nf_L$ with n varying from -3 to 0, the latter corresponding to the main peak. The comparison to the optical image of the crack provided in Fig. 6.7a confirms that all the nonlinear photoacoustic images show sensitivity to the crack presence. The first nonlinear sidelobe image (at $f_H - f_L$, Fig. 6.7d) displays a 40 dB amplitude dynamics over the 2D scanned area, and the second ($f_H - 2f_L$, Fig. 6.7c) and third ($f_H - 3f_L$, Fig. 6.7b) ones have a 35 dB amplitude dynamics. This is a demonstration that the technique is highly sensitive because, despite the strong variations of crack nonlinearity along its length, the crack is always detected. At the opposite, the main peak (Fig. 6.7e) varies by less than 15 dB and the crack cannot, similarly to earlier discussed 1D scans, be clearly identified through its mapping. It can also be observed that the second ($n = -2$) and third ($n = -3$) nonlinear sidelobes display nonlinearities in a similar area around the crack but in a much narrower region when compared with the image corresponding to the first ($n = -1$)

nonlinear sidelobe. These two nonlinear sidelobes also display similar amplitude over the whole image.

These observations can be understood by considering the differences between classic analytic and non-classic non-analytic nonlinearities [29, 81]. In the following, nonlinearities are considered classic analytic if they are described by a stress/strain relationship that is indefinitely differentiable without introducing discontinuities in the result. Conversely, non-classic non-analytic nonlinearities contain discontinuities in the stress/strain relationship or in one of their derivatives. The present method is based on non-classic non-analytic nonlinearity of the crack [78]. However, classic analytic ones can originate from the bulk of the sample and from the crack, for example from stationary (not breathing) contacts between the crack faces. Classic analytic quadratic nonlinearity initiates the processes of cascade frequency-mixing. This means that a nonlinear sidelobe n is generated from the previous one, $n-1$, so it necessarily has reduced amplitude compared to the previous nonlinear sidelobe [24]. Higher order classic nonlinearities provide an opportunity to generate higher order sidelobes without cascade processes. For example, the cubic nonlinearity leads to direct excitation of the second order sidelobe. However, the strength of the higher order classic nonlinearities progressively diminishes with their order. As a result, in the case of classic nonlinearities, the amplitude of the sidelobes always drastically falls down with the increasing order of the sidelobe [81]. This fast fall with increasing sidelobe order n makes it very difficult to detect higher sidelobes (for $n \geqslant 2$) due to their low amplitude. This also makes higher sidelobes being detected only in places with high nonlinearities but quickly fade with distance from the crack. Thus, such nonlinearities also make crack images narrower with the increasing sidelobe order [29, 81]. Conversely, for non-classic non-analytic nonlinearities, all nonlinear sidelobes are generated at once [78, 81, 99]. The amplitude of the nth sidelobe is proportional to the amplitude of the acoustic field at frequency f_H and the spectral component at frequency nf_L of a periodic non-analytical function describing modulation of the crack rigidity [81]. The spectral components of this function exhibit an overall tendency to diminish slowly with n (proportional to $1/n$). However, the fingerprint of the non-classical nonlinearity of the breathing crack is in the possible non-monotonous spectrum of the modulation function and its spectacular dependence on the parameters of the crack and the strength of its thermoelastic loading. Due to the described process of the frequency-mixing caused by non-analytical nonlinearity of the crack, there is no fixed relation between the amplitude of a nonlinear sidelobe n and $n-1$: a sidelobe n can have a greater amplitude than sidelobe $n-1$ or vice versa, and crack images realised with non-classic non-analytic nonlinearities in general do not become narrower with the increasing sidelobe order [78, 81]. In the various 1D and 2D scans achieved with the presented technique, the following observations have been documented:

- the crack image obtained from the first nonlinear sidelobe (Fig. 6.7d) is broader than the ones obtained from higher nonlinear sidelobes (Fig. 6.7b, c, and Fig. 6.8);
- nonlinear sidelobe n (with $n > 1$) having greater amplitude than the one associated with $n = 1$ have never been reported [81–83].

These are indications that classic analytic nonlinearities could influence the results for $n = 1$. However, for higher nonlinear sidelobes ($n \geqslant 2$):

- the second and third nonlinear sidelobes images are similar both in amplitude and width (Figs. 6.7b, c and 6.8);
- nonlinear sidelobes are detected up to the 5th sidelobe in the spectrum presented in Fig. 6.5b, and up to $n = 10$ in Ref. [83];
- nonlinear sidelobe n having greater amplitude than the $n - 1$ one has been reported (Figs. 6.8b, 6.13 and Refs. [81–83]).

These are experimental evidence that the non-classic non-analytic nonlinearity from the crack breathing is dominating in the reviewed imaging technique (at least for $n > 1$). In summary, the crack is detected with a better dynamics analyzing the first nonlinear sidelobes, but higher order nonlinear sidelobes can offer a better spatial resolution of the crack.

Variations in the nonlinearities over the scan can be observed in different parts of the cracks (see, for example, Fig. 6.8a–d) as well as when the scan is achieved at the same location but with different pump powers (see Fig. 6.8c, e and 6.8d, f). This indicates that the amplitude of the frequency-mixed spectral component depends on crack properties and on experimental parameters (as they both influence the crack breathing). Therefore, provided a model to estimate the nonlinear frequency amplitudes as a function of such properties exists, one could extract some information on the crack properties. This will be the aim of the work presented in Sect. 6.3.

6.2.5 *Spatial Resolution of the Crack Images*

Until now, we have always considered the nonlinear frequency-mixing components to be generated when a crack is present within the laser beams common focus spot, and to be absent otherwise. However, one can notice in Fig. 6.6, for example, the presence of these components over 400 µm, which is much larger than the laser foci diameters (and the distance between the crack faces). This means that these nonlinear mixed-frequencies are also generated when both beams are focused nearby the crack. Therefore, the spatial resolution of the technique does not correspond to the beam diameter only. To better understand how the spatial resolution is influenced by the crack and the experimental parameters, we present in Fig. 6.8 different results that are 1D scans extracted from the previous 2D scan (Fig. 6.7) as well as from a second additional 2D scan, realised under the same conditions but with a higher pump power (of roughly 50% higher). In Fig. 6.8a–d, one can compare four 1D scans at different positions ($y = 0.8$, 1.6, 2.8 and 0.3 mm, respectively). In Fig. 6.8c–f, two different 1D scans (corresponding to $y = 1.6$ and 0.3 mm) are repeated with two different pump powers ($P_{L1} = 76$ mW and $P_{L2} = 106$ mW).

The first experimental observation is relative to the amplitudes of the sidelobe images. In the images presented in Fig. 6.8, there is about a 20 dB difference between the amplitudes of the first nonlinear sidelobe image (denoted $A - 1$) and the amplitudes of the second and third nonlinear sidelobe images (A_{-2} and A_{-3}, respectively).

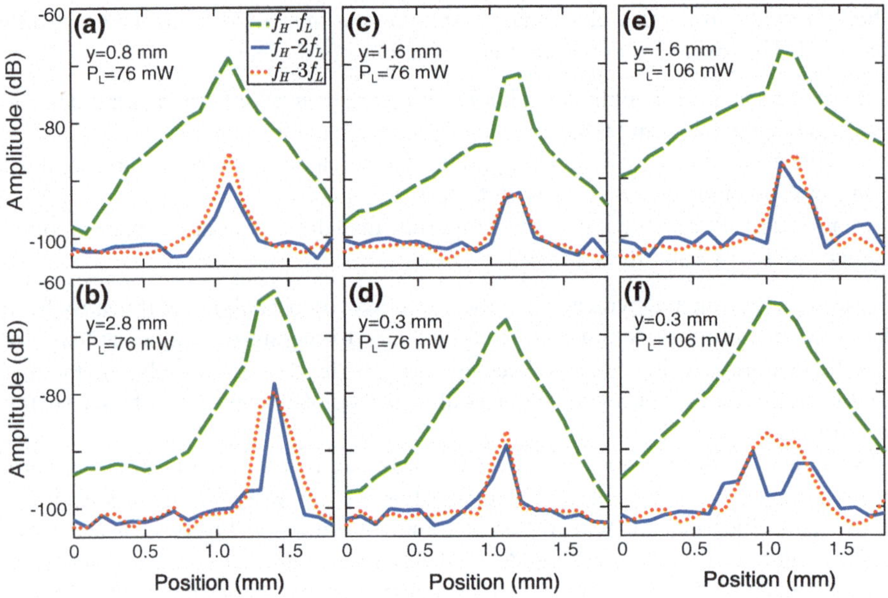

Fig. 6.8 Crack images at the first three nonlinear sidelobes taken at the positions of the crack (see Fig. 6.7) **a** $y = 0.8$ mm, **b** 2.8 mm, **c**,**e** 1.6 mm and **d**, **f** 0.3 mm with a pump power of **a**–**d** 76 mW, and **e**, **f** 106 mW, where both classic and non-classic nonlinearities contribute to first sidelobe images, while only non-classic nonlinearity contributes importantly to the second sidelobe and third sidelobe images

The same observation can be done at many other positions along the crack. These observations are consistent with the hypothesis, introduced in the previous section, of the contribution of both classic and non-classic nonlinearities to the first sidelobe, but (almost) only non-classical ones for higher sidelobes (for $n \geqslant 2$).

The second experimental observation concerns the relative widths and wings of the crack images from sidelobes of different orders. The width is here defined by a 6 dB decrease of the amplitude from its maximum while the wings size, much larger, corresponds to a 20 dB decrease. The latter is not always defined for nonlinear sidelobes associated with $n \geqslant 2$ as some cases do not exhibit a 20 dB amplitude dynamics. In most 1D images, the A_{-1} images look broader than the $A_{-2,-3}$ ones. In reality, the width (-6 dB decrease) of the first nonlinear sidelobe is regularly comparable to the one associated with the second and/or third sidelobe, and comparable to the pump beam diameter (216 μm). For example in Fig. 6.8a, the widths of A_{-1} and A_{-2} are similar (250 and 225, respectively); both being larger than the one of A_{-3} (140 μm). In Fig. 6.8b, the width of A_{-1} (235 μm) is comparable to the one of A_{-3} (220 μm) (and larger than the one of A_2 of 75 μm). The same observation can be made in other places with either the second or the third sidelobe. This indicates a predominance of the non-classical tapping/clapping nonlinearity of the crack over the classical ones on the crack. In contrast, the wings (-20 dB decrease) are significantly larger for A_{-1} images than for $A_{-2,-3}$ (when defined). In Fig. 6.8b, those of

A_{-1} are of 720 μm, to be compared with 410 and 540 μm for A_{-2} and A_{-3}, respectively. There is no, in these two 2D images, observation of similar wings between A_{-1} and $A_{-2,-3}$; the wings of A_{-1} are always broader than those of $A_{-2,-3}$ and the ratio $A_{-1}/A_{-2,-3}$ is also always larger for the wings than for the widths. These differences can be attributed to classical nonlinearities, that affect more A_{-1} due to drastic diminishing in the efficiency of the cascade frequency-mixing processes with increasing n [81]. This also demonstrates that, while classic analytic nonlinearities are present, non-classical non-analytic nonlinearities are predominant when the laser beams are focused on (or very close to) the crack.

In many locations along the crack in the image of the first sidelobe, a rather sharp peak, around the crack location, caused by non-classical nonlinearity has been observed on top of the much broader peak caused by the classic nonlinearities. This is, for example, evident in Fig. 6.8c, with a sudden ~10 dB increase. In this particular case, the widths are of 210, 240 and 290 μm for $A_{-1,-2,-3}$, respectively, demonstrating that the width of A_{-1} can be similar and even (slightly) narrower than others. This demonstrates that, in the optimal configuration, the non-classical non linearities can strongly dominate all other non-linearities. It is also possible to estimate the width due to classical nonlinearities by extrapolating the curves far away from the crack; it leads to an estimation of a −6 dB decrease for 630 μm, explaining why classical nonlinearities can degrade A_{-1} width and wing when non-classical nonlinearities do not predominate over classical ones. There are even particular points along the crack, where the classic nonlinearity is so strong that it importantly degrades the spatial resolution of the A_{-1} images relative to the one attainable in A_{-2} and A_{-3} images. This can be observed, for example, in Fig. 6.8d at $y = 0.3$ mm where the widths are $A_{-1} = 270$ μm but $A_{-2} = 160$ and $A_{-3} = 145$ μm. This has also been observed in other parts of the crack. However, in the different scan observations, there is no evidence of cases where the smooth nonlinearities dominate over the contribution of non-classic nonlinearity for the sidelobes $n \geqslant 2$. The balance between the contributions of the two nonlinearities to the A_{-1} images is delicate and depends on the distance between the crack faces (and thus on the thermal loading induced by the laser beams) and its rigidity (and thus on the position along the crack), as well as on the possible deviation from the classic nonlinearity with increasing stationary thermo elastic loading.

Conversely, for other cases, the non-classic non-analytic nonlinearity can control the complete form of the sidelobes including their wings. The scan presented in Fig. 6.6a is a good example of such observation where the amplitude evolution of $A_{-1,-2,-3}$, are all similar. In Fig. 6.6a all the sidelobe images have similar widths (235, 240 and 215 μm) and all can be interpreted as being generated due to non-classic non-analytic parametric interaction process. Besides, the amplitudes of all three sidelobes in Fig. 6.6a are comparable, which is an essential feature of the non-classic mechanism of their generation.

Thus, in general, this nonlinear frequency-mixing imaging from high order sidelobes, $n = \pm 2, \pm 3, \ldots$, could be advantageous in terms of spatial resolution in the regime of tapping/clapping crack, because the images at the first sidelobes $A_{\pm 1}$ are the most influenced ones by the smooth classic nonlinearity. In the absence of crack breathing, this classic nonlinearity would provide under particular conditions, importantly broader images. A thorough investigation of the roles of various nonlinearities, as well as of the evolution of the crack rigidities along its length has been done in Ref. [81].

Another important observation is the role of the pump power. We already demonstrated that the pump power needs to be high enough to generate crack tapping ($P \geqslant P_o$) but low enough to avoid a completely closed crack ($P < P_{cc}$, see Sect. 6.2.2). In Fig. 6.8c, e, the same 1D scan is repeated with two different pump powers ($P_{L1} = 76$ and $P_{L2} = 106$ mW). One can see that the crack is detected in both cases (thus $P_o \leqslant P_{L1} < P_{L2} < P_{cc}$). However, the evolutions of $A_{-1,-2,-3}$ are not proportional between the two scans. In Fig. 6.8e, the amplitudes of $A_{-1,-2,-3}$ are larger than in Fig. 6.8c but in different proportions: +6 dB for $A_{-2,-3}$ but only +4 dB for A_{-1}. Classical nonlinearities, that affect the wings size, are also more importantly increased on A_{-1} (by ~7 dB). As a result, the width associated with A_{-1} is degraded from 210 to 285 μm while the ones of $A_{-2,-3}$ are improved to 235 and 215 μm, respectively (from 240 and 290 μm initially). In some other cases, the spatial resolution of some nonlinear sidelobes is considerably degraded for the scan corresponding to the higher pump power. In Fig. 6.8d, f, for example, the widths evolve with increasing power from 270 to 390 μm for A_{-1} and from 150 to 430 μm for A_{-3}. Furthermore, the evolution of A_{-2} is drastically modified in the higher power scan and the amplitude goes through a minimum at the location of the crack, which, alone, may not allow its localisation. These observations are linked to the modifications of the crack breathing regime by the pump power. With the increase in the pump power, the crack spends more time in the closed state, leading to the modifications of the non-analytical crack width modulation function and non-monotonous variations in the amplitudes of its spectral components. More detailed explanations can be found in Sect. 6.3. The observation of the multiple peaks in the nonlinear mixed-frequencies images of a single crack (as observed for $n = 2$ in Fig. 6.8f) is the manifestation of the effect of crack phantoms that has been experimentally reported and theoretically interpreted in Ref. [81].

To better investigate the role of the pump and the probe beams, two different experiments have been performed on a different crack and with improved focusing (both beam radii are of ~20 μm). In these experiments, the setup is similar to the one in Fig. 6.4b but is modified so that one beam can be moved relative to the other and the sample. In the first experiment, the probe beam is focused on the crack and fixed while the pump beam is performing 1D scan perpendicularly to the crack direction (x axis in Fig. 6.4b) at different pump powers. Results are displayed in Fig. 6.9a. For low powers ($P < 40$ mW), the crack is hardly detectable because the pump is too low to make the crack breathe. However, a slight increase in the nonlinear sidelobe amplitude is noticed. It can be attributed to partial contacts, i.e., to the creation of few contacts between the crack faces. The crack starts to be detected for $P_o \simeq 40$ mW. For the most important powers, $P \geqslant 100$ mW, the crack is not localisable anymore as nonlinear mixed-frequencies are generated all over the scan with almost the same efficiency. This can be directly linked with the fact that the thermoelastic field generated by the

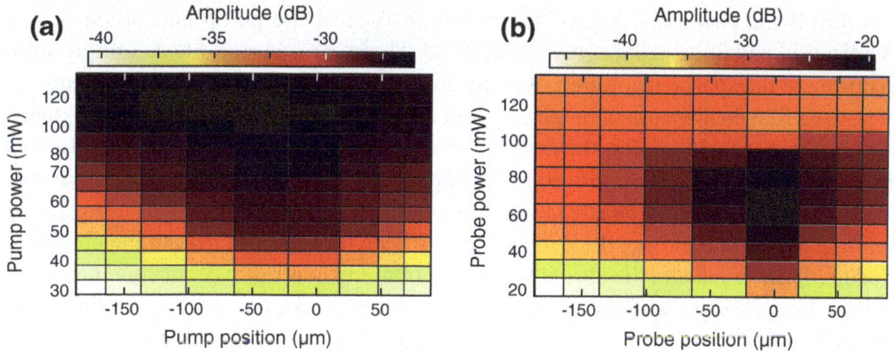

Fig. 6.9 **a–b** Colour encoded presentation of the 1D nonlinear frequency-mixing images at the first fundamental left sidelobe ($f_H - f_L$) obtained **a** by focusing the probe laser beam on the crack and by scanning with the pump laser beam of increasing power, and **b** by focusing the pump laser beam on the crack and by scanning with the probe laser beam of increasing power. Our images indicate that the crack is situated between the measurement positions at -50 and $0\,\mu$m

pump is large enough to provide crack breathing even if the pump is not focused on the crack. In this case, the breathing can be achieved even when the pump beam is focused 175 μm far from the crack -corresponding to ~8 times the pump radius. Thus, the spatial resolution can be importantly affected by the pump power.

In the second experiment, the situation is inverted: the pump is fixed on the crack with a power of 50 mW (slightly larger than $P_o = 40$ mW previously observed) while the probe beam performs 1D scans across the crack at different powers. From Fig. 6.9b, it is immediately clear that the magnitudes of the first sidelobe in the position $x = 0$ (when pump and probe beams are both focused in the close vicinity of the crack) is a strongly non-monotonous function of probe power P_H. It has been observed that the widths (at -6 dB) of A_{-1} in the different scans are approximately the same (and equal to ~120 μm), and are only weakly dependent on the probe power for $P_H \leqslant 60$ mW. These observations are consistent with the hypothesis that the pump power of $P_L \geqslant 50$ mW is sufficient to initiate the tapping of the crack in the examined crack position (this is indicated by the comparable amplitudes of the $A_{-2,-3}$ images, not presented here). Also, they are consistent with the fact that the influence of the heating caused by the probe laser on the crack closure is relatively weak at these probe powers. However, the increase in the probe power above $P_H \simeq 60$–70 mW leads to the progressively increasing role of the stationary stresses induced by the probe laser beam (see Sect. 6.1.3). This reduces the distance between the crack faces and the pump power required to achieve crack breathing. The non-classical frequency mixing phenomenon is thus expected to occur even when the probe beam is further from the crack which is observable by the broadening of the images with increasing probe power. Finally, at probe powers $P_H \geqslant 90 - 100$ mW the nonlinear sidelobes amplitudes drastically reduce. This is an indication of a closed crack, due to the high probe power, as it was already observed in the experiments with high pump power (Sect. 6.2.2).

From these two experiments, the possible roles of the pump and probe beams localisation and powers are better understood. There is no a priori ideal configuration to optimize the spatial resolution over the complete image of the crack as the interval of powers that ensures crack breathing depends on the crack rigidity and the distance between its faces. However, the imaging could potentially take advantage of the fact that the average heating produced by both pump and probe beams could reduce the crack width in order to reach tapping/clapping regime with the oscillating pump heating. At the same time, the overheating by both lasers could lead to the complete closure of the crack and the abrupt loss in the imaging contrast. By varying the power of both lasers focused on the crack, it is possible to make the crack breathing from an open to a closed crack via tapping or clapping regimes. This means that one can tune the time the crack spent in the closed state over a pump period depending on the laser powers. This dependence on the controllable laser powers can be used to our advantage to extract information on the crack properties.

6.3 Towards Quantitative Evaluation of Local Crack Parameters

In the previous experiments, it has been shown that while the crack breathing is a necessary condition to nonlinear frequency-mixing, the amplitude of these new frequencies is not proportional to the beam powers but rather is non-monotonous (see for example the two identical 1D scans at two different pump powers in Fig. 6.8). A better understanding of the role of the crack breathing in the nonlinear mixed-frequency generation is thus necessary to interpret these results.

It has already been demonstrated that the contact between two faces of a solid, or, in our case, of crack faces, is an efficient source of nonlinearities in the case of a monochromatic excitation at f generating higher harmonics at nf [33, 73, 100]. Besides, the amplitudes of the frequency spectrum components that could result from these nonlinearities as a function of the vibration excitation have also been shown to be highly non-monotonous in the presence of crack (or contact) tapping/clapping. For particular loadings, a given higher harmonic can even have an amplitude minimum when the crack breathes while the other harmonics could have an important amplitude [100]. These minima, which seem similar to the observation of the frequency-mixed component at $f_H - 2f_L$ in Fig. 6.8f, are very interesting as they are associated with a peculiar loading. Due to their dependence on the crack properties and the excitation parameters, their detection should open a way to resolve these crack properties.

The following last section is devoted to a brief presentation of the theoretical model developed to explain the nonlinear frequency-mixing processes taking place due to the crack and the prediction of the amplitude of the generated sidelobes as a function of the parameters (loading force, pump modulation frequency, etc.). More details can be found in Refs [78, 83]. Afterwards, an experiment analysing the nonlinear sidelobe amplitudes as a function of the pump beam power is described. It proves their non-monotonous behaviours. Finally, a comparison between these experimental observations and the developed theoretical model is achieved, demonstrating the possibility to extract several crack parameters.

6.3.1 Theoretical Model

In the following, the pump and probe laser beams are considered to be co-focused on the crack. Figure 6.10a is a 3D representation of the situation, where the laser beams are assumed to be centred on the crack. The cross section of the experimental geometry (Fig. 6.10a) in the plane $(x, 0, z)$ is presented in Fig. 6.10b. We are primarily interested in describing the crack motion in this cross section passing through the center of the co-focused pump and probe beams. The problem of thermal conduction inside the material is three dimensional, because of the three dimensional distribution of laser intensity, controlled by the radius of the laser beams and the optical penetration depth ℓ. However, to simplify the analysis and obtain the insightful predictions, the problem can be reduced to a one dimensional model independent of z coordinate if we describe the role of the 3D thermal conductivity in the saturation of the temperature growth by an effective relaxation time and neglect thermal conduction into air. Thus, the light penetration length and the cylindrical symmetry of the laser beams are not explicitly considered in the equations of this simplified 1D model but are taken into account implicitly via a thermal relaxation time τ_T that will be introduced later. The two crack faces, assumed parallel, can move along x, and u represents the mechanical displacement along this direction. Finally, F represents the force due to the interactions between the opposite faces of the crack and acting on a unit surface area at the crack face. The force depends on the distance between the crack faces, which can be modified by laser heating.

The model is based on one dimensional inhomogeneous equation which describes the thermoelastic generation of mechanical motion [78],

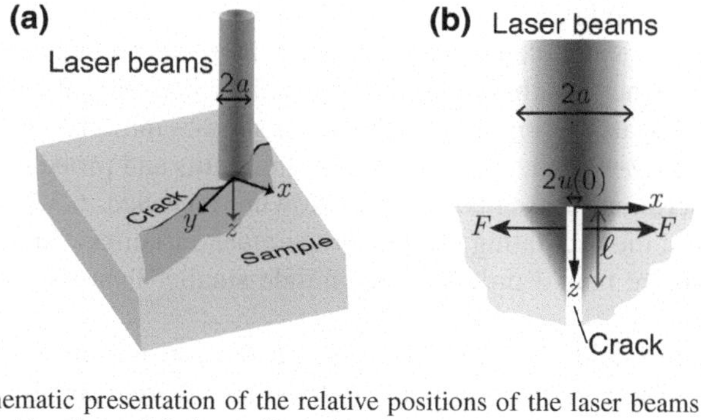

Fig. 6.10 **a** Schematic presentation of the relative positions of the laser beams of radius a, and the crack. **b** Considered simplified one dimensional geometry and position of the laser beams of radius a relatively to the crack. In the figure, the light penetration length is ℓ, the force is F and the distance between the crack faces is $2u(0)$. **b** Reprinted from [83] with permission from Elsevier

$$\frac{\partial^2 u}{\partial x^2} - \frac{1}{c^2}\frac{\partial^2 u}{\partial t^2} = \frac{K\beta}{\rho c^2}\frac{\partial T}{\partial x}, \tag{6.2}$$

with the boundary condition at $x = 0$ (the crack position),

$$\frac{\partial u(x=0)}{\partial x} - \frac{K\beta}{\rho c^2} T(x=0) = -\frac{F[2u(x=0)]}{\rho c^2}, \tag{6.3}$$

and a one dimensional equation for heat transport initiated by light absorption,

$$\frac{\partial T}{\partial t} + \frac{T}{\tau_T} = \chi \frac{\partial^2 T}{\partial x^2} + \frac{I}{\ell \rho c_p} g(t)\Psi(x), \tag{6.4}$$

where c is the velocity of the longitudinal waves, K the bulk modulus of the glass, β the bulk thermal expansion coefficient (see Table 6.1), τ_T a characteristic temperature relaxation time, I the absorbed part of laser intensity, $g(t)$ the modulation function of laser intensity in time ($g(t) = [1 + \cos(\omega t)]/2$) and $\Psi(x)$ the laser intensity distribution in space on the sample surface, assumed gaussian ($\Psi(x) = e^{-(x/a)^2}$). The factor 2 in Eq. 6.3 refers to the two crack faces that are in symmetric motion relative to $x = 0$. The introduced thermal relaxation time τ_T is accounting for the 3D character of the thermal diffusion and avoids the temperature divergence when the heat conduction equation in the cylindrical symmetry is reduced to the 1D form by assuming its homogeneity over the depth (over the z coordinate). The relaxation time is evaluated by equalising the maximum temperature rise predicted by the 3D heat equation (Eq. 6.1) with the one predicted by its 1D approximation (Eq. 6.4), under the same conditions [83]. To this thermal relaxation time is also associated a thermal relaxation frequency $\omega_T = 1/\tau_T$. From the system (Eqs. 6.2–6.4) one can see that unknown variables are the mechanical displacement u along x, the interaction force F and the temperature rise T.

The crack is assumed to be heated by two beams 100% intensity modulated. The pump beam is modulated at the cyclic frequency ω_L ($= 2\pi f_L$) and produces a constant heating and a modulated one (see Sect. 6.1.3). The probe beam is modulated at the cyclic frequency ω_H high enough to consider that only its averaged power contributes to materials heating and thermoelastic stresses initiating the crack motion. At a few tens of kHz with the experimental beam radius and probe power that were used in the experiment presented in the following (see Sec. 6.3.2), and with the sample properties, the oscillating sample temperature variations can be evaluated of only ~0.05 K (more than 3 orders of magnitude smaller than the constant part of ~60 K).

The temperature field can be solved with Eq. 6.4 and introduced into Eqs. 6.2–6.3 for the case of two monochromatic modulations (at frequencies f_L and f_H) of gaussian laser radiations $Ig(t)$. The crack faces motion can then be described by the equation [78]:

$$\frac{\partial u(0)}{\partial t} - \frac{F[2u(0)]}{\rho c} = -I_H A_H(0) - I_L A_L(0) - I_L|A_L(\omega_L)|\cos[\omega_L t - \varphi(\omega_L)], \tag{6.5}$$

with φ the phase of the pump beam, A the amplitude terms of the (constant and oscillating) pump and (constant) probe beams, and where $u(0)$ and $A_{H,L}(0)$ correspond, here and after, to $u(x=0)$ and $A_{H,L}(\omega=0)$, respectively. The amplitude term is defined as:

$$A(\omega) = -\frac{K\beta}{\ell\kappa\rho c^2}\frac{i\omega}{\left(\frac{\omega}{c}\right)^2 + \frac{\omega_T - i\omega}{\chi}}\left[\frac{i\omega/c}{\sqrt{\frac{\omega_T - i\omega}{\chi}}}\widehat{\Psi}\left(\sqrt{\frac{\omega_T - i\omega}{\chi}}\right) + \widehat{\Psi}\left(-i\frac{\omega}{c}\right)\right], \tag{6.6}$$

with i the imaginary number ($i^2 = -1$), and $\widehat{\Psi}(p)$ the Laplace transform of the laser distribution on the sample surface [78]. Equation 6.5 relates the crack faces mechanical motion $u(0)$ to the constant thermoelastic loading by the probe beam ($I_H A_H(0)$, abbreviated $I_H A_H(0)$ in the following) and to both constant and oscillating thermoelastic loading by the pump beam ($I_L A_L(0)$ and $I_L|A_L(\omega_L)|\cos[\omega_L t - \varphi(\omega_L)]$, respectively).

In order to solve these equations, it is necessary to model the dependence of the force F acting between the crack faces on the crack width. As discussed earlier, from the experimental observations, it follows that the crack can be either in open or closed state and in transitional states (when some contacts are created but insufficiently to consider the crack to be closed, see Sect. 6.1.3). Figure 6.11a is a realistic representation of the evolution of the crack faces displacement as a function of an external loading [35, 96]: the initial distance between the crack faces h_0 diminishes with the increasing force F until the crack reaches a critical width h_c associated with the force F_c that closes it, the crack faces 'jump' into contact and the crack rigidity suddenly drastically increases. The crack then remains closed until the force decreases down to F_o when the crack opens and its rigidity decreases (Fig. 6.11b). The difference between the forces F_c and F_o is due to quasistatic adhesion hysteresis. In the present simplest model, the relation between F and $2u(0)$ (Fig. 6.11a) is simplified to a quasi-linear approximation, displayed in Fig. 6.11c, for which the motions of the crack in the open and closed states are linear (a constant rigidity η is considered for each state) and the hysteresis is not taken into account ($F_o = F_c = F_i$). The two possible states of the crack, open or closed, are subscripted o and c hereafter, respectively. The two rigidities $\eta_{o,c}$ can be combined with the sample characteristics (acoustic impedance) into cyclic relaxation frequencies: $\omega_{o,c} = 2\eta_{o,c}/(\rho c)$. From physical considerations, it is expected that the crack is much softer in its open state where there is a finite distance between the crack faces than in its closed state. As it is assumed that the rigidity of the crack is piecewise constant, the crack exhibits linear elasticity both in open and closed states, while the overall nonlinearity of the crack is due to the transition from one state to another in the process of loading (bimodular elasticity). The interaction force between the crack faces is therefore approximated by:

$$F[2u(0)] = \begin{cases} -\eta_o[2u(0) - h_o] & \text{if } h_i \leqslant 2u(0) < \infty, \\ -\eta_c[2u(0) - h_c] & \text{if } 0 \leqslant 2u(0) \leqslant h_i. \end{cases} \quad (6.7)$$

The transition from the open to the closed state happens when the distance between the crack faces reaches h_i (see Fig. 6.11c). Thus, h_i corresponds to the minimum (maximum) crack thickness in the open (closed) state (point E in Fig. 6.11c). The distance difference $h_o - h_i$ ($= [A; E]$ in Fig. 6.11c) corresponds to the crack width, i.e., the difference of the distances between the crack faces in the absence of thermoelastic loading (h_o) and when they are in contact (h_i). From physical considerations, if the thermoelastic loading formally leads to $u(0) \to 0$, one should account in theory for the nonlinear growth of crack rigidity up to its infinite value which would correspond to disappearance of the crack (because it is a crack of infinite rigidity which would not scatter the acoustic wave in the considered model). Therefore, in

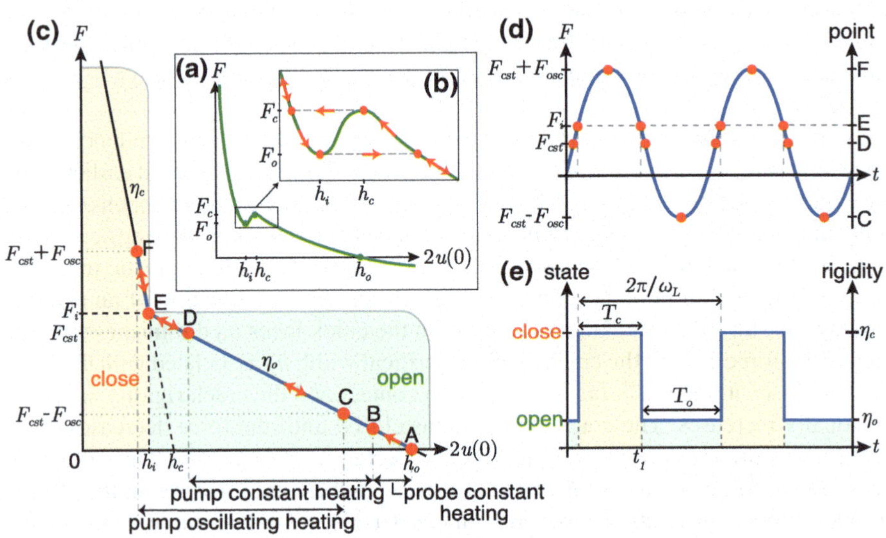

Fig. 6.11 **a** Qualitative presentation of the dependence of the interaction force F between the crack faces on the width $2u(0)$ of the crack. **b** Zoom of **a** in the crack state transition zone in the presence of hysteresis. **c** Considered piece-wise linear interaction force F between the crack faces as a function of the distance $2u(0)$ between them. **d** Evolution of the sinusoidal force, associated with the thermoelastic wave, loading the crack as a function of the time when crack breathing occurs. **e** Schematic representation of the crack state and rigidity evolution as a function of time. **c**, **d** Point A corresponds to the crack width without any loading ($2u(0) = h_o$ and $F = 0$), point B represents the reduced distance between the crack faces because of the average heating by the probe light. Point C is the maximum distance between the crack faces under crack loading (when the pump loading is minimum), point D corresponds to the distance between the crack faces under the constant heating from the probe and the pump beams, E represents the transition point when the crack switches from open to closed state and vice versa ($2u(0) = h_i$) and F is the minimum distance between the crack faces under crack loading (when the pump loading is maximum)

the piece-wise linear model presented in Fig. 6.11c, the distance between the crack faces is always assumed to be strictly positive.

The forces required to close and open the crack are deduced from Eq. 6.7 and are equal to $F = \eta_{o,c}(h_{o,c} - h_i)$. In the presence of hysteresis, as developed in [78], these two forces F_o and F_c required to close and open the crack, respectively, are not equal (see Fig. 6.11b).

The distance $2u(0) = h_o$ (point A in Fig. 6.11c) corresponds to the situation before the start of laser action ($F = 0$). The average heating caused by the probe beam is assumed to reduce this distance but insufficiently to close the crack (point B in Fig. 6.11c). The thermoelastic stress induced by the pump can be decomposed into a constant heating corresponding, with addition of the average heating caused by the probe, to the total average heating (point D in Fig. 6.11c, reached for a loading force F_{cst}) and an oscillating part -relative to this average level- that has both positive and negative parts (Sect. 6.1.3). The distance between the crack faces evolves from a maximum when the oscillating force is minimum (point C in Fig. 6.11c, associated with the loading force $F_{\text{cst}} - F_{\text{osc}}$), to a minimum when the oscillating force is maximum (point F in Fig. 6.11c, reached for $F_{\text{cst}} + F_{\text{osc}}$). Note that the modulation of the pump beam takes place at so low frequency that the temperature variations and stresses induced by this pump beam are quasi-stationary and follow the temporal variations in the intensity, so point C, and F are reached when the pump intensity is null and maximum, respectively. The difference between points B and C is due to the minimum temperature rise from the pump beam (Fig. 6.1b). For an infinitely low frequency f_L points B and C would be identical. The transition between the open and closed states happens when the distance h_i is reached (point D in Fig. 6.11c), corresponding to a loading force of F_i. Over a pump period, the distances C→E and E→F are covered during time duration T_o and T_c, respectively, where $T_{o,c}$ is the time spent by the crack in open and closed state, respectively (and with $T_o + T_c = 2\pi/\omega_L$, see Fig. 6.11d,e). By modifying the pump beam intensity or the pump beam modulation frequency, both the constant and the modulated heating are modified (points C, D and F), affecting the maximum and minimum distances between the crack faces, as well as T_o and T_c. It is now possible to reconsider the different possible regimes of crack motion. These four regimes are schematically summarised as a function of the minimum, average, and maximum loading in Fig. 6.12. If the loading is insufficient to induce crack breathing, point F remains in the open state ($F_{\text{cst}} + F_{\text{osc}} < F_i$). At the opposite, if point C is in the closed state, the crack remains closed ($F_{\text{cst}} - F_{\text{osc}} > F_i$). If the crack is breathing, then it is tapping if the average heating (point D) is in open state ($F_{\text{cst}} < F_i$) and it is clapping if point D is in the closed state ($F_{\text{cst}} > F_i$).

It was shown in Ref. [78] that by introducing the piece-wise linear relation (Fig. 6.11) between F and u in the model provides the opportunity to estimate the normalised amplitude A_n of the nonlinear sidelobe n as a function of the pump cyclic frequency ω_L and the time spent by the crack in the closed state during a pump period T_c:

$$|A_n| = \left| \frac{1}{n} \sin\left[n\frac{1}{2}\omega_L T_c \right] \right|. \tag{6.8}$$

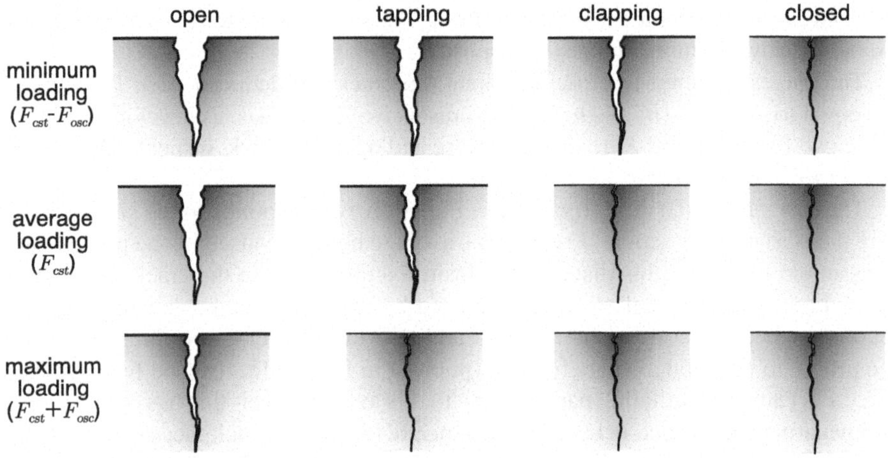

Fig. 6.12 Evolution of the crack motion regime from left to right: in open state, tapping and clapping regimes of crack breathing, and in closed state from top to bottom: for the minimum loading ($F = F_{\rm cst} - F_{\rm osc}$) corresponding to point C in Fig. 6.11, the average loading ($F = F_{\rm cst}$) corresponding to point D in Fig. 6.11 and for the maximum loading ($F = F_{\rm cst} + F_{\rm osc}$) corresponding to point F in Fig. 6.11

The first conclusion from Eq. 6.8 is that if the crack remains open ($T_c = 0$) or closed ($T_c = 2\pi/\omega_L$), these nonlinear mixed-frequencies are not generated. This confirms earlier observations (Fig. 6.5c). In addition, one can see that by changing the pump power (or the pump frequency), the amplitudes of the nonlinear sidelobes evolve non-monotonously because of the sine function; this also agrees with earlier observations (Fig. 6.8c–f). For the first nonlinear sidelobe associated with $n = 1$, the maximum is reached at the tapping/clapping transition ($T_o = T_c = \pi/\omega_L$), as suggested in Sect. 6.2.2 when discussing the results presented in Fig. 6.5c. For other nonlinear sidelobes (associated with $n > 1$), a zero-amplitude can be observed not only in absence of crack breathing but also when $T_c = (m2\pi)/(n\omega_L)$ (with $m = 1, 2, \ldots$ and $m \leqslant n$). In Fig. 6.8f, as the evolutions of $A_{-1,-3}$ demonstrate crack breathing, the decrease of A_{-2} corresponds to (or is close to) the minima for $n = 2$. This minima occurs when $T_o = T_c = \pi/\omega_L$, and corresponds to the clapping/tapping transition. For the third nonlinear sidelobe, the minima happen when the crack spends $1/3$ or $2/3$ of its time in the closed state. A similar conclusion can be made for higher sidelobes (e.g., for $n = 4$, it corresponds to the crack being closed for $1/4$, $1/2$ or $3/4$ of the time over the pump period).

It now remains to relate the time spent in the closed state T_c to the loading force F. Implementing the piecewise linear approximation Eq. 6.7 for the crack width evolution in open and closed states in Eq. 6.5 leads to linear equations coupled at the transitional time between the closed and open states that can be resolved analytically [83]:

$$2u(0) = C_{o,c}e^{-\omega_{o,c}t} + h_{o,c} + \\ -\frac{2}{\omega_{o,c}} I_H A_H(0) - 2I_L|A(\omega_L)| \left(\frac{\cos\left[\omega_L t - \varphi(\omega_L) + \phi_{o,c}\right]}{\sqrt{\omega_{o,c}^2 + \omega_L^2}} + \frac{\Gamma}{\omega_{o,c}} \right), \tag{6.9}$$

with $C_{o,c}e^{-\omega_{o,c}t}$ the solution of the homogeneous differential Eq. 6.5 (written for the variable $2u(0) - h_{o,c}$), $\phi_{o,c} = \arctan\left(\omega_L/\omega_{o,c}\right)$ and with $\Gamma = A_L(0)/|A(\omega_L)|$, the ratio between the constant and the modulated thermoelastic loading induced by the pump laser.

In the absence of tapping (for small pump amplitude), Eq. 6.9 can be rewritten by omitting the "homogeneous" solution. When $2u(0) = h_i$, the clapping starts. At this particular moment t_1 (Fig. 6.11e), when the minimum force loading that initiates clapping is reached, the cosine term equals 1 and $I_L = I_o$ (with I_o the maximum intensity that the crack can support in open state). This can be introduced into Eq. 6.9 which is first developed for the particular time t_1 where it can be written for the open and closed states of the crack (Fig. 6.11e). By comparing these two equations, the ratio I_c/I_o (where I_c is the maximum intensity that the crack can support in the closed state) can be resolved. It follows:

$$\frac{I_c}{I_o} = \frac{\sqrt{\omega_c{}^2 + \omega_L{}^2}\left(\omega_o{}^2 + \Gamma\sqrt{\omega_o{}^2 + \omega_L{}^2}\right)}{\sqrt{\omega_o{}^2 + \omega_L{}^2}\left(\omega_c{}^2 + \Gamma\sqrt{\omega_c{}^2 + \omega_L{}^2}\right)}. \tag{6.10}$$

Equation 6.9 can also be developed for the particular times $t_1 - T_c$ and $t_1 + T_o$ for a closed and an open crack state, respectively (see Fig. 6.11e). By correctly adapting these four equations developed from Eq. 6.9 (for the particular times $t_1 - T_c$ and t_1 in closed state and t_1 and $t_1 + T_o$ in open state), it is possible to get rid of the integration constants and to rewrite the equations in the matrix form [83],

$$\begin{bmatrix} A_{11} & A_{12} \\ A_{21} & A_{22} \end{bmatrix} \cdot \begin{bmatrix} \cos[\omega_L t_1 - \varphi(\omega_L) - \phi_c] \\ \sin[\omega_L t_1 - \varphi(\omega_L) - \phi_c] \end{bmatrix} = \begin{bmatrix} (I_c/I_L - B_c)\left(1 - e^{-\omega_c T_c}\right) \\ (I_o/I_L - B_o)\left(1 - e^{-\omega_o T_o}\right) \end{bmatrix}, \tag{6.11}$$

where:

$$\begin{cases} A_{11} = \omega_c\left(1 - \cos[\omega_L T_c]e^{-\omega_c T_c}\right) / \left(\omega_c + \Gamma\sqrt{\omega_c{}^2 + \omega_L{}^2}\right), \\ A_{12} = -\omega_c \sin[\omega_L T_c]e^{-\omega_c T_c} / \left(\omega_c + \Gamma\sqrt{\omega_c{}^2 + \omega_L{}^2}\right), \\ A_{13} = \omega_o\left(\cos[\Delta\phi + \omega_L T_o] - \cos[\Delta\phi]e^{-\omega_o T_o}\right) / \left(\omega_o + \Gamma\sqrt{\omega_o{}^2 + \omega_L{}^2}\right), \\ A_{14} = \omega_o\left(\sin[\Delta\phi]e^{-\omega_o T_o} - \sin[\Delta\phi + \omega_L T_o]\right) / \left(\omega_o + \Gamma\sqrt{\omega_o{}^2 + \omega_L{}^2}\right), \\ B_{o,c} = \Gamma\sqrt{\omega_{o,c}{}^2 + \omega_L{}^2} / \left(\omega_{o,c} m_L + \Gamma\sqrt{\omega_{o,c}{}^2 + \omega_L{}^2}\right), \end{cases} \tag{6.12}$$

with $\Delta\phi = \arctan(\omega_L/\omega_c) - \arctan(\omega_L/\omega_o)$. One can notice that A_{11}, A_{12}, A_{21}, A_{22}, B_c and B_o are function of $\omega_{o,c,L}$, Γ and $T_{o,c}$ only. By substituting Eq. 6.10 into Eq. 6.11 and by squaring the developed equations, it is possible to get rid of the cosine and sine terms. It is finally possible to obtain an equation of the form:

$$C_1\left(\frac{I_L}{I_o}\right)^2 - 2C_2\frac{I_L}{I_o} + C_3 = 0, \tag{6.13}$$

where C_1, C_2 and C_3 are, again, functions of $\omega_{o,c,L}$, Γ and $T_{o,c}$ only. Their explicit definitions can be found in Ref. [83]. Solving Eq. 6.13 leads to two possible roots: $(I_L/I_o)_\pm = \left(C_2 \pm \sqrt{C_2{}^2 - 4C_1C_3}\right)/(2C_1)$. It can be demonstrated that the root $(I_L/I_o)_-$ is the correct solution for $T_c \in [0; t_0]$ while the solution $(I_L/I_o)_+$ is applied for the cases $T_c \in [t_0; 2\pi/\omega_L]$, with t_0 verifying the equation $C_1(t_0 > 0) = 0$. The time t_0 corresponds to the maximum time the crack can spend in the closed state during a single pump period if we consider only the oscillating part of the laser and neglect the constant heating part ($\Gamma = 0$).

By solving Eq. 6.13, the time spent by the crack in the closed state over a pump period, T_c, can be evaluated as a function of loading parameters (I_L, ω_L, Γ) and crack parameters (I_o, ω_o, ω_c). Loading parameters I_L, ω_L are set by the experimenter. P_o (and thus I_o) is obtained from the experiments. Parameters $\omega_{o,c}$ and Γ (that remains constant for a given experimental configuration) can then be extracted by comparison of the theoretical predictions with the experimental results. Once their values are estimated, several parameters can be assessed from the previous equations: the intensity I_c, the crack rigidities in open and closed states $\eta_{o,c}$, the distances $(h_{o,c} - h_i)$, and the force F_i to locally close the crack.

6.3.2 *Evolution of the Nonlinear Sidelobes Amplitude with the Loading*

Equation 6.8 provides a prediction of the nonlinear mixed-frequencies amplitudes, as a function of the pump modulation frequency ω_L and the time T_c spent by the crack in the closed state over a pump period. By modifying the pump loading intensity I_L without modifying the other parameters, the temperature rise will differ (see Sect. 6.1.3), and the force F_i required to close/open the crack will be reached at another moment. Points C, D and F in Fig. 6.11 will then differ. Thus, modifying I_L affects the time T_c which, in turns, modifies the amplitude A_n (Eq. 6.8). Furthermore, we demonstrated with Eq. 6.13 the possibility to predict the evolution of T_c as a function of the loading I_L, and thus, to obtain the amplitude of each nonlinear sidelobe using Eq. 6.8 -as long as the other experimental parameters of the laser radiation (ω_L and Γ) and crack ones ($\omega_{o,c}$, and I_o) are known. With identical experimental parameters but different cracks or different locations along the crack, the parameters I_o, $\omega_{o,c}$ differ, and, thus, so will T_c. This effect can be observed in Fig. 6.8a–d for which the amplitudes, widths, and wings of the three nonlinear sidelobes are all different between the scans. Similarly, for the same location but with a different loading intensity I_L, the amplitude A_n is also modified, explaining the differences between Fig. 6.8c and e as well as those between 6.8d and f. With a different pump cyclic frequency ω_L, the amplitude of the different nonlinear sidelobes is also modified [88], as predicted by Eq. 6.8.

It is important to point out that while Eq. 6.8 allows an estimation of the nonlinear sidelobes amplitude as a function of T_c and ω_L, the solutions of this equation are not unique, but multivalued. Therefore, to one measured amplitude A_n can correspond several set time durations T_c (and/or cyclic pump frequencies ω_L if the latter is also unknown). Having only as many measurements (with different loadings) as unknown variables is then insufficient to deduce crack parameters. Another important point from Eq. 6.8 is the possibility to have amplitude minima occurring for a sidelobe n at a particular loading (while the crack is breathing), as it has been discussed earlier. This can be used to our advantage to deduce what is the time T_c for a given loading [78, 83]. As the first mixed-frequency ($n = 1$) cannot have a minimum if the crack breathes, its observation ensures that a crack remains detectable by this technique. Besides, there is no single T_c for which all minima of higher order cancel simultaneously (e.g. for $n = 2$ it occurs when the crack spends 1/2 of its time in the closed state over the pump period while for $n = 3$, these minima are reached when the crack spends 1/3 and 2/3 of its time in the closed state over the pump period). For a breathing crack, at least some of the nonlinear mixed-frequencies will thus remain detectable at all time. For the particular loading where $T_c = T_o = \pi/\omega_L$, the crack spends half of the time in each state and all even nonlinear sidelobes have a minimum amplitude. This corresponds to the transition from tapping regime ($T_o \geqslant T_c$) to clapping regime ($T_o \leqslant T_c$). The tapping regime can be reached only if the stationary heating affects the crack breathing as the crack cannot spend more than half time in the closed state with a (sinusoidal) oscillating heating only (the latter situation corresponding mathematically to $\Gamma = 0$). Finally, the crack can remain completely closed, despite the modulation heating, only if $\Gamma > 1$.

To experimentally observe the evolution of the nonlinear sidelobe amplitudes as a function of the pump loading, we first achieved a 1D scan (as in Sect. 6.2.3) with the setup presented in Fig. 6.4b. For this experiment, the probe spot radius is $a_H = 34\,\mu\text{m}$, the modulation frequency is $f_H = 24.9$ kHz and the power is $P_H = 35$ mW. The pump spot has a radius of $a_L = 36\,\mu\text{m}$, a modulation frequency $f_L = 1.5$ Hz and a power P_L of 60 mW. Once the crack is detected, the sample is positioned so that the laser beams foci are centred on it. Then, the power of the pump beam is slowly increased (by modifying the AOM input voltage) from 0.5 up to 142.3 mW with 94 steps (while the other parameters remain constant). By increasing the pump power, both the modulated and stationary thermoelastic stresses from the pump beam increase (points C, D, and F in Fig. 6.11c–e) which leads to progressive closure of the crack. For each pump amplitude, the spectrum is recorded. For low pump powers, the crack is not supposed to breathe as it remains open. For high pump powers, and if $\Gamma > 1$, the crack is supposed to remain closed, and the crack does not breathe either (Fig. 6.12). It is important to keep in mind that while the amplitude of the detected

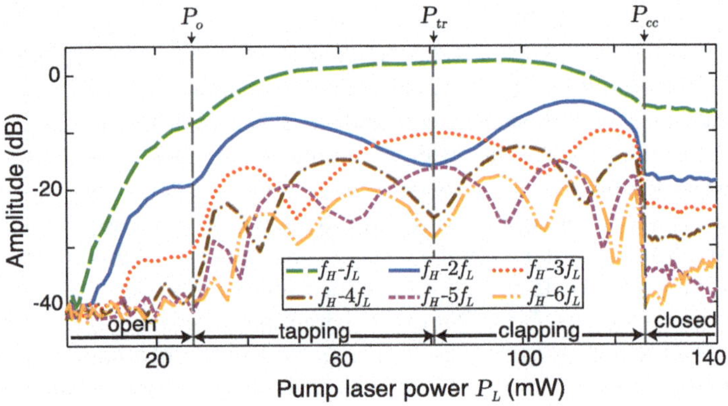

Fig. 6.13 Dependence of the amplitude of the photoacoustic signal of the first six nonlinear left sidelobes $f_H - nf_L$ ($n = 1 - 6$) on the pump laser power loading the crack. P_o indicates the critical pump power when the clapping starts, P_{tr} indicates the transition from clapping to tapping regime and P_{cc} the complete closing of the crack. Reprinted from [83] with permission from Elsevier

frequency-mixed spectral components is proportional to the probe beam power - which thus should be high-, the latter induces constant heating that reduces the crack width. Depending on the crack width, this can be either an advantage (by reducing the apparent distance between the crack faces to the pump beam and allowing crack breathing by the latter) or a drawback (by closing a thin crack and preventing its breathing, and thus its detection).

Figure 6.13 presents the evolution of the first six left nonlinear sidelobes (i.e. $f_H - nf_L$ with $n = 1 - 6$). For low pump powers, only the first sidelobe is generated, followed, as the pump power increases, by the second and then the third one. At $P_L = 28.1$ mW, a sudden increase of the three first nonlinear sidelobes ($n = 1 - 3$) amplitudes and the apparition of the three successive higher order nonlinear sidelobes ($n = 4 - 6$) are observed. This power threshold corresponds to the initiation of the crack breathing (and is referred after as $P_o = 28.1$ mW, see Fig. 6.13). Below P_o, the theory predicts no generation of the nonlinear sidelobes (Eq. 6.8 and Ref. [78]). The presence of the first three nonlinear sidelobes before this limit is attributed to smooth analytic nonlinearity of the crack (see Sect. 6.2.4). The decrease in their amplitude with the increasing order n of the sidelobe leading to undetectable sidelobes of the order higher than 3, and the monotonous increase of the sidelobe amplitudes with increasing pump power (for $P_L < P_o$) are clear indications that they are generated due to smooth analytical nonlinearity and in particular due to quadratic acoustic nonlinearity. Moreover, if in the probed region of the crack there are multiple tapping/clapping contacts between the crack faces, which are exhibiting different thresholds for the start (or the end) of breathing, their local non-analytic nonlinearities could be statistically smoothed to produce a contribution to effective smooth nonlinearity. Thus, our observations below P_o do not exclude the possibility of the existence of local contacts between the crack faces, especially when P_o is approached, but rather indicate the absence of global (on average) closing/opening of the complete tested part of the crack.

Several minima are observed when the pump power continuously increases above P_o. For $P_L = 80.7$ mW (indicated by P_{tr} in Fig. 6.13), the even sidelobes all exhibit a minimum, corresponding to the tapping-clapping transition ($T_o = T_c = \pi/\omega_L$). Furthermore, one can notice two minima of the third sidelobe at 50.9 and 105.0 mW corresponding to the crack spending 1/3 and 2/3 of the time in the closed state over a pump period, respectively. The observation of the clapping regime indicates that the stationary thermoelastic stresses induced by the stationary component of the absorbed optical power from the probe and the pump beams influence our results, mathematically corresponding to $\Gamma > 0$. For $P_L = 126.5$ mW (denoted as P_{cc} in the following), all the nonlinear sidelobes have a sudden important amplitude decrease: this corresponds to the complete closing of the crack when the oscillating thermoelastic stress cannot open the crack which remains closed during the whole pump period due to stationary heating. This indicates that, for this experiment, $\Gamma > 1$. For larger pump power ($P_L > P_{cc}$), the nonlinear mixed-frequency amplitudes are almost constant. One can notice the clear change in their amplitude evolutions as a function of the loading. The presence of remaining nonlinearities can be attributed, again, to classical nonlinearities related to the contact between the rough crack faces, and possible evolution in the number and quality of the contacts between the crack faces even if the crack remains closed. The full report of the minima of the first six nonlinear sidelobes and their associated pump powers is presented in Table 6.2.

As discussed in the second part (Sect. 6.2.4), and as deducible from Eq. 6.8, a nonlinear sidelobe n (for $n \geqslant 2$) can have a larger or smaller amplitude A_n than the one of $n - 1$ depending on the loading level provided by the pump and the probe beams. This is again evident in Fig. 6.13 in many locations. For example, the amplitude of the third nonlinear sidelobe A_3 is smaller than the one of the second sidelobe A_2 except for $P_L \in [68; 92]$ mW. Similar observations can be done with other nonlinear sidelobes but the first one, such as $A_3 < A_4$ for $P_L \in [47; 62]$ and [97; 108] mW, or even $A_3 < A_5$ for $P_L \in [47; 55]$ and [104; 108] mW (and $A_{4,5} < A_3$ otherwise). Again, the measured amplitude of the first nonlinear sidelobe A_1 is always larger than the one of all the other sidelobes for all the different powers (Fig. 6.13). While this result could be interpreted as due to classical analytic nonlinearities, it can ac-

Table 6.2 Pump power associated with the minima of the first six nonlinear sidelobes

Sidelobe	Pump power (mW)				
1	–	–	–	–	–
2	80.7	–	–	–	–
3	50.9	105.0	–	–	–
4	42.7	80.7	114.1	–	–
5	38.5	62.5	97.1	117.4	–
6	35.7	50.9	80.7	105.0	120.8

Reprinted from [83] with permission from Elsevier

tually be demonstrated from Eq. 6.8 that A_1 is always larger than A_2 in the case of the considered non-classical non-analytic nonlinearity as well.

It is worth mentioning that in the case of a closed crack (non-tapping/non-clapping), local contacts between the crack faces, i.e., contact Hertzian nonlinearity, could importantly contribute to the smooth analytic nonlinearity [81, 101]. For $P_o < P_L < P_{cc}$, the non-monotonous amplitude evolution of the nonlinear mixed-frequencies, the presence of minima and the possibility for a nonlinear sidelobe n to have a greater amplitude than the one of $(n - 1)$, all are indicators, once again, that non-classical nonlinearities overcome classical nonlinearities, as expected. The amplitudes of all six sidelobes are all comparable which is also a clear signature of non-classic mechanism in their generation.

6.3.3 Extraction of Crack Parameters

From the previously described experiment, the powers required to initiate and end the crack breathing (P_o and P_{cc}, respectively) as well as the power associated with 15 other minima on the first six sidelobes are identified (Table 6.2) and can be converted into the dimensionless value P_L/P_o to be comparable with the ratio I_L/I_o used in the theoretical model (Sect. 6.3.1) (as $P_L \propto I_L$). Theoretical predictions have been evaluated for many configurations varying Γ, ω_o, and ω_c. For each case, the number of minima, their positions and the ratio I_{cc}/I_o (with I_{cc} the intensity for complete closing of the crack) are evaluated and compared with the experimental results. Only theoretical cases with both the correct number of minima and a correct ratio (within 1%) compared to experimental results (Table 6.2 and Fig. 6.13) are considered [83]. Then, the discrepancy between the theoretical and experimental position of each minima is estimated. This discrepancy is weighted by $1/n$ for the nonlinear sidelobe n. This makes the observed minima of nonlinear sidelobes associated with $n = 2$ and 3 more important than those with $n = 5, 6$. This choice is made to take into account the $1/n$ amplitude decrease factor in Eq. 6.8.

From all the considered theoretical cases fulfilling the required conditions, the case with the lowest discrepancy between theoretical and experimental results is obtained for $\Gamma = 1.47$, $\omega_o = 8.77$ Hz and $\omega_c = 100$ kHz. The associated theoretical evolutions are presented in Fig. 6.14. Considering the cases within a 1% error discrepancy from the best case, the parameters evolve as follow: $\Gamma \in [1.47; 1.49]$, $\omega_o \in [8.05; 9.93]$ Hz, and $\omega_c \in [2.0; 100]$ kHz. The variation of Γ and ω_o are small, indicating a good parameter estimation. At the opposite, ω_c variations cover nearly two orders of magnitude leading to a poor estimation. The value of 100 kHz being the maximum value considered in the fit, a more reasonable estimation is $\omega_c \geqslant 2$ kHz. This is due to the resulting inequality $\omega_c \gg \omega_{o,L}$ which indicates that the rigidity of the closed crack is at least about 4 orders of magnitude larger than of the open one.

Comparison between the best theoretical evolution (Fig. 6.14) and the experimental one (Fig. 6.13) first demonstrates that the beginning of tapping and complete closing of the crack clearly appear in both cases. The 15 minima previously observed

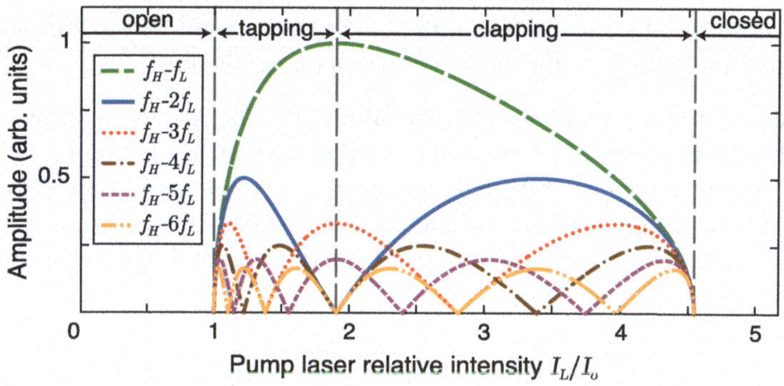

Fig. 6.14 Theoretical evolutions of the first six nonlinear sidelobes as a function of the pump laser relative intensity I_L/I_o after the fit with the experimental results ($\omega_o = 8.77$ Hz, $\omega_c = 100$ kHz and $\Gamma = 1.47$). Reprinted from [83] with permission from Elsevier

are all modelled. The possibility to observe, depending on the loading, $A_n < A_{n-1}$ or $A_{n-1} < A_n$, except for $A_1 < A_2$, is also visible in the theoretical fit, as in the experiment. Evolution of the nonlinear sidelobe amplitudes associated with $n = 2$ and 3, for example, present similar changes. However, the minima positions are still imperfect. For instance, the transition from tapping to clapping regime occurs experimentally roughly at $(P_o + P_{cc})/2$, while it occurs around $(I_o + I_{cc})/4$ with the considered parameters fit. Similar shifts can be observed for some other minima. Several reasons can be listed to explain this imperfect fit:

- the smooth analytic nonlinearities are not taken into account in the theoretical model. If, as proposed, they are responsible for the nonlinear sidelobes apparition at powers lower than P_o, and are present at least for $n = 1$ (see Sect. 6.2.4), their influence should not be neglected at higher pump powers either, especially on the low-order nonlinear sidelobes (because of the cascade frequency-mixing processes);
- the model assumes a bi-modular elasticity of the crack (Fig. 6.11c) while the crack rigidity is, in reality, not linear piecewise but should continuously increase with the crack closing, especially with a more realistic modelling of a crack (Fig. 6.11a);
- the hysteresis between F_o and F_c (Fig. 6.11b) has been neglected affecting the nonlinear sidelobe amplitudes, although strong hysteresis could probably also affect the number of minima that are well reproduced in the current case. Demodulation/rectification of the oscillating thermoelastic stress [35, 96] is also neglected in the present model;
- the possible asymmetry of the pump laser beam position relative to the crack position is also ignored. When the heating is asymmetric relatively to the crack position, closing/opening of the crack starts to influence the heat transport and temperature distribution. The heat transport across the crack could be modelled by introducing the concept of a thermal resistance which in general depends on the distance between the crack faces [78, 88, 95, 102]. The dependence of the thermal

resistance on the distance between the crack faces could be smooth analytic or abrupt non-analytic, as in the case of crack opening/closing.

Nevertheless, it is worth emphasising that values of Γ, ω_o, and ω_c are now extracted, and give the correct number of minima as well as similar I_{cc}/I_o and P_{cc}/P_o magnitudes. With these values, several crack parameters can be estimated. In the following, the first result corresponds to the estimation from the best fit while uncertainties are estimated from the 1% variation of the error from this best fit. Firstly, open and closed rigidities are obtained:

$$\eta_o = \frac{\rho c \omega_o}{2} = 66.0\,\text{MN.}\,\text{m}^{-3}(60.2 \leqslant \eta_o \leqslant 74.7), \tag{6.14a}$$

$$\eta_c = \frac{\rho c \omega_c}{2} = 752\,\text{GN.}\,\text{m}^{-3}(\eta_c \geqslant 13.7). \tag{6.14b}$$

Then, the pump power used to reach P_o being known, I_o can be deduced, and I_c follows from Eq. 6.10:

$$I_o = (1-r)\frac{P_o}{\pi a^2} = 6.14\,\text{MW.}\,\text{m}^2, \tag{6.15a}$$

$$I_c = 5.35\,\text{MW.}\,\text{m}^2(5.27 \leqslant I_c \leqslant 5.46). \tag{6.15b}$$

It is now possible, with Eq. 6.6, to estimate the distance between the crack faces reduced by the probe constant heating. In other words, the distance between points B and E in Fig. 6.11a can be evaluated. It follows:

$$h_o - \frac{I_H A_H(0)}{\omega_o} - h_i = 103.8\,\text{nm}\,(93.3 \leqslant h_o - I_H A_H(0)/\omega_o - h_i \leqslant 112.7). \tag{6.16}$$

The estimation of the reduction in the distance between the crack faces induced by the probe beam (term proportional to $I_H A_H(0)$, corresponding to the change from point A to point B in Fig. 6.11c), is necessary in order to find the total displacement of the crack faces between their position in the absence of loading and when the faces are in contact. It can be demonstrated that the amplitude term $A_H(0)$ can be estimated to be equal to $\Gamma|A(\omega_L)|(a_H\omega_T(a_L))/(a_L\omega_T(a_H))$ [83], and $I_H(0)$ can be determined similarly as in Eq. 6.15a. It follows that the complete crack width (distance $[A; E]$ in Fig. 6.11c) is equal to:

$$h_o - h_i = 146.8\,\text{nm}\,(131.3 \leqslant h_o - h_i \leqslant 160.3). \tag{6.17}$$

Finally, from Eq. 6.7, the force F_i, required to close and open the crack, can now be estimated:

$$F_i = 9.7\,\text{N.}\,\text{m}^{-2}\,(9.6 \leqslant F_i \leqslant 9.9). \tag{6.18}$$

The distance $(h_o - h_i)$ of a few hundreds of nanometers is of the same order of magnitude as the crack width measured in similar cracks with AFM microscopy [71]. This illustrates the accuracy of this newly developed method. To the authors' knowledge, the other locally evaluated parameters (crack rigidities, the force required to close the crack) cannot be obtained by other methods, preventing more comparison. The distance $(h_c - h_i)$ (Fig. 6.11c) can also be deduced similarly (it follows $h_c - h_i \leqslant 12.9$ pm), but this does not give any extra information about the crack. This distance is associated with the closed state (see Fig. 6.11c) and is required to calculate F_c in presence of hysteresis between the forces F_o and F_c that are required to close and open the crack, respectively (which, as previously said, has not been considered here).

The obtained crack rigidities can also be compared with the one of the glass sample. A possible way is to evaluate an equivalent Young elastic modulus in open and closed state. The elastic modulus can be estimated by the force exerted on the sample per unit area multiplied by the ratio of the original length of the object divided by the length change. It follows:

$$E_o = F_i \frac{2a}{h_o - h_i} = 4.7\,\text{kPa}\,(4.3 \leqslant E_o \leqslant 5.4). \tag{6.19}$$

The equivalent Young modulus in closed state E_c cannot be precisely determined due to the uncertainty on ω_c but can be estimated to be at least ~4 orders of magnitude higher than E_o. The glass Young modulus, $E_g = K(1 - \nu) = 40.16$ GPa, with $\nu = 0.22$ the Poisson ratio, is of seven orders of magnitude higher than the open effective elastic moduli of the crack. As it is clear that the weakest area of the sample remains, by far, the crack (even when closed), this estimation could be used to determine a higher limit on the previous estimations in closed states (η_c, $h_c - h_i$, and E_c).

6.4 Conclusion

In this chapter, we demonstrated the possibility to detect and characterise cracks with a novel technique. It is based:

- on the modulation of the distance between the crack faces, oscillating between closed crack (faces in contact) and open crack (no contact between the faces), by a local thermoelastic stress generated by a laser;
- on the laser generation of an ultrasonic probe wave in the vicinity of the modulated crack;
- on the detection of the resulting nonlinear frequency-mixing (sidelobes around the probe frequency) originating from the modulation of the ultrasonic wave reflection/transmission by the crack via crack breathing.

When the laser beams are focused far from a crack, this nonlinear mixing process is hardly detectable, while in the vicinity of a crack, an efficient nonlinear frequency mixing process takes place. The detection of these new mixed frequencies is related to the presence of a crack within or near the laser foci on the sample. The method offers the advantages to combine a non-contact optical excitation with high amplitude dynamics. The detection of nonlinear frequency-mixing, not generated without this crack breathing, is also very attractive as it avoids false alarms from other nonlinear phenomena, especially when monitoring the higher order frequency-mixed sidelobes $n \geqslant 2$ which can not be (efficiently) generated by nonlinear cascade processes. The possibility of an all-optical setup is demonstrated by the use of a laser vibrometer or of the deflectometry technique, adapted for the required frequency range. It is also demonstrated that this technique has important capabilities concerning spatial resolution.

Furthermore, with the use of the developed theoretical model, this method can be used to extract local crack parameters. The experimentally observed evolution of the amplitude of the nonlinear sidelobes as a function of the pump power exhibits clear non-monotonous behaviour, as predicted by the theory. The evidence of minima, for particular pump powers and sidelobes, is demonstrated and their positions as a function of the loading are studied. The theoretical model provides the evolution of the nonlinear sidelobes, as a function of the pump power, considering a dynamic motion of the crack faces and the constant heating due to both pump and probe beams. A fit between the experimental results and the theoretical ones is possible. The estimations of the crack mechanical relaxation frequencies and of the ratio between the constant and the modulated loading are deduced from that fit. With these information, it is then possible to estimate several crack parameters, including some for the first time to the authors' knowledge. These parameters are the open and closed rigidities $\eta_{o,c}$, the intensities $I_{o,c}$, the force F_i required to close and open the crack, and the displacement of the crack faces induced by the pump loading, as well as the one required to close the crack.

This nonlinear frequency-mixing photoacoustic technique is very promising both regarding its detection sensitivity and its characterisation possibilities. The two dimensional images of the crack have been achieved and led to a better understanding of the spatial resolution of the method. While this has not been reported yet, the images at -6 dB could potentially be narrower than the laser excitation focused beams. Repeating 2D images with more focused laser beams should help improving the spatial resolution as the non-classical non-analytical nonlinearities are generated only in the crack vicinity. Such experiment will also help to better understand and discriminate the role of the classical analytical nonlinearities and of the non-classical non-analytical ones in the region of the crack.

Furthermore, the local crack properties can be evaluated by comparison with the model. As the 2D images showed sensitivity to the local crack properties, repeating such experiments with diffraction-limited focused laser beams will open a way to sub-micrometric mapping of the crack localisation, thickness and rigidity.

However, the fit between theoretical and experimental results is not yet perfect and the theoretical modeling could be improved in several ways. In particular, the asymmetric laser heating of the crack and the hysteresis between F_o and F_c could be added (as it was done in Ref. [88] for the quasistatic state). A realistic 3D description of the light absorption region, and, eventually, a full 3D theoretical model, would also be of high interest.

References

1. C.B. Scruby, L.E. Drain, *Laser Ultrasonics Techniques & Applications* (Adam Hilger, New York, 1990)
2. V.E. Gusev, A.A. Karabutov, *Laser Optoacoustics* (American Institute of Physics, Maryland, 1993)
3. D. Royer, E. Dieulesaint, *Elastic Waves in Solids II: Generation, Acousto-Optic Interaction, Applications* (Springer, Berlin, 2000)
4. W. Bai, G.J. Diebold, Moving photoacoustic sources: Acoustic waveforms in one, two, and three dimensions and application to trace gas detection. J. Appl. Phys. **125**(6), 060902 (2019)
5. K.-H. Lin, G.-W. Chern, C.-T. Yu, T.-M. Liu, C.-C. Pan, G.-T. Chen, J.-I. Chyi, S.-W. Huang, P.-C. Li, C.-K. Sun, Optical piezoelectric transducer for nano-ultrasonics. IEEE Trans. UFFC **52**(8), 1404–1414 (2005)
6. P.-A. Mante, Y.-R. Huang, S.-C. Yang, T.-M. Liu, A.A. Maznev, K.-K. Sheu, C.-K. Sun, THz acoustic phonon spectroscopy and nanoscopy by using piezoelectric semiconductor heterostructures. Ultrasonics **56**, 52–65 (2015)
7. A. Huynh, B. Perrin, A. Lemaître, Semiconductor superlattices: A tool for terahertz acoustics. Ultrasonics **56**, 66–79 (2015)
8. C. Mechri, P. Ruello, J.M. Breteau, M.R. Baklanov, P. Verdonck, V. Gusev, Depth-profiling of elastic inhomogeneities in transparent nanoporous low-k materials by picosecond ultrasonic interferometry. Appl. Phys. Lett. **95**(9), 091907 (2009)
9. A. Steigerwald, Y. Xu, J. Qi, J. Gregory, X. Liu, J.K. Furdyna, K. Varga, A.B. Hmelo, G. Lüpke, L.C. Feldman, N. Tolk, Semiconductor point defect concentration profiles measured using coherent acoustic phonon waves. Appl. Phys. Lett. **94**(11), 111910 (2009)
10. A.M. Lomonosov, A. Ayouch, P. Ruello, G. Vaudel, M.R. Baklanov, P. Verdonck, L. Zhao, V.E. Gusev, Nanoscale noncontact subsurface investigations of mechanical and optical properties of nanoporous low-k material thin film. ACS Nano **6**(2), 1410–1415 (2012). PMID: 22211667
11. V.E. Gusev, P. Ruello, Advances in applications of time-domain brillouin scattering for nanoscale imaging. Appl. Phys. Rev. **5**(3), 031101 (2018)
12. A. Vertikov, M. Kuball, A.V. Nurmikko, H.J. Maris, Time resolved pump-probe experiments with subwavelength lateral resolution. Appl. Phys. Lett. **69**(17), 2465–2467 (1996)
13. J.M. Atkin, S. Berweger, A.C. Jones, M.B. Raschke, Nano-optical imaging and spectroscopy of order, phases, and domains in complex solids. Adv. Phys. **61**(6), 745–842 (2012)
14. H.J. Maris, Picosecond ultrasonics. Sci. Am. **278**(1), 86–89 (1998)
15. K.-H. Lin, C.-M. Lai, C.-C. Pan, J.-I. Chyi, J.-W. Shi, S.-Z. Sun, C.-F. Chang, C.-K. Sun.: Spatial manipulation of nanoacoustic waves with nanoscale spot sizes. Nat. Nanotechnol. **2**, 704 EP –, 10 (2007)
16. P.-A. Mante, C.-C. Chen, Y.-C. Wen, H.-Y. Chen, S.-C. Yang, Y.-R. Huang, I.-J. Chen, Y.-W. Chen, V. Gusev, M.-J. Chen, J.-L. Kuo, J.-K. Sheu, and C.-K. Sun. Probing hydrophilic interface of solid/liquid-water by nanoultrasonics. Sci. Rep. **4**, 6249 EP –, 09 (2014)
17. J.D.G. Greener, A.V. Akimov, V.E. Gusev, Z.R. Kudrynskyi, P.H. Beton, Z.D. Kovalyuk, T. Taniguchi, K. Watanabe, A.J. Kent, A. Patanè, Coherent acoustic phonons in van der Waals nanolayers and heterostructures. Phys. Rev. B **98**, 075408 (2018)
18. C. Li, V. Gusev, E. Dimakis, T. Dekorsy, M. Hettich.: Broadband photo-excited coherent acoustic frequency combs and mini-brillouin-zone modes in a MQW-SESAM structure. Appl. Sci. **9**(2), (2019)

19. S.A. Akhmanov, V.E. Gusev, Laser excitation of ultrashort acoustic pulses: New possibilities in solid-state spectroscopy, diagnostics of fast processes, and nonlinear acoustics. Sov. Phys. Usp. **35**(3), 153–191 (1992)
20. P. Ruello, V.E. Gusev, Physical mechanisms of coherent acoustic phonons generation by ultrafast laser action. Ultrasonics **56**, 21–35 (2015)
21. C. Thomsen, H.T. Grahn, H.J. Maris, J. Tauc, Surface generation and detection of phonons by picosecond light pulses. Phys. Rev. B **34**, 4129–4138 (1986)
22. H.T. Grahn, H.J. Maris, J. Tauc, Picosecond ultrasonics. IEEE J. Quantum Electron. **25**(12), 2562–2569 (1989)
23. L.K. Zarembo, V.A. Krasil'nkov, Nonlinear phenomena in the propagation of elastic waves in solids. Sov. Phys. Usp. **13**(6), 778–797 (1971)
24. O.V. Rudenko, S.I. Soluyan, *Theoretical Foundations of Nonlinear Acoustics* (Consultants Bureau, New York, 1977)
25. V.E. Nazarov, L.A. Ostrovsky, I.A. Soustova, A.M. Sutin, Nonlinear acoustics of micro-inhomogeneous media. Phys. Earth Planet. Int. **50**, 65–73 (1988)
26. Robert A. Guyer, Paul A. Johnson, Nonlinear mesoscopic elasticity: evidence for a new class of materials. Phys. Today **52**(4), 30–36 (1999)
27. M. Born, K. Huang, *Dynamical Theory of Crystal Lattices* (Clarendon Press, Oxford, 1954)
28. V. Gusev, V. Tournat, B. Castagnède, Nonlinear acoustic phenomena in micro-inhomogeneous materials, in *Materials and Acoustics Handbook*, ed. by M. Bruneau, C. Potel (Wiley, London, New York, 2009), pp. 433–472
29. V. Tournat, V. Gusev, B. Castagnède, Non-destructive evaluation of micro-inhomogeneous solids by nonlinear acoustic methods, in *Materials and Acoustics Handbook*, ed. by M. Bruneau, C. Potel (Wiley, London, New York, 2009), pp. 473–504
30. P.P. Delsanto (ed.), *Universality of Nonclassical Nonlinearity* (Springer, New York, 2006)
31. K.E.A. Van Den Abeele, P.A. Johnson, A. Sutin, Nonlinear elastic wave spectroscopy (NEWS) techniques to discern material damage, part I: Nonlinear wave modulation spectroscopy (NWMS). Res. Nondestruct. Eval. **12**(1), 17–30 (2000)
32. V.Y. Zaitsev, L.A. Matveev, A.L. Matveyev, Elastic-wave modulation approach to crack detection: Comparison of conventional modulation and higher-order interactions. NDT & E Int. **44**(1), 21–31 (2011)
33. O. Buck, W.L. Morris, J.M. Richardson, Acoustic harmonic generation at unbonded interfaces and fatigue cracks. Appl. Phys. Lett. **33**, 371–373 (1978)
34. A. Novak, M. Bentahar, V. Tournat, R. El Guerjouma, L. Simon, Nonlinear acoustic characterization of micro-damaged materials through higher harmonic resonance analysis. NDT & E Int. **45**, 1–8 (2012)
35. A. Moussatov, V. Gusev, B. Castagnède, Self-induced hysteresis for nonlinear acoustic waves in cracked material. Phys. Rev. Lett. **90**(12), 124301 (2003)
36. B.A. Korshak, I.Y. Solodov, E.M. Ballad, DC effects, sub-harmonics, stochasticity and "memory" for contact acoustic non-linearity. Ultrasonics **40**(1–8), 707–713 (2002)
37. A. Moussatov, B. Castagnède, V. Gusev, Observation of non linear interaction of acoustic waves in granular materials: demodulation process. Phys. Lett. A **283**, 216–223 (2001)
38. L. Fillinger, V. Zaitsev, V. Gusev, B. Castagnède, Wave self-modulation in an acoustic resonator due to self-induced transparency. Europhys. Lett. **76**, 229–235 (2006)
39. L. Fillinger, V.Y. Zaitsev, V.E. Gusev, B. Castagnède, Self-modulation of acoustic waves in resonant bars. J. Sound Vibr. **318**, 527–548 (2008)
40. V. Zaitsev, V. Gusev, B. Castagnède, Luxemburg-Gorky effect retooled for elastic waves: a mechanism and experimental evidence. Phys. Rev. Lett. **89**(10), 105502 (2002)
41. V.Y. Zaitsev, V. Nazarov, V. Gusev, B. Castagnède, Novel nonlinear-modulation acoustic technique for crack detection. NDT&E Int. **39**, 184–194 (2006)
42. P.B. Nagy, Fatigue damage assessment by nonlinear ultrasonic materials characterization. Ultrasonics **36**(1–5), 375–381 (1998)
43. A.M. Sutin, Nonlinear acoustic nondestructive testing of cracks. J. Acoust. Soc. Am. **99**(4), 2539–2574 (1996)

44. Y. Zheng, R.G. Maev, I.Y. Solodov, Nonlinear acoustic applications for material characterization: A review. Can. J. Phys. **77**(12), 927–967 (1999)
45. K.-Y. Jhang, Nonlinear ultrasonic techniques for nondestructive assessment of micro damage in material: A review. Int. J. Precis. Eng. Manuf. **10**(1), 123–135 (2009)
46. M.W. Sigrist, Laser generation of acoustic waves in liquids and gases. J. Appl. Phys. **60**(7), R83–R122 (1986)
47. S.A. Akhmanov, V.E. Gusev, A.A. Karabutov, Pulsed laser optoacoustics: Achievements and perspective. Infrared Phys. **29**(2–4), 815–838 (1989)
48. A.A. Karabutov, V.T. Platonenko, O.V. Rudenko, B.A. Chupryna, Experimental investigation of shock-front formation in solid (in russian). Mosc. Univ. Phys. Bull. **25**(3), 89–91 (1984)
49. Y. Yasumoto, A. Nakamura, R. Takeuchi, Developments in the use of acoustic shock pulses in the study of elastic properties of solids. Acta Acust. United Acust. **30**(5), 260–267 (1974)
50. V.N. In'kov, E.B. Cherepetskaya, V.L. Shkuratnik, A.A. Karabutov, V.A. Makarov, Ultrasonic laser spectroscopy of mechanic-acoustic nonlinearity of cracked rocks. J. Appl. Mech. Tech. Phys. **46**, 452–457 (2005)
51. M. Li, A.M. Lomonosov, Z. Shen, H. Seo, K.-Y. Jhang, V.E. Gusev, C. Ni, Monitoring of thermal aging of aluminum alloy via nonlinear propagation of acoustic pulses generated and detected by lasers. Appl. Sci. **9**(6), (2019)
52. O.L. Muskens, J.I. Dijkhuis, High amplitude, ultrashort, longitudinal strain solitons in sapphire. Phys. Rev. Lett. **89**(28), 285504 (2002)
53. A. Bojahr, M. Herzog, D. Schick, I. Vrejoiu, M. Bargheer, Calibrated real-time detection of nonlinearly propagating strain waves. Phys. Rev. B **86**, 144306 (2012)
54. C. Klieber, V.E. Gusev, T. Pezeril, K.A. Nelson, Nonlinear acoustics at ghz frequencies in a viscoelastic fragile glass former. Phys. Rev. Lett. **114**, 065701 (2015)
55. H.-Y. Hao, H.J. Maris, Experiments with acoustic solitons in crystalline solids. Phys. Rev. B **64**, 064302 (2001)
56. P.J.S. van Capel, E. Péronne, J.I. Dijkhuis, Nonlinear ultrafast acoustics at the nano scale. Ultrasonics **56**, 36–51 (2015)
57. A.A. Kolomenskii, A.M. Lomonosov, R. Kuschnereit, P. Hess, V.E. Gusev, Laser generation and detection of strongly nonlinear elastic surface pulses. Phys. Rev. Lett. **79**, 1325–1328 (1997)
58. A. Lomonosov, P. Hess, Effects of nonlinear elastic surface pulses in anisotropic silicon crystals. Phys. Rev. Lett. **83**, 3876–3879 (1999)
59. A. Lomonosov, V.G. Mikhalevich, P. Hess, EYu. Knight, M.F. Hamilton, E.A. Zabolotskaya, Laser-generated nonlinear rayleigh waves with shocks. J. Acoust. Soc. Am. **105**(4), 2093–2096 (1999)
60. A.M. Lomonosov, P. Hess, A.P. Mayer, Observation of solitary elastic surface pulses. Phys. Rev. Lett. **88**, 076104 (2002)
61. P. Hess, Surface acoustic waves in materials science. Phys. Today **55**(3), 42–47 (2002)
62. P. Hess, A.M. Lomonosov, A.P. Mayer, Laser-based linear and nonlinear guided elastic waves at surfaces (2D) and wedges (1D). Ultrasonics **54**, 39–55 (2014)
63. S. Choi, H. Seo, K.-Y. Jhang, Noncontact evaluation of acoustic nonlinearity of a laser-generated surface wave in a plastically deformed aluminum alloy. Res. Nondestr. Eval. **26**, 13–22 (2015)
64. H. Seo, J. Jun, K.-Y. Jhang, Assessment of thermal aging of aluminum alloy by acoustic nonlinearity measurement of surface acoustic waves. Res. Nondestruct. Eval. **28**(1), 3–17 (2017)
65. T. Stratoudaki, R. Ellwood, S. Sharples, M. Clark, M.G. Somekh, I.J. Collison, Measurement of material nonlinearity using surface acoustic wave parametric interaction and laser ultrasonics. J. Acoust. Soc. Am. **129**(4), 1721–1728 (2011)
66. R. Ellwood, T. Stratoudaki, S.D. Sharples, M. Clark, M.G. Somekh, Determination of the acoustoelastic coefficient for surface acoustic waves using dynamic acoustoelastography: An alternative to static strain. J. Acoust. Soc. Am. **135**(3), 1064–1070 (2014)

67. H. Xiao, P.B. Nagy, Enhanced ultrasonic detection of fatigue cracks by laser-induced crack closure. J. Appl. Phys. **83**(12), 7453–7460 (1998)
68. Z. Yan, P.B. Nagy, Thermo-optical modulation for improved ultrasonic fatigue crack detection in Ti-6Al-4V. NDT&E Int. **33**, 213–223 (2000)
69. Z. Yan, P.B. Nagy, Thermo-optical modulation of ultrasonic surface waves for NDE. Ultrasonics **40**(1–8), 689–696 (2002)
70. C.-Y. Ni, N. Chigarev, V. Tournat, N. Delorme, Z.-H. Shen, V.E. Gusev, Probing of laser-induced crack modulation by laser-monitored surface waves and surface skimming bulk waves. J. Acoust. Soc. Am. 131(3) (2012)
71. C.-Y. Ni, N. Chigarev, V. Tournat, N. Delorme, Z.-H. Shen, V.E. Gusev, Probing of laser-induced crack closure by pulsed laser-generated acoustic waves. J. Appl. Phys. **113**(1), 014906 (2013)
72. V. Tournat, C. Ni, N. Chigarev, N. Delorme, Z. Shen, V. Gusev, Probing of crack breathing by pulsed laser-generated acoustic waves. Proc. Meet. Acoust. **19**(030081), (2013)
73. N. Krohn, R. Stoessel, G. Busse, Acoustic non-linearity for defect selective imaging. Ultrasonics **40**(1–8), 633–637 (2002)
74. I.Y. Solodov, J. Wackerl, K. Pfleiderer, G. Busse, Nonlinear self-modulation and subharmonic acoustic spectroscopy for damage detection and location. Appl. Phys. Lett. **84**(26), 5386–5388 (2004)
75. S.-H. Park, J. Kim, K.-Y. Jhang, Relative measurement of the acoustic nonlinearity parameter using laser detection of an ultrasonic wave. Int. J. Precis. Eng. Manuf. **18**(10), 1347–1352 (2017)
76. G. Grégoire, V. Tournat, D. Mounier, V. Gusev, Nonlinear photothermal and photoacoustic processes for crack detection. Eur. Phys. J. Spec. Top. **153**, 313–315 (2008)
77. N. Chigarev, J. Zakrzewski, V. Tournat, V. Gusev, Nonlinear frequency-mixing photoacoustic imaging of a crack. J. Appl. Phys. 106(036101) (2009)
78. V. Gusev, N. Chigarev, Nonlinear frequency-mixing photoacoustic imaging of a crack: Theory. J. Appl. Phys. **107**(124905) (2010)
79. J. Zakrzewski, N. Chigarev, V. Tournat, V. Gusev, Combined photoacoustic-acoustic technique for crack imaging. Int. J. Thermophys. **31**(1), 199–207 (2010)
80. S. Mezil, N. Chigarev, V. Tournat, V. Gusev, All-optical probing of the nonlinear acoustics of a crack. Opt. Lett. **36**(17), 3449–3451 (2011)
81. S. Mezil, N. Chigarev, V. Tournat, V. Gusev, Two dimensional nonlinear frequency-mixing photo-acoustic imaging of a crack and observation of crack phantoms. J. Appl. Phys. **114**(174901) (2013)
82. V.E. Gusev, N. Chigarev, S. Mezil, V. Tournat, All-optical nonlinear frequency - mixing acoustics of cracks. Proc. Meet. Acoust. **19**(1), 030079 (2013)
83. S. Mezil, N. Chigarev, V. Tournat, V.E. Gusev, Evaluation of crack parameters by a nonlinear frequency-mixing laser ultrasonics method. Ultrasonics **69**, 225–235 (2016)
84. S. Mezil, N. Chigarev, V. Tournat, V. Gusev, Review of a nonlinear frequency-mixing photo-acoustic method for imaging a crack. J. JSNDI **66**(12), 589–592 (2017)
85. C. Bakre, P. Rajagopal, K. Balasubramaniam, Nonlinear mixing of laser generated narrowband rayleigh surface waves. AIP Conf. Proc. **1806**(1), 020004 (2017)
86. Y. Liu, S. Yang, X. Liu, Detection and quantification of damage in metallic structures by laser-generated ultrasonics. Appl. Sci. **8**(5) (2018)
87. P. Liu, J. Jang, S. Yang, H. Sohn, Fatigue crack detection using dual laser induced nonlinear ultrasonic modulation. Opt. Lasers Eng. **110**, 420–430 (2018)
88. S. Mezil, *Nonlinear optoacoustics method for crack detection & characterization*. Ph.D. thesis, LUNAM Université, Université du Maine (2012)
89. Y. Ohara, K. Takahashi, Y. Ino, K. Yamanaka, T. Tsuji, T. Mihara, High-selectivity imaging of closed cracks in a coarse-grained stainless steel by nonlinear ultrasonic phased array. NDT & E Int. **91**, 139–147 (2017)
90. O.B. Wright, K. Kawashima, Coherent phonon detection from ultrasfast surface vibrations. Phys. Rev. Lett. **69**(11), 1668–1671 (1992)

91. J.E. Rothenberg, Observation of the transient expansion of heated surfaces by picosecond photothermal deflection spectroscopy. Opt. Lett. **13**(9), 713–715 (1988)
92. V. Gusev, A. Mandelis, R. Bleiss, Theory of second harmonic thermal wave generation: 1D geometry. Int. J. Thermophys. **14**, 321 (1993)
93. V. Gusev, A. Mandelis, R. Bleiss, Theory of strong photothermal nonlinearity from sub-surface non-stationary ("breathing") cracks in solids. Appl. Phys. A **57**, 229–233 (1993)
94. V. Gusev, A. Mandelis, R. Bleiss, Non-linear photothermal response of thin solid films and coatings. Mat. Sci. Eng. B **26**, 111 (1994)
95. V. Gusev, A. Mandelis, R. Bleiss, Theory of combined acousto-photo-thermal spectral decomposition in condensed phases: parametric generation of thermal waves by a non-stationary ("breathing") sub-surface defect. Mat. Sci. Eng. B **26**, 121 (1994)
96. V.E. Gusev, B. Castagnède, A.G. Moussatov, Hysteresis in response of nonlinear bistable interface to continuously varying acoustic loading. Ultrasonics **41**, 643–654 (2003)
97. I.Y. Solodov, B.A. Korshak, Instability, chaos, and "memory" in acoustic-wave-crack interaction. Phys. Rev. Lett. **88**(1), 014303 (2001)
98. I. Y. Solodov, B. A. Korshak, K. Pfleiderer, J. Wackerl, and G. Busse. Nonlinear ultrasonics inspection and NDE using subharmonic and self-modulation modes. *WCU*. Paris, France (2003)
99. V. Tournat, C. Inserra, V. Gusev, Non-cascade frequency-mixing processes for elastic waves in unconsolidated granular materials. Ultrasonics **48**(6), 492–497 (2008)
100. I.Y. Solodov, N. Krohn, G. Busse, CAN: an example of nonclassical acoustic nonlinearity in solids. Ultrasonics **40**, 621–625 (2002)
101. K.L. Johnson, *Contact Mechanics*, 2nd edn. (Cambridge University Press, Cambridge, Angleterre, 1985)
102. G.C. Wetzel Jr., J.B. Spiecer, Nonlinear effects in photothermal-optical-beam-deflection imaging. Can. J. Phys. **64**, 1269 (1986)

Index

K.-Y. Jhang et al. (eds.), *Measurement of Nonlinear Ultrasonic Characteristics*, Springer Series in Measurement Science and Technology,
https://doi.org/10.1007/978-981-15-1461-6

The manufacturer's authorised representative in the EU is Springer Nature Customer Service Centre GmbH, Europaplatz 3, 69115 Heidelberg, Germany. If you have any concerns regarding our products, please contact ProductSafety@springernature.com

Printed and bound by CPI Group (UK) Ltd, Croydon, CR0 4YY
25/09/2026
02222901-0005